Pitman Research Notes in Mathematics Series

Submission of proposals for consideration
Suggestions for publication, in the form of outlines and representative samples, are invited by the Editorial Board for assessment. Intending authors should approach one of the main editors or another member of the Editorial Board, citing the relevant AMS subject classifications. Alternatively, outlines may be sent directly to the publisher's offices. Refereeing is by members of the board and other mathematical authorities in the topic concerned, throughout the world.

Preparation of accepted manuscripts
On acceptance of a proposal, the publisher will supply full instructions for the preparation of manuscripts in a form suitable for direct photo-lithographic reproduction. Specially printed grid sheets are provided and a contribution is offered by the publisher towards the cost of typing. Word processor output, subject to the publisher's approval, is also acceptable.

Illustrations should be prepared by the authors, ready for direct reproduction without further improvement. The use of hand-drawn symbols should be avoided wherever possible, in order to maintain maximum clarity of the text.

The publisher will be pleased to give any guidance necessary during the preparation of a typescript, and will be happy to answer any queries.

Important note
In order to avoid later retyping, intending authors are strongly urged not to begin final preparation of a typescript before receiving the publisher's guidelines and special paper. In this way it is hoped to preserve the uniform appearance of the series.

Longman Scientific & Technical
Churchill Livingstone Inc.
1560 Broadway
New York, NY 10036, USA
(tel (212) 819-5453)

Longman Scientific & Technical
Longman House
Burnt Mill
Harlow, Essex, UK
(tel (0279) 26721)

Titles in this series

52 Wave propagation in viscoelastic media
F Mainardi
53 Nonlinear partial differential equations and their applications: Collège de France Seminar. Volume I
H Brezis and J L Lions
54 Geometry of Coxeter groups
H Hiller
55 Cusps of Gauss mappings
T Banchoff, T Gaffney and C McCrory
56 An approach to algebraic K-theory
A J Berrick
57 Convex analysis and optimization
J-P Aubin and R B Vintner
58 Convex analysis with applications in the differentiation of convex functions
J R Giles
59 Weak and variational methods for moving boundary problems
C M Elliott and J R Ockendon
60 Nonlinear partial differential equations and their applications: Collège de France Seminar. Volume II
H Brezis and J L Lions
61 Singular systems of differential equations II
S L Campbell
62 Rates of convergence in the central limit theorem
Peter Hall
63 Solution of differential equations by means of one-parameter groups
J M Hill
64 Hankel operators on Hilbert space
S C Power
65 Schrödinger-type operators with continuous spectra
M S P Eastham and H Kalf
66 Recent applications of generalized inverses
S L Campbell
67 Riesz and Fredholm theory in Banach algebra
B A Barnes, G J Murphy, M R F Smyth and T T West
68 Evolution equations and their applications
F Kappel and W Schappacher
69 Generalized solutions of Hamilton-Jacobi equations
P L Lions
70 Nonlinear partial differential equations and their applications: Collège de France Seminar. Volume III
H Brezis and J L Lions
71 Spectral theory and wave operators for the Schrödinger equation
A M Berthier
72 Approximation of Hilbert space operators I
D A Herrero
73 Vector valued Nevanlinna Theory
H J W Ziegler
74 Instability, nonexistence and weighted energy methods in fluid dynamics and related theories
B Straughan
75 Local bifurcation and symmetry
A Vanderbauwhede
76 Clifford analysis
F Brackx, R Delanghe and F Sommen
77 Nonlinear equivalence, reduction of PDEs to ODEs and fast convergent numerical methods
E E Rosinger
78 Free boundary problems, theory and applications. Volume I
A Fasano and M Primicerio
79 Free boundary problems, theory and applications. Volume II
A Fasano and M Primicerio
80 Symplectic geometry
A Crumeyrolle and J Grifone
81 An algorithmic analysis of a communication model with retransmission of flawed messages
D M Lucantoni
82 Geometric games and their applications
W H Ruckle
83 Additive groups of rings
S Feigelstock
84 Nonlinear partial differential equations and their applications: Collège de France Seminar. Volume IV
H Brezis and J L Lions
85 Multiplicative functionals on topological algebras
T Husain
86 Hamilton-Jacobi equations in Hilbert spaces
V Barbu and G Da Prato
87 Harmonic maps with symmetry, harmonic morphisms and deformations of metrics
P Baird
88 Similarity solutions of nonlinear partial differential equations
L Dresner
89 Contributions to nonlinear partial differential equations
C Bardos, A Damlamian, J I Díaz and J Hernández
90 Banach and Hilbert spaces of vector-valued functions
J Burbea and P Masani
91 Control and observation of neutral systems
D Salamon
92 Banach bundles, Banach modules and automorphisms of C*-algebras
M J Dupré and R M Gillette
93 Nonlinear partial differential equations and their applications: Collège de France Seminar. Volume V
H Brezis and J L Lions
94 Computer algebra in applied mathematics: an introduction to MACSYMA
R H Rand
95 Advances in nonlinear waves. Volume I
L Debnath
96 FC-groups
M J Tomkinson
97 Topics in relaxation and ellipsoidal methods
M Akgül
98 Analogue of the group algebra for topological semigroups
H Dzinotyiweyi
99 Stochastic functional differential equations
S E A Mohammed

100 Optimal control of variational inequalities
V Barbu
101 Partial differential equations and dynamical systems
W E Fitzgibbon III
102 Approximation of Hilbert space operators. Volume II
C Apostol, L A Fialkow, D A Herrero and D Voiculescu
103 Nondiscrete induction and iterative processes
V Ptak and F-A Potra
104 Analytic functions – growth aspects
O P Juneja and G P Kapoor
105 Theory of Tikhonov regularization for Fredholm equations of the first kind
C W Groetsch
106 Nonlinear partial differential equations and free boundaries. Volume I
J I Díaz
107 Tight and taut immersions of manifolds
T E Cecil and P J Ryan
108 A layering method for viscous, incompressible L_p flows occupying R^n
A Douglis and E B Fabes
109 Nonlinear partial differential equations and their applications: Collège de France Seminar. Volume VI
H Brezis and J L Lions
110 Finite generalized quadrangles
S E Payne and J A Thas
111 Advances in nonlinear waves. Volume II
L Debnath
112 Topics in several complex variables
E Ramírez de Arellano and D Sundararaman
113 Differential equations, flow invariance and applications
N H Pavel
114 Geometrical combinatorics
F C Holroyd and R J Wilson
115 Generators of strongly continuous semigroups
J A van Casteren
116 Growth of algebras and Gelfand–Kirillov dimension
G R Krause and T H Lenagan
117 Theory of bases and cones
P K Kamthan and M Gupta
118 Linear groups and permutations
A R Camina and E A Whelan
119 General Wiener–Hopf factorization methods
F-O Speck
120 Free boundary problems: applications and theory, Volume III
A Bossavit, A Damlamian and M Fremond
121 Free boundary problems: applications and theory, Volume IV
A Bossavit, A Damlamian and M Fremond
122 Nonlinear partial differential equations and their applications: Collège de France Seminar. Volume VII
H Brezis and J L Lions
123 Geometric methods in operator algebras
H Araki and E G Effros
124 Infinite dimensional analysis–stochastic processes
S Albeverio
125 Ennio de Giorgi Colloquium
P Krée
126 Almost-periodic functions in abstract spaces
S Zaidman
127 Nonlinear variational problems
A Marino, L Modica, S Spagnolo and M Degiovanni
128 Second-order systems of partial differential equations in the plane
L K Hua, W Lin and C-Q Wu
129 Asymptotics of high-order ordinary differential equations
R B Paris and A D Wood
130 Stochastic differential equations
R Wu
131 Differential geometry
L A Cordero
132 Nonlinear differential equations
J K Hale and P Martinez-Amores
133 Approximation theory and applications
S P Singh
134 Near-rings and their links with groups
J D P Meldrum
135 Estimating eigenvalues with *a posteriori/a priori* inequalities
J R Kuttler and V G Sigillito
136 Regular semigroups as extensions
F J Pastijn and M Petrich
137 Representations of rank one Lie groups
D H Collingwood
138 Fractional calculus
G F Roach and A C McBride
139 Hamilton's principle in continuum mechanics
A Bedford
140 Numerical analysis
D F Griffiths and G A Watson
141 Semigroups, theory and applications. Volume I
H Brezis, M G Crandall and F Kappel
142 Distribution theorems of L-functions
D Joyner
143 Recent developments in structured continua
D De Kee and P Kaloni
144 Functional analysis and two-point differential operators
J Locker
145 Numerical methods for partial differential equations
S I Hariharan and T H Moulden
146 Completely bounded maps and dilations
V I Paulsen
147 Harmonic analysis on the Heisenberg nilpotent Lie group
W Schempp
148 Contributions to modern calculus of variations
L Cesari
149 Nonlinear parabolic equations: qualitative properties of solutions
L Boccardo and A Tesei
150 From local times to global geometry, control and physics
K D Elworthy

151 A stochastic maximum principle for optimal control of diffusions
U G Haussmann

152 Semigroups, theory and applications. Volume II
H Brezis, M G Crandall and F Kappel

153 A general theory of integration in function spaces
P Muldowney

154 Oakland Conference on partial differential equations and applied mathematics
L R Bragg and J W Dettman

155 Contributions to nonlinear partial differential equations. Volume II
J I Díaz and P L Lions

156 Semigroups of linear operators: an introduction
A C McBride

157 Ordinary and partial differential equations
B D Sleeman and R J Jarvis

158 Hyperbolic equations
F Colombini and M K V Murthy

159 Linear topologies on a ring: an overview
J S Golan

160 Dynamical systems and bifurcation theory
M I Camacho, M J Pacifico and F Takens

161 Branched coverings and algebraic functions
M Namba

162 Perturbation bounds for matrix eigenvalues
R Bhatia

163 Defect minimization in operator equations: theory and applications
R Reemtsen

Contributions to nonlinear partial differential equations
VOLUME II

J I Díaz & P L Lions (Editors)

Universidad Complutense de Madrid / Université de Paris-Dauphine

Contributions to nonlinear partial differential equations

VOLUME II

Longman Scientific & Technical

Copublished in the United States with
John Wiley & Sons, Inc., New York

Longman Scientific & Technical
Longman Group UK Limited
Longman House, Burnt Mill, Harlow
Essex CM20 2JE, England
and Associated Companies throughout the world.

Copublished in the United States with
John Wiley & Sons, Inc., 605 Third Avenue, New York, NY 10158

First published 1987

AMS Subject Classifications: (main) 35J65, 35K60, 35L60
(subsidiary) 49B60, 35P30, 58F14

ISSN 0269–3674

British Library Cataloguing in Publication Data
Contributions to nonlinear partial
differential equations : volume II.—
(Pitman research notes in mathematics
series, ISSN 0269–3674; 155)
1. Differential equations, Partial
2. Differential equations, Nonlinear
I. Díaz, J. I. II. Lions, P.L.
515.3′53 QA374

ISBN 0-582-44267-2

Library of Congress Cataloging-in-Publication Data
(Revised for volume 2)

Contributions to nonlinear partial differential
equations.

(Research notes in mathematics ; 89,)
Proceedings of an international meeting held in Madrid on Dec. 14–17, 1981.
Vol. 2 contains the proceedings of the Second Franco-Spanish Colloquium on Nonlinear Partial Differential Equations held in Paris in December 1985, edited by J.I. Díaz and P.L. Lions.
Includes bibliographical references.
1. Differential equations, Partial — Addresses, essays, lectures. 2. Nonlinear operators — Addresses, essays, lectures. I. Bardos, C. (Claude), 1940– .
II. Díaz, J.I. III. Lions, P.L. (Pierre Louis)
IV. Franco-Spanish Colloquium on Nonlinear Differential Equations (2nd : 1985 : Paris, France) V. Series.
VI. Series: Pitman research notes in mathematics series ;
89, etc.
QA377.C763 1983 515.3′53 83-8112
ISBN 0-273-08595-6 (pbk. : v. 1)
ISBN 0-470-20810-4 (pbk. : v. 2)(USA only)

Printed and bound in Great Britain by
Biddles Ltd, Guildford and King's Lynn

Contents

Preface

This volume contains the proceedings of the *Second Franco-Spanish Colloquium on Nonlinear Partial Differential Equations* held in Paris in December 1985.

As for the first colloquium, held in Madrid in December 1981, the aim was to bring together French and Spanish workers in the field of partial differential equations. (The proceedings of the first meeting are contained in volume 89 of the *Research Notes* series, edited by C. Bardos, A. Damlamian, J.I. Díaz and J. Hernández).

The invited contributors survey the state of the art and indicate new directions of developments in the treatment of genuine nonlinear partial differential equations related to realistic problems in physics or engineering.

The organization of the meeting was made possible by the cooperation programme of the French and Spanish governments, who supported the 'action intégrée' involving the universities of Paris-Dauphine and the Complutense of Madrid, and the scientific programme of the French Embassy in Madrid. We would also like to thank Project No. 3308/83 of the CAICYT (Spain) and the French Minister of Foreign Affairs for their support.

Finally, we express our gratitude to Longman Scientific & Technical for their technical support in the publication and distribution of this volume.

J.I. Díaz
P.L. Lions
Madrid & Paris,
October 1986.

List of contributors

M. ARIAS
Departamento de Ecuaciones Funcionales, Universidad de Granada, Granada, Spain.

G. BARLES
CEREMADE, Université Paris-Dauphine, Paris, France.

L. BARTHELEMY
Equipe de Mathématiques de Besançon, Université de Franche-Comté, Besançon, France.

A. BENSOUSSAN
Université Paris-Dauphine and INRIA, Paris, France.

J. BLAT
Departament de Matemàtiques, Universitat Illes Balears, Palma de Mallorca,Spain.

H. BREZIS
Départment de Mathématiques, Université Pierre et Marie Curie, Paris, France.

J. CARRILLO
Departamento de Matématica Aplicada, Universidad Complutense de Madrid, Spain.

T. CAZENAVE
Laboratoire d'Analyse Numérique, Université Pierre et Marie Curie, Paris,France.

M. CHIPOT
Départment de Mathématiques, Université de Metz, Metz, France.

G. DÍAZ
Facultad de Matemáticas, Universidad Complutense de Madrid, Madrid, Spain.

J. ESQUINAS
Departamento de Matematica Aplicada, Universidad Complutense de Madrid, Madrid, Spain.

M.J. ESTEBAN
Laboratoire d'Analyse Numérique, Université Pierre et Marie Curie, Paris, France.

E. FERNÁNDEZ-CARA
Universidad de Sevilla, Spain.

J. FRAILE
Departamento de Matematica Aplicada, Universidad Complutense de Madrid, Madrid, Spain.

J. HERNÁNDEZ
Departamento de Matemáticas, Universidad Autónoma, Madrid, Spain.

J.M. LASRY
CEREMADE, Université Paris-Dauphine, Paris, France.

P.L. LIONS
CEREMADE, Université Paris-Dauphine, Paris, France.

J. LÓPEZ-GÓMEZ
Departamento de Matematica Aplicada, Universidad Complutense de Madrid, Madrid, Spain.

P. MARTINEZ-AMORES
Departamento de Ecuaciones Funcionales, Universidad de Granada, Granada, Spain.

Y. MEYER
CEREMADE, Université Paris-Dauphine, Paris, France.

X. MORA
Departament de Matemàtiques, Universitat Autònoma de Barcelona, Barcelona, Spain.

J-M. MOREL
CEREMADE, Université Paris-Dauphine, Paris, France.

C. MORENO
Universidad Autónoma, Madrid, Spain.

R. ORTEGA
Departamento de Ecuaciones Funcionales, Universidad de Granada, Granada, Spain.

L. OSWALD
Laboratoire d'Analyse Numérique, Université Pierre et Marie Curie, Paris, France.

I. PERAL ALONSO
División de Matemáticas, Facultad de Ciencias, Universidad Autónoma, Madrid, Spain.

B. PERTHAME
Ecole Normale Superieure, 45, rue D'ulm, Paris, France.

M. PIERRE
Département de Mathématiques, Université de Nancy I, Vandoeuvre-les-Nancy, France.

J. SABINA
ETSI de Montes, Universidad Politécnica de Madrid, Madrid, Spain.

JUAN L. VAZQUEZ
Departamento de Matemáticas, Universidad Autónoma, Madrid, Spain

L. VÁZQUEZ
Departamento de Física Teórica, Facultad de Ciencias Físicas, Universidad Complutense, Madrid, Spain.

J.M. VEGAS
Departamento de Matematica Aplicada, Universidad Complutense, Madrid, Spain.

L. VERON
Département de Mathématiques, Faculté des Sciences, Parc de Grandmont, Tours, France.

M ARIAS, P MARTINEZ–AMORES & R ORTEGA

Some notes on a forced semilinear wave equation

This note is a preliminary communication of the results in [1].

We consider the existence problem of weak doubly 2π-periodic solutions of the semilinear wave equation

$$u_{tt} - u_{xx} + g(u) = f(t,x), \quad (t,x) \in R^2, \tag{1}$$

where f is a given 2π-periodic function in t and x and g is a continuous non-decreasing function.

By a weak solution of this problem we understand a function $u \in L^\infty(J)$, $J = (0,2\pi) \times (0,2\pi)$, such that

$$(u,\Box\phi) + (g(u),\phi) = (f,\phi), \quad \phi \in \mathcal{D} \tag{2}$$

where $\Box = \frac{\partial^2}{\partial t^2} - \frac{\partial^2}{\partial x^2}$, $(.,.)$ is the inner product of $L^2(J)$ and $\mathcal{D}$ is the class of functions of $C^2(\bar{J})$ which are 2π-periodic in t and x (we identify, whenever needed, a function on J to its doubly-periodic extension to R^2).

A function $u \in L^\infty(J)$ verifies (2) if and only if u verifies

$$Au + g(u) = f. \tag{3}$$

Here A is the abstract realization in $L^2(J)$ of the wave operator with doubly-periodic conditions (see [1]).

We now present our result.

<u>THEOREM</u> Assume that the following conditions are verified.

(a) $g(-\infty) > -\infty$, $g(+\infty) = +\infty$.

(b) $f \in L^\infty(J)$ admits a decomposition in the form $f = f^* + f^{**}$ with $f^* \in R(A) \cap L^\infty(J)$, $f^{**}(t,x) \geqq g(-\infty) + \delta$, a.e. $(t,x) \in J$, for some $\delta > 0$.

Then there exists at least one solution of (3) in $L^\infty(J)$.

This result has been motivated by the work of Bahri and Brezis [2] (see also [3]) on the Dirichlet-periodic problem for (1). Their corresponding condition on g is

(a') $|g(u)| \leq \gamma|u| + c$, γ, c constants with $\gamma < |\lambda_{-1}|$.

Here λ_{-1} is the first negative eigenvalue of the operator $\square$ when it acts on functions satisfying the boundary conditions.

We remark that condition (a') does not allow the crossing of g and the eigenvalues of $\square$ different of $\lambda_0 = 0$.

Also, the growth of g must be of linear type at most and limited by γ. Although, as indicated before, [2] deals with the Dirichlet-periodic problem, the results in [2] can be translated to the doubly-periodic case.

In contrast with (a'), condition (a) allows that g may cross other eigenvalues besides λ_0 or grow arbitrarily in the positive direction. However, g must be bounded in the negative direction, which is not required by (a').

Another result related to ours is given by Ward in [4], where g is not allowed to interact with λ_0 and verifying a condition of linear growth of type (a') with arbitrary γ.

The idea of the proof of the Theorem is simple. For details, see [1]. We consider the sequence of truncated equations

$$Au + g_n(u) = f, \quad n = 1,2,... \tag{4}$$

where $g_n(u) = \min[g(u),g(n)]$, and we apply the results of [2] to each one of them to prove the existence of a solution $u_n \in L^\infty(J)$ of (4) for large n. Then we obtain a L^∞-bound of u_n independent of n, implying therefore that $g_n(u_n) = g(u_n)$ and, thus, u_n is a solution of (3) for large n. The method used to get the bounds is based in [2] combined with estimates similar to those of [4].

<u>Some remarks</u>

1. Some nonlinearities verifying (a) are $\alpha u^+ (\alpha \geq 1)$, u^+u^2, e^u. The results in [2], [4] do not apply to these examples. However, we cannot study a nonlinear term of the type $\alpha u^+ - \beta u^-$ $(0 < \alpha, \beta < 1)$, that can be studied from the results in [2] or [4].

2. Condition (b) was first formulated in [2,3] and is a sharp condition for

the solvability of (3). In fact, when g is strictly increasing then (b) characterizes the solvability of (3). However, when $g(-\infty) = g(u), u \leq c$ for some c, (b) is only sufficient. A necessary condition for the solvability is

$$f \in L^{\infty}(J) \text{ admits a decomposition in the form } f^{*} + f^{**} \text{ with}$$

$$f^{*} \in R(A) \cap L^{\infty}(J),\ f^{**}(t,x) \geq g(-\infty),\ \text{ a.e. } (t,x) \in J.$$

We do not know whether this last condition is also sufficient in this case or not.

3. Our proof is based in the existence of positive functions in the kernel of A. Therefore it is not possible to adapt the proof to the Dirichlet-periodic problem for equation (1). However, it seems possible to obtain some results for this last problem in $(0,2\pi) \times (0,\pi)$ when the resonance is at $\lambda_1 = 1$ since $\phi_1(t,x) = \sin x$ belongs to $N(A-\lambda_1 I)$ and is positive.

References

1. Arias, M., Martinez-Amores, P. and Ortega, R. Doubly periodic solutions of a forced semilinear wave equation. To appear in Proc. Amer. Math. Soc.
2. Bahri, A. and Brezis, H. Periodic solutions of a nonlinear wave equations. Proc. Roy. Soc. Edimburg. 85 (1980), 313-320.
3. Brezis, H. Periodic solutions of nonlinear vibrating strings and duality principles. Bull. Amer. Math. Soc. 8 (1983), 409-426.
4. Ward, J. A wave equation with a possible jumping nonlinearity. Proc. Amer. Math. Soc. 92 (1984), 209-214.

The authors wish to acknowledge the support from CAICYT, Ministerio de Educación y Ciencia, Spain.

M. Arias, P. Martinez-Amores and R. Ortega
Departamento de Ecuaciones Funcionales
Universidad de Granada
18071 Granada
Spain.

G BARLES & P L LIONS

Remarks on existence and uniqueness results for first–order Hamilton–Jacobi equations

Introduction

We consider here existence and uniqueness results for bounded uniformly continuous viscosity solutions of

$$H(x,u,Du) = 0 \text{ in } \Omega \quad ; \quad u = \phi \text{ on } \partial\Omega \tag{1}$$

that we call the Dirichlet problem for Hamilton-Jacobi equations and

$$\begin{cases} \frac{\partial u}{\partial t} + H(x,t,u,Du) = 0 \quad \text{in} \quad \Omega \times]0,T[\\ u(x,0) = u_o(x) \text{ in } \Omega \; ; \; u(x,t) = \phi(x,t) \text{ on } \partial\Omega \times]0,T[\end{cases} \tag{2}$$

that we call the Cauchy problem for Hamilton-Jacobi equations. In both cases, Ω is any open subset of $\mathbb{R}^N$, H is at least a continuous function, ϕ and u_o are given continuous functions (boundary conditions). When $\Omega = \mathbb{R}^N$ there is no boundary condition: we require only u to be bounded.

The results proved here are based upon the notion of viscosity solution introduced by M.G. Crandall and P.L. Lions [3]. For a complete presentation of this notion of solution and of its properties, the reader can refer also to [3], M.G. Crandall, L.C. Evans and P.L. Lions [4], P.L. Lions [16].

Existence results for bounded uniformly continuous viscosity solutions of (1) or (2) are essentially obtained under two types of assumptions: one concerns the behaviour of H in x (cf. P.L. Lions [16, 17]; P.E. Souganidis [18], G. Barles [1], H. Ishii [11, 12] and M.G. Crandall and P.L. Lions [6]), the other concerns the behaviour of H in p (essentially $H \to +\infty$ if $|p| \to +\infty$) (cf. P.L. Lions [16, 17], G. Barles [2]). In this work, we give results under these two types of assumptions.

In the first part, we treat the case of the Dirichlet problem. One of the main remarks in [1] was that, when we are interested in viscosity solutions in $BUC(\bar{\Omega})$(*), we may assume that H is bounded for u bounded. Using this remark, we are able to relax the classical assumption on the modulus of continuity of H in x by assuming that it depends strongly in H. Under this relaxed assumption, we obtain existence and comparison results in $BUC(\bar{\Omega})$. Moreover, we show that this result contains also all the existence results obtained under the assumption $H \to +\infty$ if $|p| \to +\infty$.

In the second part we consider the case of the Cauchy problem. As in the first part, we can relax the assumption on the modulus of continuity of H in x but in a less general way than in the first part. Moreover, as we cannot prove a comparison result under this relaxed assumption, we assume that the viscosity subsolution $\underline{u}$ and supersolution $\bar{u}$ (the existence of which is a classical assumption) satisfy $\underline{u} \leq \bar{u}$. Then, we show that there exists a unique viscosity solution u in $BUC(\bar{\Omega} \times [0,T])$ which satisfies $\underline{u} \leq u \leq \bar{u}$. Finally, as this result does not contain all the results obtained by assuming $H \to \infty$ if $|p| \to +\infty$, we give an extension of the results proved by P.L. Lions [16, 17] under this type of assumption.

I. Existence and uniqueness results for the Dirichlet problem

We will use the following assumptions

$$\left.\begin{array}{l} H \in C(\bar{\Omega} \times \mathbb{R} \times \mathbb{R}^N) \\ \text{For each } R > 0 \text{, there is a modulus } n_R \text{ such that} \\ \text{for all } x, y \in \bar{\Omega},\ |r| \leq R,\ p, q \in B_R \\ |H(x,r,p) - H(y,r,q)| \leq n_R(|x-y| + |p-q|) \end{array}\right\} \quad (3)$$

$$\left.\begin{array}{l} \forall R < \infty,\ \exists \gamma_R > 0 \text{ such that} \\ H(x,t,p) - H(x,s,p) \geq \gamma_R(t-s) \\ \text{for } x \in \bar{\Omega},\ -R \leq s \leq t \leq R,\ p \in \mathbb{R}^N \end{array}\right\} \quad (4)$$

(*) $BUC(\bar{\mathcal{O}})$ = {u bounded uniformly continuous on $\bar{\mathcal{O}}$}

$\forall\, R < \infty$, there exists a modulues m_R and $\phi_R \in C(\mathbb{R}^2)$ such that

$$\left.\begin{aligned} &H(x,t,\lambda(x-y)) - H(y,t,\lambda(x-y)) \geqq \\ &\qquad \geqq -m_R(|x-y| + \lambda|x-y|^2)\cdot\phi_R(H(x,t,\lambda(x-y));\ H(y,t,\lambda(x-y)) \\ &\text{for } x,y \in \bar{\Omega},\ \lambda \geqq 0 \text{ and } -R \leqq t \leqq R. \end{aligned}\right\} \qquad (5)$$

REMARK I.1: The assumption (5) is a weaker requirement on H than the classical one in which $\phi_R \equiv 1$. For example, (5) contains Hamiltonians like

$$H(x,t,p) = b(x)|(Ap/p)|^m + t - f(x)$$

where $b,f \in BUC(\bar{\Omega})$, $b \geqq \eta \geqq 0$, $p \in \mathbb{R}^N$, $m > 0$ and A is a $N \times N$ matrix. It is easy to see that we can take

$$\phi_R(z_1,z_2) = \max\,(|z_1|,|z_2|) + R + \|f\|_{L^\infty(\bar{\Omega})}.$$

Our results are the following:

THEOREM I.1: Let H satisfy (3), (4), (5).

COMPARISON: Let v and $w \in BUC(\bar{\Omega})$ be respectively viscosity sub and supersolution of (1) and of

$$H(x,w,Dw) + f = 0 \text{ in } \Omega.$$

Then

$$\|(v-w)^+\|_{L^\infty(\bar{\Omega})} \leqq \max\left(\frac{\|f^+\|_{L^\infty(\bar{\Omega})}}{\gamma}\ ;\ \|(v-w)^+\|_{L^\infty(\partial\Omega)}\right) \qquad (6)$$

where $\gamma = \gamma_R$ defined in the assumption (4) with $R = \max\left(\|v\|_{L^\infty(\bar{\Omega})}\ ;\ \|w\|_{L^\infty(\bar{\Omega})}\right)$.

EXISTENCE: Assume that there exists $\underline{u}$ and $\bar{u} \in BUC(\bar{\Omega})$ respectively sub and supersolution of (1) with $\underline{u} = \bar{u} = \phi$ on $\partial\Omega$, then there exists a unique viscosity solution u of (1) in $BUC(\bar{\Omega})$.

COROLLARY I.1: If $\Omega = \mathbb{R}^N$, under assumptions (3), (4), (5) and if there exists $M \geq 0$ such that

$$H(x,M,0) \geq 0 \text{ and } H(x,-M,0) \leq 0, \quad \forall x \in \mathbb{R}^N \tag{7}$$

then there exists a unique viscosity solution u in $BUC(\mathbb{R}^N)$ of

$$H(x,u,Du) = 0 \quad \text{in} \quad \mathbb{R}^N. \tag{8}$$

REMARK I.2: These results extends those obtained previously by P.E. Souganidis [18], H. Ishii [11, 12] and the authors [1], [17].

REMARK I.3: It is worth noticing that our existence results contains all the existence results obtained under the assumptions (3), (4) and

$$H(x,t,p) \to +\infty \quad \text{when } |p| \to +\infty \quad \text{uniformly for } x \in \bar{\Omega} \tag{9}$$

and $|t| \leq R$, $\forall R < \infty$.

In this case, it is easy to see that (5) holds with, for example:

$$\phi_R(z_1,z_2) = [\max(|z_1|, |z_2|)]^2 + 1.$$

REMARK I.4: In the case when H is convex, the existence of the subsolution $\underline{u}$ is discussed in S.N. Kruzkov [13, 14, 15], W.H. Fleming [8] and P.L. Lions [16]. (In [16], a necessary and sufficient condition for the existence of $\underline{u}$ is given).

We now turn to the proof of Theorem I.1.

PROOF OF THEOREM I.1: The idea of the proof is based on the truncation argument given in [1].

Proof of the comparison result: We will denote by $a \wedge b = \text{Inf}(a,b)$ and $a \vee b = \sup(a,b)$. Let us define $\tilde{H}$ by

$$\tilde{H}(x,t,p) = -M \vee (H(x,t,p)-t) \wedge M + t.$$

The main remark of the truncation argument is that if M is greater than $R + \|f\|_{L^\infty(\Omega)}$, R defined in Theorem I.1, then v and w are still respectively viscosity sub and supersolution of

$$\tilde{H}(x,u,Du) = 0 \quad \text{in } \Omega$$

and

$$\tilde{H}(x,u,Du) + f(x) = 0 \text{ in } \Omega .$$

Moreover $\tilde{H}$ satisfies (3), (4) with the same γ_R and finally it satisfies (5) with $\phi_R \equiv 1$ (or equivalently depending only on R). So we can apply to $\tilde{H}$ the classical comparison results for BUC viscosity sub and supersolutions (cf. M.G. Crandall and P.L. Lions [3, 6], M.G. Crandall, L.C. Evans and P.L. Lions [4], P.L. Lions [16], H. Ishii [12], M.G. Crandall, H. Ishii and P.L. Lions [5]). These results yield (6).

PROOF OF THE EXISTENCE RESULT: First, let us remark that the comparison result proved above implies $\underline{u} \leq \bar{u}$. Moreover, if u is a viscosity solution of (1), the comparison result gives:

$$\underline{u} \leq u \leq \bar{u} .$$

And so, we can apply the truncation argument of [1] which implies that the problem (1) and the problem

$$\left.\begin{aligned} &\tilde{H}(x,u,Du) = 0 \quad \text{in } \Omega \\ &u = \phi \qquad\qquad \text{on } \partial\Omega \end{aligned}\right\}$$

where

$$\tilde{H}(x,t,p) = (-\bar{u}(x)) \vee (H(x,t,p)-t) \wedge (-\underline{u}(x)) + t$$

have the same viscosity solution in $BUC(\bar{\Omega})$. But now H satisfies (3), (4), (5) with $\phi_R \equiv 1$ (or depending only on R). To conclude, it suffices to extend the problem in to a problem in $\mathbb{R}^N$ as in [1] and to use the existence results of H. Ishii [11, 12] or M.G. Crandall and P.L. Lions [6]. And the proof is complete.

II. Existence results for the Cauchy problem

In this section, we obtain existence results for the Cauchy problem under the two classical types of assumptions described in the introduction. We will use the following assumptions.

$$H \text{ is bounded uniformly continuous on } \bar{\Omega} \times [0,T] \times (-R,R) \times \bar{B}_R \quad (\forall R < \infty) \tag{10}$$

$$\left.\begin{aligned} &\forall R < \infty, \ \exists \gamma_R \in \mathbb{R} \text{ such that} \\ &H(x,t,u_2,p) - H(x,t,u_1,p) \geqq \gamma_R(u_2-u_1) \\ &\text{for } x \in \bar{\Omega},\ t \in [0,T],\ -R \leqq u_1 \leqq u_2 \leqq R,\ p \in \mathbb{R}^N \end{aligned}\right\} \tag{11}$$

$\forall R < \infty$, there exists a modulus m_R and $C_R \in \mathbb{R}$ such that

$$\left.\begin{aligned} &H(x,t,u,\lambda(x-y)) - H(y,t,u,\lambda(x-y)) \geqq \\ &\qquad \geqq -m_R(|x-y| + \lambda|x-y|^2 + C_R|x-y|\cdot|H(y,t,u,\lambda(x-y))|) \\ &\text{for all } x,y \in \bar{\Omega},\ t \in [0,T],\ |u| \leqq R \end{aligned}\right\} \tag{12}$$

$$\left.\begin{aligned} &H(x,t,u,p) \to +\infty \ \text{ when } |p| \to +\infty \text{ uniformly in } x \in \bar{\Omega} \\ &t \in [0,T],\ |u| \leqq R,\ (\forall R < \infty). \end{aligned}\right\} \tag{13}$$

$$\left.\begin{aligned} &H \text{ is Lipschitz in } t \text{ for all } x \in \Omega,\ u \in \mathbb{R},\ p \in \mathbb{R}^N \text{ and} \\ &H_t(x,t,u,p) \leqq C_R^1 + C_R^2\, H(x,t,u,p),\ C_R^1 \in \mathbb{R},\ \ C_R^2 > 0, \text{ for} \\ &x \in \Omega,\ t \in\,]0,T[,\ |u| \leqq R,\ \ (\forall R < \infty),\ p \in \mathbb{R}^N \end{aligned}\right\} \tag{14}$$

REMARK II.1: The assumption (12) is a weaker requirement on H than the classical one in which $C_R \equiv 0$. For example, it involves Hamiltonians of Eikonal type like

$$H(x,t,u,p) = \Big(\sum_{i,j} a_{i,j}(x,t)p_i p_j \Big)^{m/2} + u + f(x,t)$$

where

$$a_{i,j}, f \in BUC(\bar{\Omega} \times [0,T])$$

and

$$\sum_{i,j} a_{i,j}(x,t)p_i p_j \geqq \eta|p|^2, \qquad \eta > 0.$$

We denote by $Q_T = \Omega \times]0,T[$, $\partial_0 Q_T = \bar{\Omega} \times \{0\} \cup (\partial\Omega \times [0,T])$, then we have the following results.

THEOREM II.1: Under assumptions (10), (11), (12) and if there exists $\underline{u}$ and $\bar{u} \in BUC(\bar{Q}_T)$ respectively viscosity sub and supersolution of (2) such that $\underline{u} \leqq \bar{u}$ in Q_T, $\underline{u} = \bar{u} = \phi$ on $\partial\Omega \times [0,T]$ and $\underline{u} = \bar{u} = u_0$ in $\Omega \times \{0\}$, then there exists a unique viscosity solution u of (2) in $BUC(\bar{\Omega} \times [0,T])$ which satisfies

$$\underline{u} \leqq u \leqq \bar{u} \quad \text{in} \quad \bar{Q}_T \tag{15}$$

COROLLARY II.1: If $\Omega = \mathbb{R}^N$, under assumptions (10), (11), (12) and if $u_0 \in BUC(\mathbb{R}^N)$ then there exists T_0 $(0 < T_0 \leqq T)$ which depends only on $\|u_0\|_{L^\infty(\mathbb{R}^N)}$ and $u \in BUC(\mathbb{R}^N \times [0,T_0])$ viscosity solution of

$$\left.\begin{aligned} &\frac{\partial u}{\partial t} + H(x,t,u,Du) = 0 \text{ in } \mathbb{R}^N \times (0,T_o) \\ &u(x,0) = u_0(x) \text{ in } \mathbb{R}^N \end{aligned}\right\} \tag{16}$$

Moreover if there exists $C \in \mathbb{R}$ such that $\gamma_R \geqq C$ $(\forall R < +\infty)$ then $T_0 = T$.

The assumption (12) is less general than the corresponding one in the first part for the Dirichlet problem. A consequence is that this result does not contain all the results obtained under assumptions like (13). For that case, we give the following result.

THEOREM II.2: Let H satisfy (10), (11), (13), (14). Assume that there exists $\underline{u} \in W^{1,\infty}(\bar{Q}_T)$ and $\bar{u} \in BUC(\bar{Q}_T)$ respectively viscosity sub and super-solution of (2) with $\bar{u} \geqq \underline{u} = \phi$ on $\partial\Omega \times [0,T]$ and $\bar{u} \geqq \underline{u} = u_0$ in $\Omega \times \{0\}$ then there exists a unique viscosity solution u of (2) in $W^{1,\infty}(\bar{Q}_T)$.

COROLLARY II.2: If $\Omega = \mathbb{R}^N$, under assumptions (10), (11), (13), (14) and if $u_o \in BUC(\mathbb{R}^N)$, then there exists T_o $(0 < T_o \leq T)$ which depends only on $\|u_o\|_{L^\infty(\mathbb{R}^N)}$ and $u \in BUC(\mathbb{R}^N \times [0,T_o])$ viscosity solution of (16).

REMARK II.2: In Theorem II.1, the assumption $\underline{u} \leq \bar{u}$ is needed since the assumptions on H do not imply comparison result in $BUC(\bar{Q}_T)$ if $C_R \not\equiv 0$. This result complements those obtained by P.E. Souganidis [18], H. Ishii [11, 12], M.G. Crandall and P.L. Lions [6] and the authors [1], [16, 17], by assuming roughly speaking that $C_R \equiv 0$. The result of Theorem II.2 extends existence results obtained by the authors by assuming $C_R^2 \equiv 0$ (cf. [2], [16,17]).

REMARK II.3: As in Remark I.5, it is worth mentioning that, when H is convex, the existence of $\underline{u}$ is discussed in W.H. Fleming [7, 8, 9], A. Friedman [10], S.N. Kruzkov [13, 14, 15] and P.L. Lions [16]. (In [16], a necessary and sufficient condition for the existence of $\underline{u}$ is given).

Now we prove Theorem II.1:

PROOF OF THEOREM II.1: The proof is based upon the truncation method of [1]. We search a viscosity solution of (2) satisfying (15). As in [1], we can reduce to the case when $\gamma_R \equiv 1$ and then we can consider the Hamiltonian H defined by

$$\tilde{H}(x,t,u,P) = (-\bar{u}(x,t)) \vee (p_{n+1} + H(x,t,u,p)-u) \wedge (-\underline{u}(x,t)) + u \qquad \text{if } t \leq T$$

and

$$\tilde{H}(x,t,u,P) = \tilde{H}(x,T,u,P) \qquad \text{if } t \geq T$$

where $P = (p,p_{n+1})$. Extending for $t \geq T$, $\underline{u}$ and $\bar{u}$ by respectively $\underline{u}(x,T)$ and $\bar{u}(x,T)$, the shape of $\tilde{H}$ implies that the extensions - still denoted by $\underline{u}$ and $\bar{u}$ - are respectively viscosity sub and supersolution in $Q_\infty = \Omega \times]0, +\infty[$. Moreover $\tilde{H}$ satisfies (10) and (11); it suffices to verify (12). From [1], we already know $\tilde{H}$ satisfies that there exists a modulus $\tilde{m}$ such that

$$|\tilde{H}(x,t,u,P) - \tilde{H}(x,s,u,P)| \leq \tilde{m}(|t-s|(1 + |P|))$$

($\tilde{m}$ does not depend on R after the reductions, we have made on [1]). So, it suffices to examine the behaviour in x since the property above in t is stronger than (12). Let us denote by m the modulus m_R where $R = \max\left(\|\underline{u}\|_{L^\infty(Q_T)} ; \|\bar{u}\|_{L^\infty(Q_T)}\right)$. To have a property like (12) for $\tilde{H}$ is equivalent to have (12) for F where F is given by

$$F(x,t,u,P) = p_{n+1} + H(x,t,u,p).$$

Now we compute $F(x,t,u,\lambda(x-y)) - F(y,t,u,\lambda(x-y))$. It is equal to

$$H(x,t,u,\lambda(x-y)) - H(y,t,u,\lambda(x-y))$$

which is greater than

$$-m(|x-y|) + \lambda|x-y|^2 + |x-y|C\cdot|H|)$$

where $C = C_R$. Now it suffices to estimate $|H|$ by $|p_{n+1}| + |F|$ which gives

$$-m(|x-y| + \lambda|x-y|^2 + C|x-y|\cdot|p_{n+1}| + C|x-y|\cdot|F|).$$

In fact, the term $C|x-y|\cdot|p_{n+1}|$ explains the limitation for the assumption (12). And the last inequality obtained shows that F, hence $\tilde{H}$ satisfies (12), so by Theorem I.1 there exists a viscosity solution u in $BUC(\bar{Q}_\infty)$ of

$$\left.\begin{aligned} &\tilde{H}(x,t,u,(Du, \tfrac{\partial u}{\partial t})) = 0 \quad \text{in } Q_\infty \\ &u = \underline{u} \text{ on } \partial_0 Q \quad . \end{aligned}\right\} \qquad (17)$$

Moreover since $\underline{u} \leqq u \leqq \bar{u}$, u is also a viscosity solution of (2). Finally u is the unique viscosity solution of (2) satisfying $\underline{u} \leqq u \leqq \bar{u}$ because a viscosity solution of (2) is a viscosity solution of (17) at least when we have $\underline{u} \leqq u \leqq \bar{u}$ and uniqueness holds for (17). This ends the proof.

We do not give the proof of Corollary II.1 which is exactly the same as the corresponding result in [1].

PROOF OF THEOREM II.2: If u is a viscosity solution of (2) in $W^{1,\infty}(\bar{Q}_T)$ then by comparison results for the Cauchy problem (see [3], [4], [15] ... etc.), we have

$$\underline{u} \leqq u \leqq \bar{u} .$$

To have the second inequality, it is a classical remark that to have a comparison result under (10), (11), it suffices that one of the functions we want to compare is in $W^{1,\infty}(\bar{Q}_T)$. Then by classical reductions, we may assume that γ_R, C_R^1, C_R^2 do not depend on R. We call them γ, C^1, C^2. Now we make the change of variable $t = \frac{1}{C^2} \text{Log}\,(1+C^2 s)$ and we consider

$$v(x,s) = u\left(x, \frac{1}{C^2} \text{Log}\,(1+C^2 s)\right) \quad (\text{resp. } \underline{v}, \bar{v} \ldots) .$$

The Hamiltonian H is changed in F defined by

$$F(x,s,u,p) = \frac{1}{1+C^2 s} \, H\left(x, \frac{\text{Log}\,(1+C^2 s)}{C^2}, u, p\right).$$

And if we compute F_s, we have

$$F_s(x,s,u,p) = H_t\left(x, \frac{\text{Log}\,(1+C^2 s)}{C^2}, u,p\right) (1+C^2 s)^{-2} +$$

$$- \frac{C^2}{(1+C^2 s)^2} \, H\left(x, \frac{\text{Log}\,(1+C^2 s)}{C^2}, u,p\right)$$

Then using (14)

$$F_s \leqq C^1 (1+C^2 s)^{-2} \leqq C^1 .$$

Therefore by result of P.L. Lions [17] there exists a unique v in $W^{1,\infty}(\bar{Q}_T)$ viscosity solution of

$$\left.\begin{aligned} &v_s + F(x,s,v,Dv) = 0 \quad \text{in} \quad Q_s \\ &v = \underline{v} \quad \text{on } \partial_0 Q_s \end{aligned}\right\}$$

where $S = (\exp(C^2T)-1)\cdot(C^2)^{-1}$ and the associated u is the unique viscosity solution of (2) in $W^{1,\infty}(\bar{Q}_T)$.

We do not give the proof of Corollary II.2 which is an easy adaptation of the ideas of [1].

Bibliography

1. G. Barles: Existence results for first order Hamilton-Jacobi Equations. Ann. I.H.P. Anal. non lin., vol. 15 (1984).
2. G. Barles: Remarks on existence results for first order Hamilton-Jacobi Equations. Ann. I.H.P. Anal. non lin., 2 (1985), 21-33.
3. M.G. Crandall and P.L. Lions: Viscosity solutions of Hamilton-Jacobi Equations. Trans. AMS, 277, (1983), p. 1-42.
4. M.G. Crandall, L.C. Evans and P.L. Lions: Some properties of viscosity solutions of Hamilton-Jacobi Equations. Trans. AMS <u>282</u> (1984), p. 487-502.
5. M.G. Crandall, H. Ishii and P.L. Lions: Uniqueness of viscosity solutions revisited, (to appear).
6. M.G. Crandall and P.L. Lions: On existence and unqiueness of solutions of Hamilton-Jacobi Equations. To appear in Non Linear Analysis T.M.A..
7. W.H. Fleming: The Cauchy problem for a non linear partial differential equation. J. Diff. Equations, 5 (1969) p. 515-530.
8. W.H. Fleming: Non linear partial differential equations - probabilistic and games theoretic method. In Problems in Non Linear Analysis. CIME ed. Cremonese, Rome (1979).
9. W.H. Fleming: The Cauchy problem for degenerate parabolic equations. J. Math. Rech. 13 (1964) p. 987-1008.
10. A. Friedman: The Cauchy problem for first order partial differential equations. Indiana Univ. Math. J. 23 (1973) p. 27-40.
11. H. Ishii: Remarks on the existence of viscosity solutions of Hamilton-Jacobi equations, Bull. Facul. Sci. Eng. Chuo University, 26 (1983), 5-24.
12. H. Ishii: Existence and Uniqueness of solutions of Hamilton-Jacobi Equations, preprint.
13. S.N. Kruzkov: Generalized solutions of non linear first order equations and certain quasilinear parabolic equations. Vestnik Moscow Univ. Ser. I Math. Rech. 6 (1964) p. 67-74 (in Russian).

14. S.N. Kruzkov: Generalized solutions of Hamilton-Jacobi Equations of Eikonal type. USSR Sbornik, 27, (1975) p. 406-446.

15. S.N. Kruzkov: Generalized solutions of first order non linear equations in several independent variables. I. Mat. Sb. 70 (112), (1966), p. 394-415, II. Mat. Sb. (NS) 72 (114) (1967) p. 93-116 (en russe).

16. P.L. Lions: Generalized Solutions of Hamilton-Jacobi Equations; Pitman, London (1982).

17. P.L. Lions: Existence results for first order Hamilton-Jacobi Equations. Richerche Mat. Napoli, 32, (1983), 1-23.

18. P.E. Souganidis: Existence of viscosity solutions of Hamilton-Jacobi Equations. J. Diff. Eq., 56, (1985) 345-390.

G. Barles and P.L. Lions
CEREMADE
Université Paris IX - Dauphine
Place de Lattre de Tassigny
75775 Paris CEDEX 16
France.

L BARTHELEMY

Application de la théorie des operateurs sous–potentiels au contrôle stochastique

Introduction

Nous nous intéressons au problème suivant correspondant à différentes situations de contrôle stochastique. Etant donnés $\{N_t^i, t > 0\}_{i \in I}$ une famille de semi-groupes sous-markoviens sur un espace X de fonctions boréliennes bornées, $(f^i)_{i \in I}$ une famille d'éléments de X, $(M_j)_{j \in J}$ une famille d'applications de X dans X et $\lambda > 0$, trouver $u \in X$, solution maximum de

$$\left.\begin{array}{l} u \leqq \int_0^t e^{-s} N_{\lambda s}^i . f^i \, ds + e^{-t} N_{\lambda t}^i . u \quad \forall t > 0 \quad \forall i \in I \\ u \leqq M_j u \quad \forall j \in J. \end{array}\right\} \quad (1)$$

A. Bensoussan et M. Robin ont étudié ce problème dans [2] en utilisant une méthode de discrétisation et ont montré des résultats d'existence sous certaines hypothèses sur les données, card I fini et card J = 1.

Dans cet article, nous montrons des résultats d'existence plus généraux sous des hypothèses assez faibles, en utilisant la théorie des opérateurs sous-potentiels de type accrétif développée dans [1].

Dans la section I, nous précisons le problème et donnons les résultats; dans la section II, nous faisons le lien entre un semi-groupe sous-markovien et les opérateurs sous-potentiels au sens de [1]; enfin, dans la section III, nous donnons la démonstration des résultats.

I. Position du problème - Résultats

Nous nous plaçons dans l'espace $X = L^\infty(E, \mathcal{B})$ des fonctions à valeurs réelles, boréliennes bornées définies sur un espace polonais E; X est muni de la norme uniforme. X est un espace de Banach réticulé σ-complet[(1)] pour l'ordre: $u \leqq v$

---

(1) Un espace vectoriel récticulé X est σ-complet si pour toute suite (u_n) majorée dans X, $\sup u_n$ existe dans X (cf. [4], ch. 12).

si et seuiement si $u(x) \leqq v(x) \quad \forall x \in E$.

Nous rappelons (cf. [3]) qu'une famille $\{N_t,\ t > 0\}$ est un <u>semi-groupe de noyaux sous-markovien mesurable</u> sur X si elle vérifie:

$$(N_1)\begin{cases} \text{(a)} \quad N_t \in L(X,X) \quad \forall t > 0, \\ \qquad u \leqq v \Rightarrow N_t u \leqq N_t v \quad \forall t > 0 \\ \qquad N_t 1 \leqq 1 \quad \forall t > 0^{(1)} \\ \text{(b)} \quad N_{t+s} = N_t \circ N_s \quad \forall t,s > 0 \\ \text{(c)} \quad u_n \searrow u^{(2)} \Rightarrow N_t u_n \searrow N_t u \quad \forall t > 0 \\ \text{(d)} \quad (t,x) \in]0,\infty[\times E \to N_t(x,u)^{(3)} \text{ est mesurable par} \\ \qquad \text{rapport à la tribu produit } L \otimes B \text{ sur }]0,\infty[\times E, \text{ où } L \\ \qquad \text{est la tribu de Lebesgue; pour tout } u \in X. \end{cases}$$

Nous utiliserons des applications $M : X \to X$ verifiant

(M_1) M est une contraction croissante,

(M_2) $u_n \searrow u \Rightarrow Mu_n \searrow Mu$,

(M_3) il existe une constante $c_o \leqq 0$ telle que $c_o \leqq Mc_o$ dans X.

<u>EXEMPLES</u>: Les applications M suivantes vérifient (M_1) (M_2) (M_3),

(1) $M = Id_X$

(2) Si $\psi \in X$, M définie par $Mu = \psi$, $\forall u \in X$, avec $c_o \leqq -\|\psi\|$

(3) Si E est un ouvert de $\mathbb{R}^N$ et $k \in X^+$, M définie par

$$Mu(x) = k(x) + \inf_{\substack{y \in E \\ y \geqq x}} u(y), \text{ avec } c_o \leqq 0.$$

---

(1) (a) signifie que N_t est une contraction linéaire croissante, $\forall t > 0$.

(2) $u_n \searrow u$ signifie que: u_n, $u \in X$ et $u_n(x) \searrow u(x)$, $\forall x \in E$.

(3) Pour tout $u \in X$, tout $x \in E$ et tout $t > 0$, on note $(N_t u)(x) = N_t(x,u)$.

Nous avons le resultat général suivant:

THÉORÈME 1: Soient I, J $\subset \mathbb{N}$ non vides, $\{N_t^i, t > 0\}_{i\in I}$ une famille de semi-groupes de noyaux sous-markoviens mesurables sur X et $(M_j)_{j\in J}$ une famille d'applications de X dans X vérifiant (M_1) (M_2) (M_3) avec la même constante c_o, alors,

(i) Pour tout $\lambda > 0$ et tout $F = (f^i) \in \ell^\infty(I,X)$, le problème

$$\tilde{I}(\lambda,F)\begin{cases} u \in X \\ u(x) \leqq \int_0^t e^{-s} N_{\lambda s}^i(x,f^i)ds + e^{-t} N_{\lambda t}^i(x,u) \ \forall x \in E \quad \text{p.p. } t > 0 \quad \forall i \in I \\ u \leqq M_j u \quad \forall j \in J \end{cases}$$

admet une solution maximum notée $\delta_\lambda F$

(ii) $\|(\delta_\lambda F - \delta_\lambda \hat{F})^+\| \leqq \sup_{i\in I} \|(f^i - \hat{f}^i)^+\| \quad \forall\lambda > 0 \ \forall F = (f^i),\ \hat{F} = (\hat{f}^i) \in \ell^\infty(I,X)$

(iii) $(\delta_\lambda)_{\lambda>0}$ est une famille résolvante, c'est-à-dire vérifie

$$\delta_\lambda F = \delta_\mu \left(\frac{\mu}{\lambda} F + \frac{\lambda-\mu}{\lambda} \delta_\lambda F\right) \quad \forall\lambda, \mu > 0 \quad \forall F \in \ell^\infty(I,X). \tag{2}$$

Donnons maintenant le résultat plus particulier répondant au problème initial (1). Pour cela, introduisons des notations et des hypothèses supplémentaires.

Si $\{N_t, t > 0\}$ est un semi-groupe de noyaux sous-markovien mesurable sur X, pour tout $\lambda > 0$ et tout $f \in X$, posons:

$$J_\lambda f = \int_0^\infty e^{-s} N_{\lambda s}.f\, ds \ ^{(1)}. \tag{3}$$

---

(1) Pour tout $0 < t \leqq \infty$, $\int_0^t e^{-s} N_{\lambda s}.f\, ds$ est l'élément $v \in X$, d'après (N_1,d), défini par $v(x) = \int_0^t e^{-s} N_{\lambda s}(x,f)ds \ \forall x \in E$; $J_\lambda = \frac{1}{\lambda} V_{1/\lambda}$ où $V_\alpha(x,f) = \int_0^\infty e^{-\alpha s} N_s(x,f)ds \quad x \in E \quad f \in X$; avec la terminologie de [3] V_α est la résolvante du semi-groupe $\{N_t, t > 0\}$.

LEMME 1: $\{N_t, t > 0\}$ etant un semi-groupe de noyaux sous-markovien mesurable sur X et J_λ étant définie par (3) pour tout $\lambda > 0$, alors

(i) $(J_\lambda)_{\lambda>0}$ est une famille résolvante, c'est-à-dire vérifie

$$J_\lambda f = J_\mu \left(\frac{\mu}{\lambda} f + \frac{\lambda-\mu}{\lambda} J_\lambda f\right) \quad \forall \lambda,\mu > 0 \quad f \in X \tag{4}$$

(ii) Pour tout $\lambda > 0$, $J_\lambda : f \in X \to J_\lambda f \in X$ est une contraction croissante

(iii) $f_n \searrow f \Rightarrow J_\lambda f_n \searrow J_\lambda f \quad \forall \lambda > 0.$

Ces propriétés se démontrent aisément en utilisant (N_1,a,c) pour (ii) et (iii). (4) se montre en utilisant le théorème de Fubini et la relation

$$N_\tau \, . \int_0^t e^{-s} N_{\lambda s}.f \, ds = \int_0^t e^{-s} N_{\tau+\lambda s}.f \, ds \quad \forall \tau, t > 0 \quad f \in X. \tag{5}$$

Si C (resp. S) désigne l'ensemble des fonctions continues (resp. semi-continues supérieurement (s.c.s.)) bornées de E dans $\mathbb{R}$, nous utiliserons les hypothèses (N_2) et (N_3) suivantes:

(N_2) J_λ envoie C dans C pour tout $\lambda > 0$,

(N_3) $t \in]0,\infty[\to N_t(x,u)$ est s.c.s. à droite $\forall x \in E \;\; \forall u \in C$.

Enfin si M est une application de X dans X, nous aurons à supposer

(M_4) M envoie C dans S.

REMARQUE 1: Si dans l'exemple 2 $\psi \in S$ et dans l'exemple 3 $k \in S^+$, alors $Mu \in S \;\; \forall u \in S$.

THÉORÈME 2: Les hypothèses et les notations étant les mêmes que dans le théorème 1, nous supposerons de plus que les semi-groupes $\{N_t^i, t > 0\}$ vérifient (N_2) et (N_3) pour tout $i \in I$ et les applications M_j vérifient (M_4) pour tout $j \in J$.

Si $F = (f^i) \in \ell^\infty(I,X)$ et si $J_\lambda^i f^i = \int_0^\infty e^{-s} N_{\lambda s}^i . f^i \, ds \in S$ pour tout $i \in I$ et $\lambda > 0$ (1),

alors pour tout $\lambda > 0$, $\delta_\lambda F \in S$ et est solution (maximum) de

$$I(\lambda,F) \begin{cases} u \in X \\ u \leq \int_0^t e^{-s} N_{\lambda s}^i . f^i + e^{-t} N_\lambda^i . u \quad \forall t > 0 \ \forall i \in I \\ u \leq M_j u \quad \forall j \in J. \end{cases}$$

REMARQUE 2: Ce résultat recouvre les résultats d'existence démontrés dans [2].

II. Opérateurs sous-potentiels associés a un semi-groupe de noyaux sous-markovien mesurable

Nous allons tout d'abord rappeler certaines définitions et résultats de [1].

DEFINITION 1: Nous dirons qu'une famille $(J_\lambda)_{\lambda>0}$ d'applications de X dans X vérifie la propriété (R) si

$$(R) \begin{cases} \text{(a)} \ (J_\lambda)_{\lambda>0} \text{ est une famille résolvante, c'est-à-dire vérifie} \\ \quad \text{(4) pour tout } \lambda, \mu > 0 \ f \in X, \\ \text{(b) Pour tout } \lambda > 0, J_\lambda \text{ est une contraction croissante de } X, \\ \text{(c)} \ f_n \searrow f \Rightarrow J_\lambda f_n \searrow J_\lambda f. \end{cases}$$

REMARQUE 3: Une telle famille engendre un opérateur A dont les résolvantes $(I + \lambda A)^{-1}$ coïncident pour tout $\lambda > 0$ avec J_λ; il est défini par

$A = \{(J_\lambda f, \frac{f-J_\lambda f}{\lambda}); \lambda > 0 \ f \in X\}$.

La propriété (b) signifie que A est un opérateur m-T-accrétif de X.

---

(1) Cette hypothèse est vérifiée si pour tout $i \in I$ il existe $\lambda > 0$ tel que $J_\lambda^i f^i \in C$. En effet d'après (4)

$J_\lambda^i f^i = J_\mu^i (\frac{\mu}{\lambda} f^i + \frac{\lambda-\mu}{\lambda} J_\lambda^i f^i) = \frac{\mu}{\lambda} J_\mu^i f^i + \frac{\lambda-\mu}{\lambda} J_\mu^i J_\lambda^i f^i$ et compte tenu de (N_2) on en deduit que $J_\mu^i f^i \in C \ \ \forall \mu > 0$.

DEFINITION 2: Soit $(J_\lambda)_{\lambda>0}$ une famille d'applications de X dans X vérifiant (R). Un graphe $\Gamma \subset X \times X$ est un opérateur sous-potentiel (au sens de [1]) de résolvante J_λ si:

(S_1) $v \leqq \hat{v} \Rightarrow \Gamma(v) \subset \Gamma(\hat{v})$

(S_2) $u_n \searrow u,\ v_n \searrow v,\ u_n \in \Gamma(v_n) \Rightarrow u \in \Gamma(v)$

(S_3) Pour tout $\lambda > 0$ et tout $f \in X$

$$J_\lambda f = \max \{u \in X;\ u \in \Gamma(\frac{f-u}{\lambda})\}.$$

Nous dirons aussi que Γ est un opérateur sous-potentiel de générateur A, où A est l'opérateur engendré par $(J_\lambda)_{\lambda>0}$.

Nous avons les résultats suivants, démontrés dans [1] mais qu'il est aisé de vérifier.

LEMME 2 (définitions, propriétés cf. [1]): Soient $(J_\lambda)_{\lambda>0}$ une famille d'applications de X dans X vérifiant (R) et A l'opérateur engendré par cette famille, alors

(1) $\Gamma_A = \{(v,u) \in X \times X;\ u \leqq J_\lambda(u+\lambda v) \quad \forall\lambda > 0\}$ est le plus grand opérateur sous-potentiel de générateur A;

(2) Pour tout $u \in S(A) = \bigcup_{v \in X} \Gamma_A(v)$, il existe $R_A u \in S(A)$ tel que

(i) $J_\lambda(u+\lambda v) \searrow R_A u$ lorsque $\lambda \searrow 0$, pour tout v tel que $u \in \Gamma_A(v)$

(ii) $u \in \Gamma_A(v) \Rightarrow R_A u \in \Gamma_A(v)$

(iii) L'application $R_A : u \in S(A) \to R_A u \in S(A)$ est une contraction croissante de S(A) dans S(A) vérifiant $R_A \circ R_A = R_A$.

(3) On note $S^r(A) = \{u \in S(A);\ R_A u = u\} = R_A S(A)$

(i) $D(A) \subset S^r(A)$

(ii) $u \in S^r(A)$ si et seulement si $u \in S(A)$ et il existe $\{u_n\} \subset D(A)$ tel que $u_n \searrow u$

(4) $\Gamma_A^r = \{(v,u) \in \Gamma_A;\ R_A u = u\}$ est le plus petit opérateur sous-potentiel de générateur A.

Nous fixons maintenant un semi-groupe de noyaux sous-markovien mesurable $\{N_t, t > 0\}$ sur X. La famille $(J_\lambda)_{\lambda>0}$ définie par (3) vérifie d'après le lemme 1 la propriété (R); nous allons caractériser les opérateurs sous-potentiels Γ_A et Γ_A^r, définis dans le lemme 2, qui lui sont associés.

PROPOSITION:

(1) Posons $N_o = id_X$

(a) Soit $u \in S(A)$ il y a équivalence des propriétés suivantes:

(i) $t \to N_t(x,u)$ est s.c.s. en 0 $\forall x \in E$,

(ii) $t \to N_t(x,u)$ est s.c.s. sur $[0,\infty[$ $\forall x \in E$,

(iii) $t \to N_t(x,u)$ est continue à droite sur $[0,\infty[$ $\forall x \in E$,

(b) $S^r(A) = \{u \in S(A)$; vérifiant les propriétés équivalentes précédentes$\}$

(c) $\Gamma_A^r = \{(v,u) \in X \times X;\ u \leqq \int_o^t e^{-s} N_{\lambda s}.(u+\lambda v)ds + e^{-t} N_{\lambda t}.u$

$\forall t \geqq 0$ $\forall \lambda > 0$ et $t \to N_t(x,u)$ est continue à droite sur $[0,\infty[$

$\forall x \in E\}$

(2) $u \in S(A) \Rightarrow N_t(x,R_A u) = N_t(x,u)$ $\forall x \in E$ p.p. $t > 0$

$\Gamma_A = \{(v,u) \in X \times X;\ u(x) \leqq \int_o^t e^{-s} N_{\lambda s}(x,u+\lambda v)ds + e^{-t} N_{\lambda t}(x,u)$

$\forall x \in E$ $\forall \lambda > 0$ p.p. $t > 0\}$

(3) $u \in \Gamma_A(\frac{f-u}{\lambda})$ si et seulement si

$$u(x) \leqq \int_o^t e^{-s} N_{\lambda s}(x,f)ds + e^{-t}N_{\lambda t}(x,u) \quad \forall x \in E \quad \text{p.p. } t > 0.$$

REMARQUE 4: D'après (3) on a en fait

$\Gamma_A^r = \{(v,u) \in X \times X;\ \exists \lambda > 0$ tel que $u \leqq \int_o^t e^{-s}N_{\lambda s}(u+\lambda v)ds + e^{-t}N_{\lambda t}.u$ $\forall t \geqq 0$ et $t \to N_t(x,u)$ est continue à droite sur $[0,\infty[$ $\forall x \in E\}$

$\Gamma_A = \{(v,u) \in X \times X;\ \exists \lambda > 0$ tel que

$u(x) \leqq \int_o^t e^{-s} N_{\lambda s}(x,u+\lambda v)ds + e^{-t}N_{\lambda t}(x,u)$ $\forall x \in E$ p.p. $t > 0\}$.

REMARQUE 5: Si $c \leqq 0$ alors $c \in \Gamma_A(0)$: en effet d'après (N_1,a)

$$\int_0^t e^{-s} N_{\lambda s}.c\, ds \geqq (1-e^{-t})c \geqq c-e^{-t}\, N_{\lambda t}.c \quad \forall t > 0.$$

PREUVE DE LA PROPOSITION: Montrons 1) : si $u \in \Gamma_A(v)$ tel que $t \to N_t(x,u)$ est s.c.s. en 0 $\forall x \in E$ alors par definition de Γ_A et R_A,

$$u \leqq R_A u = \lim_{\lambda \searrow 0} \int_0^\infty e^{-s}\, N_{\lambda s}.(u + \lambda v)ds \leqq \int_0^\infty e^{-s} \limsup_{\tau \searrow 0} N_\tau .u \leqq u \text{ donc}$$

$u = R_A u$ et $u \in \Gamma_A^r(v)$.

Soit $u \in D(A)$; il existe $f \in X$ et $\lambda > 0$ tels que

$u = J_\lambda f = \int_0^\infty e^{-s} N_{\lambda s}.f\, ds$ et d'après (5)

$$N_t.u = \int_0^\infty e^{-s} N_{\lambda s+t}.f\, ds = e^{t/\lambda}\int_{t/\lambda}^\infty e^{-s}\, N_{\lambda s}.f\, ds \quad \forall t \geqq 0. \qquad (6)$$

On en déduit que $t \to N_t(x,u)$ est continue sur $[0,\infty[$ $\forall x \in E$.

Si $u \in S^r(A)$ d'après le lemme 2, (3) il existe $\{u_n\} \subset D(A)$ tel que $u_n \searrow u$; d'après (N_1,c) $N_t(x,u_n) \searrow N_t(x,u)$, on en déduit que $t \to N_t(x,u)$ est s.c.s. sur $[0,\infty[$ $\forall x \in E$.

D'autre part si $u \in \Gamma_A(v)$, posons $u_\lambda = J_\lambda(u+\lambda v)$. D'après l'équation résolvante (4)

$$u_\lambda = J_\mu(u + \mu v + \frac{\lambda-\mu}{\lambda}(u_\lambda - u)) \quad \forall \mu > 0, \text{ et d'après (6)} \qquad (7)$$

$$u_\lambda = \int_0^t e^{-s} N_{\mu s}(u + \mu v + \frac{\lambda-\mu}{\lambda}(u_\lambda - u))ds + e^{-t} N_{\mu t}.u_\lambda \quad \forall t > 0 \quad \mu > 0$$

donc pour $\lambda < \mu$ comme $u_\lambda - u \geqq 0$ on en déduit que

$$u_\lambda \leqq \int_0^t e^{-s}\, N_{\mu s}(u+\mu v) + e^{-t}\, N_{\mu t}.u_\lambda \quad \forall t > 0,\ \ 0 < \lambda < \mu. \qquad (8)$$

Faisons $\lambda \searrow 0$ dans (8), il vient d'après (N_1,c) et la définition de R_A,

$$R_A u \leqq \int_0^t e^{-s} N_{\mu s}(u + \mu v) + e^{-t} N_{\mu t}.R_A u \quad \forall t > 0,\ \mu > 0 \ \ u \in \Gamma_A(v) \qquad (9)$$

donc si $u \in \Gamma_A^r(v)$ on a

$$u \leq \int_0^t e^{-s} N_{\mu s}(u+\mu v) + e^{-t} N_{\mu t}.u \quad \forall t > 0, \quad \mu > 0. \tag{10}$$

On deduit de (10), de (N_1,a) et (5) que

$$N_{t_0}(x,u) \leq \liminf_{t \searrow t_0} N_t(x,u) \quad \forall t_0 \geq 0, \text{ donc si } u \in \Gamma_A^r(v),$$

$t \to N_t(x,u)$ est s.c.i. à droite sur $[0,\infty[$ $\forall x \in E$.

Nous avons donc complètement montré (1).

Montrons (2): soit $u \in \Gamma_A(v)$, reprenons l'égalité (7) et utilisons la linéarité de J_μ, il vient

$\lambda u_\lambda = \lambda u_\mu + (\lambda-\mu) J_\mu(u_\lambda - u)$ et faisons $\lambda \searrow 0$, on obtient $J_\mu(R_A u-u) = 0$, comme $R_A u \geq u$ ceci implique

$$N_t(x,R_A u) = N_t(x,u) \quad \forall x \in E \quad \text{p.p. } t > 0. \tag{11}$$

Combinons (9) (11) et $u \leq R_A u$ on obtient

$$u(x) \leq \int_0^t e^{-s} N_{\lambda s}(x,u+\lambda v)ds + e^{-t} N_{\lambda t}(x,u) \quad \forall x \in E, \lambda > 0 \quad \text{p.p. } t > 0. \tag{12}$$

Réciproquement, soit (t_n) telle que $t_n \to \infty$ et (12) est vérifiée pour t_n (x et λ etant fixés); faisons $n \to \infty$ il vient que $u(x) \leq J_\lambda(u+\lambda v)(x)$ ce qui achève de démontrer (2).

Montrons (3). On vient de voir que

$$\left.\begin{array}{l} u(x) \leq \int_0^t e^{-s} N_{\lambda s}(x,f)ds + e^{-t} N_{\lambda t}(x,u) \ \forall x \quad \text{p.p. } t > 0 \\ \Rightarrow u \leq J_\lambda f. \end{array}\right\} \tag{13}$$

Utilisons l'équation résolvante (4)

$$u \leq J_\lambda f = J_\mu\left(\frac{\mu}{\lambda} f + \frac{\lambda-\mu}{\lambda} J_\lambda f\right) = J_\mu\left(u+\mu\left(\frac{f-u}{\lambda}\right) + \frac{\lambda-\mu}{\lambda}(J_\lambda f-u)\right)$$

si $\mu \geq \lambda$ il vient que

$$u \leq J_\mu\left(u+\mu\left(\frac{f-u}{\lambda}\right)\right). \tag{14}$$

Par définition de Γ_A, $u \in \Gamma_A\left(\frac{f-u}{\lambda}\right)$ si l'on montre l'inégalité (14) pour

$0 < \mu < \lambda$. D'après l'implication (13) c'est une conséquence du

LEMME 3: Soit $\{N_t, t > 0\}$ un semi-groupe de noyaux sous-markovien mesurable. Pour $\lambda > 0$, u, f $\in$ X,

$$u(x) \leq \int_0^t e^{-s} N_{\lambda s}(x,f)ds + e^{-t} N_{\lambda t}(x,u), \quad \forall x \in E \quad \text{p.p. } t > 0$$

$$\Rightarrow$$

$$u(x) \leq \int_0^t e^{-s} N_{\mu s}(x, \frac{\mu}{\lambda} f + \frac{\lambda-\mu}{\lambda} u)ds + e^{-t} N_{\mu t}(x,u), \; \forall x \in E, \; 0 < \mu < \lambda \quad \text{p.p. } t > 0$$

PREUVE: Nous utilisons le principe de la démonstration du lemme 1.5 de [2]. Fixons x $\in$ E et posons

$$H(t) = u(x) - \int_0^t e^{-s} N_{\lambda s}(x,f)ds - e^{-t} N_{\lambda t}(x,u), \quad \forall t > 0. \tag{15}$$

Soit $\gamma > 0$; multiplions (15) par $e^{-\gamma t}$ et intégrons de 0 à T pour T > 0, il vient,

$$(1-e^{-\gamma T})u(x) = \int_0^T e^{-(1+\gamma)t} N_{\lambda t}(x,f)dt - e^{-\gamma T} \int_0^T e^{-t} N_{\lambda t}(x,f)dt$$
$$+ \gamma \int_0^T e^{-(\gamma+1)t} N_{\lambda t}(x,u)dt + \gamma \int_0^T e^{-\gamma t} H(t)dt$$

$$u(x) = \int_0^T e^{-(\gamma+1)t} N_{\lambda t}(x,f+\gamma u)dt + e^{-(\gamma+1)T} N_{\lambda T}(x,u)$$
$$+ e^{-\gamma T} H(T) + \gamma \int_0^T e^{-\gamma t} H(t)dt$$

$$u(x) = \int_0^{\frac{\lambda}{\mu}T} e^{-(\gamma+1)\frac{\mu}{\lambda}s} N_{\mu s}(x, \frac{\mu}{\lambda} f + \frac{\gamma\mu}{\lambda} u)ds + e^{-(\gamma+1)T} N_{\lambda T}(x,u)$$
$$+ e^{-\gamma T} H(T) + \gamma \int_0^T e^{-\gamma s} H(s)ds. \tag{16}$$

Posons $t = \frac{\lambda}{\mu} T$, $\gamma = \frac{\lambda-\mu}{\mu}$, $0 < \mu < \lambda$ alors (16) devient

$$u(x) = \int_0^t e^{-s} N_{\mu s}(x, \frac{\mu}{\lambda}f + \frac{\lambda-\mu}{\lambda} u)ds + e^{-t} N_{\mu t}(x,u) + \phi(t) \quad \forall t > 0 \quad (16')$$

$$\text{où } \phi(t) = e^{-\frac{\lambda-\mu}{\lambda} t} H(\frac{\mu}{\lambda} t) + \frac{\lambda-\mu}{\mu} \int_0^{\frac{\mu}{\lambda} t} e^{-\gamma s} H(s)ds$$

enfin l'hypothèse montre que $\phi(t) \leqq 0$ p.p. $t > 0$
ce qui achève de montrer le lemme.

III. Demonstration des théorèmes

Remarquons tout d'abord qu'il suffit de démontrer les théorèmes pour card $J = 1$. En effet, par hypothèse il existe une constante négative c_0 telle que $c_0 \leqq M_j c_0$ $\forall j \in J$; on en déduit d'après (M_1) que

$$M_j u \geqq c_0 - \|u-c_0\| \quad \forall j \in J \quad \forall u \in X$$

d'où $M = \inf_{j \in J} M_j$ existe, c'est une application de X dans X vérifiant (M_1) (M_2) (M_3) avec cette même constante c_0. Si de plus M_j vérifie (M_4) $\forall j \in J$, on verifie aisément que $M = \inf M_j$ vérifie aussi (M_4). Nous remplacerons désormais $(M_j)_{j \in J}$ par M.

Soit donc $\{N_t^i, t > 0\}_{i \in I}$ une famille finie ou dénombrable de semigroupes de noyaux sous-markoviens mesurables. Nous noterons J_λ^i la résolvante associée à $\{N_t^i, t > 0\}$.

Rappelons que pour $\lambda > 0$ et $f \in X$,

$$J_\lambda^i f = \int_0^\infty e^{-s} N_{\lambda s}^i . f \, ds.$$

Nous noterons A^i l'opérateur m-T-accrétif engendré par $(J_\lambda^i)_{\lambda > 0}$.

Plaçons-nous dans l'espace $\chi = \ell^\infty(I,X)$ muni de la norme $\|(u^i)\| = \sup_{i \in I} \|u^i\|_X$. C'est un espace de Banach réticulé σ-complet pour l'ordre $(u^i) \leqq (v^i)$ si et seulement si $u^i \leqq v^i$ dans X $\forall i \in I$.

Comme pour tout $i \in I$ et tout $\lambda > 0$ J_λ^i est une contraction linéaire sur X,

$$F = (f^i) \in \chi \Rightarrow (J_\lambda^i f^i) \in \chi.$$

On peut donc poser pour tout $F = (f^i) \in \chi$ et tout $\lambda > 0$

$$J_\lambda F = (J_\lambda^i f^i) \in \chi . \tag{17}$$

Nous avons alors le résultat immédiat suivant:

LEMME 4:

(1) $(J_\lambda)_{\lambda>0}$ définie par (17) vérifie la propriété (R) sur χ.

(2) Si G est l'opérateur de χ engendré par $(J_\lambda)_{\lambda>0}$, nous notons Γ_G l'opérateur sous-potentiel de χ de générateur G défini dans le lemme 2. Alors $\Gamma_G = \prod_{i\in I} \Gamma_{A^i}$, plus précisement si $U = (u^i)$

$$V = (v^i) \in \chi \quad (V,U) \in \Gamma_G \text{ si et seulement si } (v^i,u^i) \in \Gamma_{A^i} \quad \forall i \in I.$$

Nous allons utiliser des résultats de [1] que nous rappelons dans le lemme suivant,

LEMME 5 (cf. [1]): Soit $\mathcal{U}$ un sous-ensemble non vide de χ vérifiant (H)

$$(H) \begin{cases} (i) & U_n \searrow U \text{ et } U_n \in \mathcal{U} \quad \forall n \Rightarrow U \in \mathcal{U} \\ (ii) & \forall F \in \chi \quad P(F) = \max\{U \in \mathcal{U} \; ; \; U \leq F\} \text{ existe} \\ (iii) & P : F \in \chi \to P(F) \text{ est une contraction.} \end{cases}$$

Alors:

(1) $\Gamma_G^{\mathcal{U}} = \{(V,U) \in \Gamma_G ; U \in \mathcal{U}\}$ est un opérateur sous-potentiel de χ si et seulement si $\Gamma_G^{\mathcal{U}} \neq \emptyset$

(2) Si $\Gamma_G^{\mathcal{U}} \neq \emptyset$ les résolvantes $J_\lambda^{\mathcal{U}}$ sont obtenues de la manière suivante soient $\lambda > 0$ et $F \in \chi$

(i) pour $\varepsilon > 0$, il existe une unique solution $U_\varepsilon = J_\lambda^\varepsilon F$ de $U_\varepsilon \in \chi$

$$U_\varepsilon + \lambda G U_\varepsilon + \frac{\lambda}{\varepsilon}(U_\varepsilon - PU_\varepsilon) = F$$

(ii) $J_\lambda^\varepsilon F \searrow \tilde{J}_\lambda F$ dans χ lorsque $\varepsilon \searrow 0$

(iii) $J_\lambda^{\mathcal{U}} F = P(\tilde{J}_\lambda F)$.

Dans notre problème prenons

$$\mathcal{U} = \{U = (u^i) \in X \; ; \text{ il existe } u \in X \text{ tel que } u^i = u \quad \forall i \in I \text{ et } u \leqq Mu\}. \tag{18}$$

Par hypothese $C_o = (c^i) \in X$ avec $c^i = c_o \quad \forall i \in I$ est dans $\mathcal{U}$. De plus, d'après la remarque 5, $C_o \in \Gamma_G^{\mathcal{U}}(0)$.

$\mathcal{U}$ vérifie (H) (cf. [1])(1).

Nous pouvons donc appliquer le lemme 5: $\Gamma_G^{\mathcal{U}}$ est un opérateur sous-potentiel de X de résolvante $J_\lambda^{\mathcal{U}}$, nous avons

$$J_\lambda^{\mathcal{U}} F = \max \{U \in X \; ; \; U \in \Gamma_G^{\mathcal{U}} (\frac{F-U}{\lambda})\} \tag{19}$$

$$\|(J^{\mathcal{U}}F - J^{\mathcal{U}}\hat{F})^+\| \leqq \|(F-\hat{F})^+\| \quad \forall F,\hat{F} \in X \tag{20}$$

$$(J_\lambda^{\mathcal{U}})_{\lambda>0} \text{ est une famille résolvante sur } X. \tag{21}$$

Pour $F = (f^i) \in X$ et $\lambda > 0$, traduisons (19):

$J_\lambda^{\mathcal{U}} F \in \mathcal{U}$ donc il existe $\delta_\lambda F \in X$ tel que $\delta_\lambda F \leqq M(\delta_\lambda F)$ et

$$\delta_\lambda F = \max \{u \in X; \; u \in \Gamma_{A^i}(\frac{f^i - u}{\lambda}) \quad \forall i \in I, \quad u \leqq Mu\}$$

ce qui, compte tenu de la proposition, montre que $\delta_\lambda F$ est solution maximum de $\tilde{I}(\lambda,F)$.

(ii) et (iii) du théorème 1 se déduisent aisément de (20) et (21). $F = (f^i) \in X$ et $\lambda > 0$ étant fixés, notons pour simplifier $u = \delta_\lambda F$. u sera solution maximum de $I(\lambda,F)$ si pour tout $x \in E$ et $i \in I$ les inégalités

$$u(x) \leqq \int_0^t e^{-s} N_{\lambda s}^i(x,f^i)ds + e^{-t} N_{\lambda t}^i(x,u) \quad \text{p.p. } t > 0 \tag{22}$$

qu'il vérifie sont vraies pour tout $t > 0$.

---

(1) Soit $F = (f^i) \in X$; $P(F) = \max \{U \in \mathcal{U}; \; U \leqq F\}$ est défini de la manière suivante: si $f = \inf_{i \in I} f^i \in X$, $P(F) = U = (u^i)$ où pour tout i $u^i = u$ tel que $u = \searrow \lim u_n$, u_n définis par récurrence par: $u_o = f \quad u_n = u_{n-1} \wedge Mu_{n-1}$.

Pour cela il suffit que $t \to N_t^i(x,u)$ soit s.c.s. à droite sur $]0,\infty[$ $\forall x \in E$ $\forall i \in I$ et d'après l'hypothèse (N_3) il suffit de montrer que $u \in S$. D'après la construction de P rappelée précédemment, sous l'hypothèse (M_4) P envoie S^I dans S^I. Il suffit donc d'après le lemme 5 de montrer que $U_\varepsilon = J_\lambda^\varepsilon F \in S^I$. Toujours d'après ce lemme et l'équation résolvante

$$U_\varepsilon = J_\lambda(F - \frac{\lambda}{\varepsilon}(U - PU_\varepsilon)) = \frac{\varepsilon}{\lambda+\varepsilon} J_{\frac{\lambda\varepsilon}{\lambda+\varepsilon}} F + \frac{\lambda}{\lambda+\varepsilon} J_{\frac{\lambda\varepsilon}{\lambda+\varepsilon}} PU_\varepsilon = T_F U_\varepsilon .$$

Si l'on suppose que $J_\mu^i f^i \in S$ $\forall i \in I$ $\forall \mu > 0$ d'après (N_2) T_F envoie S^I dans S^I. Et l'on rappelle que $U_\varepsilon = \searrow \lim U_n$ quand $n \to \infty$, avec $U_o = J_\lambda F \in S^I$, $U_n = T_F U_{n-1}$, donc $U_\varepsilon \in S^I$. □

<u>Bibliographie</u>

1. L. Barthélemy - Ph. Bénilan: "Opérateurs sous-potentiels de type accrétif dans un espace de Banach réticulé σ-complet" (à paraître).
2. A. Bensoussan - M. Robin: "On the convergence of the discrete time dynamic programming equation for general semi-groups".
3. Meyer: "Probabilités et Potentiel" - Actualités scientifiques et industrielles. Hermann.
4. K. Yosida: "Functional Analysis" - Springer-Verlag 1980.

L. Barthélemy
Equipe de Mathématiques de Besançon
C.N.R.S. U.A. 741
Université de Franche-Comté
25030 Besançon CEDEX
France.

A BENSOUSSAN

Some remarks on ergodic control problems

Introduction

Our objective in this article is to review some techniques which have been given in the literature to solve Bellman equations of ergodic type. The presentation will by no means be exhaustive and we will just stick to the main ideas, leaving aside some particular cases. We shall rely mainly on the work of F. Gimbert [3], P.L. Lions [5] and A. Bensoussan - J. Frehse [1].

1. Setting the problem

1.1 Assumptions. Notations

Consider the 2nd order differential operator

$$A = - \sum_{i,j} \frac{\partial}{\partial x_i} a_{ij} \frac{\partial}{\partial x_j} + \sum_i a_i \frac{\partial}{\partial x_i} \tag{1.1}$$

where

$$a_{ij}(x) \text{ periodic, with period 1 in all components}^{(1)} \ (x \in \mathbb{R}^n \tag{1.2}$$

$$\sum_{i,j} a_{ij}(x)\, \xi_i \xi_j \geq \beta\, |\xi|^2, \quad \forall \xi \in \mathbb{R}^n, \quad \beta > 0$$

$$a_i \text{ periodic bounded.} \tag{1.3}$$

We consider now a function $H(x,p)$ on $\mathbb{R}^n \times \mathbb{R}^n$ such that:

$$H(x,p) \text{ is periodic in } x, \text{ Borel,} \tag{1.4}$$

$$H(x,0) \text{ bounded}$$

---

(1) This assumption is made to simplify the notation, but is by no means necessary. We denote:
$Y =]0,1[^n$

$$|H(x,p_1) - H(x,p_2)| \leqq K(1 + |p_1| + |p_2|)\ |p_1 - p_2| \tag{1.5}$$

1.2 The problem

Consider the Bellman equation ($\alpha > 0$)

$$Au_\alpha + u_\alpha = H(x, Du_\alpha) \tag{1.6}$$

u_α periodic.

We first have:

THEOREM 1.1: There exists one and only one solution of (1.6), such that $u_\alpha \in W^{2,p}(Y)$, $\forall p$, $2 \leqq p < \infty$. □

The proof is omitted since it is a classical result (cf. Ladyzhenskaya - Uralceva [4], Gilbarg - Trudinger [2]). We are interested in the behaviour of u_α as tends to 0.

2. Preliminary estimates

2.1 Estimate on αu_α

LEMMA 2.1: One has the estimate:

$$|\alpha u_\alpha(x)| \leqq K.$$

PROOF: This follows from maximum principle considerations. □

2.2 Estimate on Du_α

This estimate requires more assumptions on H. There are namely two routes, one consisting in obtaining estimates on $\|Du_\alpha\|_{L^\infty}$ and one consisting in obtaining estimates on $\|Du_\alpha\|_{L^2}$.

To obtain estimates on the L^∞ norm of the gradient we make the following additional assumptions

$$a_{ij} \in C^2,\ \ a_j \in C^1 \tag{2.2}$$

$H(x,p)$ is C^1 in x,p (2.3)

We shall use the notation $a_{ij,k}$ for $\frac{\partial a_{ij}}{\partial x_k}$, $a_{ij,ik}$ for $\frac{\partial^2 a_{ij}}{\partial x_i \partial x_k}$ and similar notation. Call also:

$$\tau = \sup_x \Sigma a_{ii}(x) \tag{2.4}$$

we make the following growth assumption on H (with respect to p)

there exists $\delta > 0$, $\delta < 2$, such that (2.5)

$$\varliminf_{p\to\infty} \{ \frac{2-\delta}{\tau} [H(x,p) + h - p_j(a_j - a_{ij,i})]^2 - 2H_x p -$$

$$- \frac{1}{\delta\beta} \sum_{ij} (a_{ij,k}p_k)^2 - 2(a_{ij,ik} - a_{j,k})p_j p_k\} > 0$$

uniformly with respect to $|h| \leqq K$ and x.

We can then state the:

<u>LEMMA 2.2</u>: Under the assumptions of §1.1 and (2.2), (2.3), (2.5) one has

$$||Du_\alpha||_{L^\infty} \leqq C \tag{2.6}$$

<u>PROOF</u>: We follow P.L. Lions [5] and F. Gimbert [3]. Let us set:

$$w_\alpha = \sum_i (u_{\alpha,i})^2$$

A calculation shows that:

$$Aw_\alpha - H_p Dw_\alpha + 2\alpha w_\alpha + 2\alpha w_\alpha + 2a_{ij} u_{\alpha,ik} u_{\alpha,jk} \tag{2.7}$$

$$= 2 H_x Du_\alpha + 2a_{ij,k} u_{\alpha,ij} + 2(a_{ij,jk} - a_{j,k})u_{\alpha,j} u_{\alpha,k}$$

Note that:

$$(a_{ij} u_{\alpha,ij})^2 \leqq \tau \; a_{ij} u_{\alpha,ik} u_{\alpha,jk}.$$

We deduce from (2.7) that:

$$Aw_\alpha - H_p\, Dw_\alpha + 2\alpha w_\alpha + \frac{2-\delta}{\tau}[H(x,Du_\alpha) - \alpha u_\alpha - u_{\alpha,j}(\alpha_j - a_{ij,i})]^2$$

$$+ \delta\beta \sum_{i,j} (u_{\alpha,ij})^2 \leq 2H_x\, Du_\alpha + 2a_{ij,k}\, u_{\alpha,k}\, u_{\alpha,ij}$$

$$+ 2\,(a_{ij,ik} - a_{j,k})\, u_{\alpha,j}\, u_{\alpha,k}$$

$$\leq 2H_x\, Du_\alpha + \delta\beta \sum_{i,j} (u_{\alpha,ij})^2 + \frac{1}{\delta\beta} \sum_{i,j} (a_{ij,k} u_{\alpha,k})^2$$

$$+ 2(a_{ij,ik} - a_{j,k})\, u_{\alpha,j}\, u_{\alpha,k}$$

If x_α is a point of maximum of w_α, one has:

$$(Aw_\alpha - H_p\, Dw_\alpha + 2\alpha w_\alpha)(x_\alpha) > 0$$

hence setting $p_\alpha = Du_\alpha(x_\alpha)$

$$\frac{2-\delta}{\tau}[H(x_\alpha,p_\alpha) - \alpha u_\alpha(x_\alpha) - p_{\alpha,j}(a_j - a_{ij,i})(x_\alpha)]^2$$

$$- 2H_x p_\alpha - \frac{1}{\delta\beta} \sum_{i,j} (a_{ij,k} p_{\alpha,k})^2 - 2(a_{ij,ik} - a_{j,k})(x_\alpha) p_{\alpha,j} p_{\alpha,k}$$

$$\leq 0.$$

If p_α does not remain bounded as $\alpha \to 0$, there will be a contradiction with the assumption (2.5). Therefore

$$w_\alpha(x_\alpha) \leq C$$

which completes the proof of the desired result. □

To obtain an L^2 estimate on Du_α we make the additional assumption:

$$H(x,p) \leq k_0(1 - |p|^2) \qquad (2.8)$$

then we can state the:

LEMMA 2.3: Under the assumptions of §1.1 and (2.8), one has:

$$\|Du_\alpha\|_{L^2} \leq C \tag{2.9}$$

PROOF: Integrate (1.6) over Y we get:

$$\sum_i \int a_i \frac{\partial u_\alpha}{\partial x_i} dx + \alpha \int_Y u_\alpha dx = \int_Y H(x,Du_\alpha)dx$$

and from (2.8):

$$\leq k_0 - k_0 \int_Y |Du_\alpha|^2 dx$$

from which one easily deduces (2.9). □

REMARK 2.1: Note that under the assumptions of Lemma 2.3, the growth condition (2.5) is satisfied, provided H_x has the same growth as H (i.e. quadratic growth).

3. Convergence

3.1 L^∞ estimates

Let us consider the function:

$$z_\alpha(x) = u_\alpha(x) - u_\alpha(x_0)$$

we can assert that:

LEMMA 3.1: Under the conditions of Lemma 2.2 or Lemma 2.3, one has:

$$\|z_\alpha\|_{L^\infty} \leq C \tag{3.1}$$

PROOF: Note that:

$$Dz_\alpha = Du_\alpha.$$

Consider first the situation of Lemma 2.2, one has $\|Dz_\alpha\|_{L^\infty} \leq C$, therefore using:

$$z_\alpha(x) = \int_0^1 Dz_\alpha(x_0 + \lambda(x - x_0))(x - x_0)d\lambda$$

one has:

$$|z_\alpha(x)| \leq C\ |x - x_o|$$

and since we need only to consider x, x_o in Y, the result (3.1) follows.

Consider now the situation of Lemma 2.3. A proof based upon the use of Green functions has been given in Bensoussan - Frehse [1]. We give here a different one, based upon an idea of P.L. Lions. Define:

$$\zeta_\alpha = \frac{z_\alpha}{\|z_\alpha\|_{L^\infty}}$$

and assume that $\|z_\alpha\|_{L^\infty} \to \infty$ for a subsequence. We deduce from (1.6) that:

$$Az_\alpha + \alpha u_\alpha = H(x, Dz_\alpha)$$

hence

$$A\zeta_\alpha + \frac{\alpha u_\alpha}{\|z_\alpha\|_{L^\infty}} = \frac{H(x, Dz_\alpha)}{\|z_\alpha\|_{L^\infty}} \tag{3.2}$$

By definition ζ_α remains in a bounded subset of L^∞, and from Lemma 2.3, $D\zeta_\alpha \to 0$ in L^2. Therefore:

$$\zeta_\alpha \to C \text{ in } H^1.$$

Using the equation (3.2), it is possible to prove additional regularity results, implying namely that:

$$\zeta_\alpha \to C \text{ in } C(\bar{Y}).$$

Since $\zeta_\alpha(x_o) = 0$, we have $C = 0$, but this contradicts the fact that $\|\zeta_\alpha\| = 1$, tends also to C. This completes the proof of the desired result. □

3.2 Main result

We can now state the main result of the theory:

THEOREM 3.1: Under the conditions of Lemma 2.2 or Lemma 2.3, one has:

$$\alpha u_\alpha(x) \to \rho \tag{3.3}$$

$$u_\alpha(x) - u_\alpha(x_0) \to z \text{ in } W^{2,p}(Y) \text{ weakly}$$

where x_0 is arbitrary in Y, ρ is a constant, and the pair z,ρ is solution of

$$Az + \rho = H(x,Dz) \tag{3.4}$$

$$z \text{ periodic.}$$

The constant ρ is uniquely defined, the function z is uniquely defined if one adds the condition $z(x_0) = 0$.

PROOF:

Uniqueness:

Since the solution z is regular, its gradient is bounded, hence the quadratic growth does not matter anymore. The uniqueness then follows from standard maximum principle arguments.

Existence:

In the case of Lemma 2.2, since the gradient is bounded, the function $H(x,Du_\alpha)$ remains bounded, hence z_α remains bounded in $W^{2,p}$. From this estimate it is easy to prove that:

$$z_\alpha \to z, \quad \alpha u_\alpha(x_0) \to \rho$$

and z,ρ is solution of (3.4).

In the case of Lemma 2.3, pick a subsequence such that

$$z_\alpha \to z \text{ in } H^1 \text{ weakly and a.e.} \tag{3.5}$$

One first shows that:

$$z_\alpha \to z \text{ in } H^1 \text{ strongly.} \tag{3.6}$$

Consider for that the function:

$$\chi_\alpha = e^{\lambda(z_\alpha - z)} - e^{-\lambda(z_\alpha - z)}$$

which is bounded (independently of and periodic). Set also:

$$\psi_\alpha = e^{\lambda(z_\alpha - z)} + e^{-\lambda(z_\alpha - z)}$$

and write the equation (1.6) as:

$$Az_\alpha + \alpha z_\alpha + \rho_\alpha = H(x, Dz_\alpha) \tag{3.7}$$

where

$$\rho_\alpha = \alpha u_\alpha(x_o).$$

Without loss of generality we may assume that:

$$\rho_\alpha \to \rho. \tag{3.8}$$

Note that from (3.5) one has:

$$\chi_\alpha \to 0 \text{ a.e.} \tag{3.9}$$

We multiply (3.7) by χ_α and integrate over Y. We obtain:

$$\begin{aligned}
&\lambda \int a_{ij} \frac{\partial}{\partial x_j}(z_\alpha - z) \frac{\partial}{\partial x_i}(z_\alpha - z)\, \psi_\alpha \, dx + \lambda \int a_{ij} \frac{\partial z}{\partial x_j} \frac{\partial}{\partial x_i}(z_\alpha - z)\psi_\alpha dx \\
&+ \int a_i \frac{\partial z_\alpha}{\partial x_i} \chi_\alpha \, dx + \int \alpha z_\alpha \chi_\alpha \, dx + \rho_\alpha \int \chi_\alpha \, dx \\
&= \int H(x, Dz_\alpha)\, \chi_\alpha \, dx \\
&\leqq K \int (1 + |Dz_\alpha|^2)\, |\chi_\alpha| \, dx \\
&\leqq 2K \int |D(z_\alpha - z)|^2 \psi_\alpha \, dx + K \int (1 + 2\,|Dz|^2\, |\chi_\alpha| \, dx.
\end{aligned} \tag{3.10}$$

We choose λ so that $\lambda\beta > 2K$. By virtue of (3.5), (3.9), (3.8) and the

L^∞ bound on z_α, we easily deduce from (3.10) that:

$$\overline{\lim_{\alpha\to}} \int |D(z_\alpha - z)|^2 \, dx = 0$$

which proves (3.6).

From the assumption (1.5) we deduce:

$$\int |H(x,Dz_\alpha) - H(x,Dz)| \, dx \leq K \int |D(z_\alpha - z)| dx \leq$$

$$\leq K \int |D(z_\alpha - z)| \, (1 + |Dz_\alpha| + |Dz|) dx$$

$$\to 0 \text{ as } \alpha \to 0.$$

Therefore, we can pass to the limit in (3.7), and obtain (3.4). The equation (3.4) implies by standard regularity arguments that $z \quad W^{2,p}(Y)$.

In fact, it is also true that z_α remains bounded in $W^{2,p}$, but this requires a different argument. The proof has been completed.

References

1. A. Bensoussan, J. Frehse, On Bellman equations of ergodic type with quadratic growth Hamiltonian, to be published.
2. D. Gilbarg, N.S. Trudinger, Elliptic Partial Differential Equations of Second Order, Springer Verlag 1977.
3. F. Gimbert, Sur quelques équations non linéaires intervenant en contrôle stochastique, Thèse 3e cycle, Paris Dauphine.
4. O.A. Ladyzhenskaya, N.N. Ural'ceva, Equations aux dérivées parielles de type elliptique, Dunod (1968).
5. P.L. Lions, Quelques remarques sur les problèmes elliptiques quasi-linéaires du 2e ordre, Journ. Anal. Math. 45, 1985, pp. 234-254.

A. Bensoussan
Université Paris IX-Dauphine
Place de Lattre de Tassigny
75775 Paris CEDEX 16
France.

J BLAT

Bifurcation to positive solutions in systems of elliptic equations

1. Introduction

The aim of this paper is to give a description of some global bifurcation techniques which have been used to prove existence results for steady-state soltuions of some kinds of reaction-diffusion systems. More detailed results and proofs appear elsewhere (1,2,3).

In order to be more precise we shall centre around the system

$$u_t = d_1\Delta u + au - a_1u^2 - a_2f(u)v$$

$$v_t = d_2\Delta v + bv - b_1v^2 + b_2f(u)v$$

where u and v are functions of x and t, $x \in \Omega \subset R^n$ (n = 1,2 or 3), Ω is a smooth bounded domain, d_1, a_1, b_1 are positive constants, a and b are real numbers which are allowed to vary as bifurcation parameters, f(u) is either u or u/(1+cu) and we impose homogeneous Dirichlet boundary conditions.

This system is used to describe the evolution of the concentration of two species denoted by u and v which diffuse in a bounded region, with a logistic growth for each species, a predator-prey interaction which is either classical or Holling-Tanner type depending on f(u), u being the prey and v the predator. The boundary conditions are supposed to model a hostile environment.

We choose this system as an example because it does not possess monotonicity properties and one has to resort to bifurcation techniques to prove satisfactory results about existence of positive solutions which are the only ones that have physical meaning (to see a comparison between the results obtained with both methods see (1)). Perhaps one should note that this system does not have variational properties.

The nonnegative steady-state solutions of that system can be of three different types: the trivial solution (∅,∅), possible semitrivial solutions (u,∅), (∅,v) with u and v positive and possible nontrivial (u,v) with both components positive. The first two types mean the extinction of one or both species while the latter would suggest the coexistence of predator and prey.

In order to obtain the semitrivial solutions one has to solve either $-d_1\Delta u = au - a_1u^2$ or $-d_2\Delta v = bv - b_1v^2$, with the corresponding boundary conditions. The single semilinear elliptic equations have been studied quite a lot in the last years (see for instance the review paper by P.L. Lions (13)) and thus the problem of finding semitrivial solutions is solved in quite a few cases for different kinds of equations. It is more interesting for its meaning to find nontrivial solutions and also more difficult in general. In the following we describe two ways of using global bifurcation techniques to obtain these solutions: one via decoupling of the system and another one of bifurcation applied directly to the system.

In the next paragraph we collect some terminology and preliminary results and the following ones deal with the two methods and a comparison between them.

2. Terminology and preliminary results

It is well known that the linear eigenvalue problem

$$-\Delta\phi = \lambda\phi \text{ in } \Omega, \ \phi = 0 \text{ on } \partial\Omega$$

has an infinite sequence of eigenvalues λ_n such that $0 < \lambda_1 < \lambda_2 \leqq \ldots$ with corresponding eigenfunctions ϕ_n and ϕ_1 can be taken as positive in Ω. Suppose that $q : \Omega \to R$ is smooth. Then the linear eigenvalue problem

$$-\Delta\phi + q\phi = \lambda\phi \quad \text{in } \Omega, \ \phi = 0 \text{ on } \partial\Omega$$

also has an infinite sequence of eigenvalues which are bounded below. We denote them by $\lambda_i(q)$ and it is known that the first eigenvalue is simple and the corresponding eigenfunctions do not change sign in Ω: $\lambda_i(q)$ is increasing with respect to q.

If 0 is not an eigenvalue of the latter problem, then we can define a corresponding solution operator K_q, i.e., $K_q f$ is the unique solution of

$$-\Delta u + qu = f \text{ in } \Omega; \quad u = 0 \text{ on } \partial\Omega$$

and K_q is the inverse of the differential operator $L_q = -\Delta + q$ associated with the respective boundary conditions. It is well known that this operator is compact on $C^1(\Omega)$ and $L^2(\Omega)$ and an isomorphism from $C^\alpha(\Omega)$ into $C^{2+\alpha}(\Omega)$.

In order to be more precise, consider the system with all the parameters fixed except for b, which we shall treat as a bifurcation parameter.

Let $v \in C^1(\Omega)$ and consider the following equation for u

$$-\Delta u = au - a_1u^2 - a_2uv \text{ in } \Omega; \quad u = \emptyset \text{ on } \partial\Omega .$$

Using the results mentioned in the previous paragraph, we know that if $a \leqq \lambda_1(a_2v)$ the equation has no positive solution whereas if $a > \lambda_1(a_2v)$, it has a unique positive solution. We define a map on $C^1(\Omega)$ by u(v) being this unique positive solution when it exists or $\emptyset$ otherwise.

<u>LEMMA 1</u>: The map u(v) is continuous on $C^1(\Omega)$ and takes bounded sets into bounded sets. Moreover if $v_1 \leqq v_2$ then $u(v_1) \geqq u(v_2)$.

Let us give a simple preliminary result:

<u>THEOREM 2</u>: If $a < \lambda_1$ and (u,v) is a nonnegative solution of the system, then

(i) $u = \emptyset$

(ii) if $b \leqq \lambda_1$, $v = \emptyset$ and if $b > \lambda_1^*$ then we have also a semitrivial solution with $v > \emptyset$.

The proof is immediate.

From now on we suppose $a > \lambda_1$. Then our system has always the solutions $(\emptyset,\emptyset)$ and $(u_a,\emptyset)$ for all values of b. We proceed with the bifurcation analysis.

Let L be the differential expression defined by

$$Lv = -\Delta v - b_2u_av.$$

We can assume without loss of generality that L is invertible (Otherwise, we replace L by L + k so that L + k is invertible and a similar analysis gives the same results). Taking K as the inverse of L with the corresponding boundary conditions and F as defined by

$$F(v) = -b_1v^2 + b_2(u(v) - u(\emptyset))v$$

clearly our system can be reduced to the resolution of the single equation

$$v = bKv + KFv.$$

As both K and KF are compact and continuous and $\|KF(v)\| = o(\|v\|)$ as v tends to $\emptyset$, the well known bifurcation results of Crandall and Rabinowitz (7) and Rabinowitz (14) can be applied to this equation. There is bifurcation from the trivial solution when $b = \lambda_1(-b_2u_a)$ and it has global character as this point is a simple characteristic value for K. As the corresponding characteristic functions do not change sign in Ω, there is a bifurcating branch of solutions lying in the positive cone which we denote by S^+.

Using a priori bounds and the fact that this characteristic value is the only one with positive characteristic functions we can prove

<u>THEOREM 3</u>: The branch S^+ is contained in $R \times P$ (where P is the positive cone) and $\{b \in R : (b,v) \in S^+\} = (\lambda_2(-b_2u_a),\infty)$.

One can prove as well that for any given fixed value of a, there exists b big enough such that there are no solutions of the system with both u and v positive, so that the branch S^+ joins with the branch $(\emptyset,v_b)$.

Similar analysis can be performed with b fixed, also for systems of competing species and other types of boundary conditions (see 1,2) for instance).

4. Existence results via global bifurcation directly

Now we consider the following system

$$-\Delta u = au - a_1u^2 - a_2uv/(1+cu)$$

$$-\Delta v = bv - b_1v^2 + b_2uv/(1+cu),$$

with the corresponding boundary conditions. One can see that this system cannot be easily decoupled: the map u(v) has sometimes two positive solutions and if we try to choose only one, the map loses the continuity property which is essential to the preceding analysis.

But we expect bifurcation to positive solutions from $(u_a,\emptyset)$ or from $(\emptyset,v_b)$ as shown by the previous paragraph. Thus we try to perform a bifurcation analysis directly on the system.

For instance, let us linearise the system around $(u_a, \emptyset)$: taking $u = u_a - U$ and $v = V$ we have

$$- \Delta U = aU - 2a_1 u_a U + a_2 u_a V/(1+cu_a) + h(a,x,U,V)$$

$$- \Delta V = bV + b_2 u_a V/(1+cu_a) + g(a,x,U,V)$$

where g and h are functions with higher order terms.

The system can be written in an equivalent form as

$$\begin{pmatrix} U \\ V \end{pmatrix} = K \begin{bmatrix} a - 2a_1 u_a & a_2 u_a/(1+cu_a) \\ b & b_2 u_a/(1+cu_a) \end{bmatrix} \begin{pmatrix} U \\ V \end{pmatrix} + \text{h.o.t.}$$

and in order to obtain nontrivial solutions we need that $V > \emptyset$ provided $U < u_a$. We look for bifurcation from $(U,V) = (\emptyset,\emptyset)$ to solutions positive in V.

The well known theorems of global bifurcation apply to equations of the form $\omega - T(a.\omega) = \emptyset$ when T is compact and continuous and its Fréchet derivative $T_\omega(a,\emptyset)$ is of the form $aK\omega$ or $aK_1\omega + K_2\omega$. But in our case the dependence of T_ω on a is more complicated and we need a slightly different version of the classical global bifurcation theorems which can be found in (3). The theorem states that if the index (local Leray-Schauder degree) of the zero solution of the linearised operator changes when some value $a_\emptyset$ is crossed by the bifurcation parameter, then this $a_\emptyset$ is a bifurcation value and the bifurcation has global character in the sense that the continuum of solutions emanating from this point either goes to infinity or to another bifurcation point. This result can be proved by using exactly the same argument as in Rabinowitz (14). The index is easily computed by investigating the eigenvalues of T_ω. The change of index corresponds to the linearised operator becoming singular and we also need that the corresponding eigenfunctions are positive, at least for the V component.

The preliminary results required are listed in the following lemmas:

LEMMA 4: All the eigenvalues of the operator $- \Delta - a + 2a_1 u_a$ are positive if $a > \lambda_1$. (A similar result holds for b,v)

LEMMA 5: The map $a \to u_a$ is a C^1 map from (λ_1,∞) to $C^{2+\alpha}(\Omega)$ and du_a/da is positive in Ω on the same interval.

LEMMA 6: Suppose (u,v) is a nontrivial solution of the system

(i) $u < u_a$ and $v < b_1^{-1}(b + b_2\, a/(a_1 + ca))$

(ii) if $b > \lambda_1$ then $v > v_b$.

In order to be precise, we treat a as a bifurcation parameter and assume that all the other constants are fixed. When $b > \lambda_1$ the decoupling technique works and it can be proved with the methods of the previous paragraph:

THEOREM 7: Suppose $b > \lambda_1$. The system has a nontrivial solution provided $a > \lambda_1(a_2 v_b)$.

Thus we assume that $b < \lambda_1$. Using the linearisation around $(u_a,\emptyset)$ we see that we expect bifurcation from $(\emptyset,\emptyset)$ to $(u_a,\emptyset)$, u_a positive, when a crosses λ_1 and afterwards the linearised operator is singular only when $b = \lambda_1(-b_2u_a/(1+cu_a))$ with positive characteristic functions, due to Lemma 5, and we expect bifurcation to positive solutions from this point.

The following result can be proved using the methods described above and the fact that $\lambda_1(-b_2u_a/(1+cu_a))$ is decreasing with respect to a and its infimum is $\lambda_1 - b_2/c$.

THEOREM 8: If $\lambda_1 - b_2/c < b < \lambda_1$ then the system has a nontrivial solution provided $b > \lambda_1(-b_2u_a/(1+cu_a))$.

The branch of positive solutions extends globally due to the a priori bounds and stays in the positive cone because only the first eigenvalue possesses positive eigenfunctions and the branch cannot go to another eigenvalue without the operator becoming singular, precisely abandoning the cone and contradicting the simplicity of this eigenvalue to positive eigenfunctions.

Similar results can be proved with this method taking b as a bifurcation parameter and other linear boundary conditions can be studied as well.

5. Discussion

The method explained in the last paragraph compares advantageously to the decoupling method: first, it is conceptually simpler; secondly, it can be used in more general conditions; thirdly, the decoupling method can become very cumbersome for systems of more than two equations, while the other method only requires linearisation and a local condition on the linearised operator.

But in some cases decoupling has some advantages because it reduces the analysis of the system to a better known analysis of a single equation. See, for instance (10) and (11). The method was used by Rothe (15) with one of the equations being linear and by Brown (4) with semilinear equations.

Related work has been done, among others, by Dancer (8) and (9) using index theory; Leung (12) using iteration methods; Conway-Gardner-Smoller (6), Cantrell and Cosner (5). This work was done with K.J. Brown at Heriot-Watt University.

References

1. Blat, J: Ph.D. Thesis, Heriot-Watt University.
2. Blat, J, Brown, K.J: Bifurcation of steady-state solutions in predator-prey and competition systems, Proc Roy Soc Edinburgh 97A (1984), 21-34.
3. Id.: Global bifurcation of positive solutions in some systems of elliptic equations, to appear in SIAM J Math Anal.
4. Brown, K.J: Spatially inhomogeneous steady-state solutions for systems of equations describing interacting populations, J. Math Anal Appl 95 (1983), 22-37.
5. Cantrell, R.S, Cosner, C: On the steady-state problem for the Volterra-Lotka competition model with diffusion, preprint.
6. Conway, E, Gardner, R, Smoller, J: Stability and bifurcation of steady-state solutions for predator-prey equations, Adv Appl Math 3 (1982), 288-334.
7. Crandall, M.G, Rabinowitz, P.H: Bifurcation from simple eigenvalues, J. Funct Anal 8 (1971), 321-340.
8. Dancer, E.N: On positive solutions of some pairs of differential equations I, Trans Amer Math Soc 284 (1984), 729-743.

9. Id.: On positive solutions of some pairs of differential equations II, J Diff Eqns.
10. de Figueiredo D.G, Mitidieri, E: A maximum principle for an elliptic system and applications to semilinear problems, preprint.
11. Hernández, J: paper in this Proceedings.
12. Leung, A: Monotone schemes for semilinear elliptic systems related to Ecology, Math Methods Appl Sci 4 (1982), 272-285.
13. Lions P.L: On the existence of positive solutions of semilinear elliptic equations, SIAM Rev 24 (1982), 441-467.
14. Rabinowitz P.H: Some global results for nonlinear eigenvalue problems, J. Funct Anal 7 (1971), 487-513.
15. Rothe, F: Some analytical results about a simple reaction-diffusion system for morphogenesis, J Math Biol 7 (1979), 375-384.

J. Blat
Facultat Ciències
Departament de Matemàtiques
Universitat Illes Balears
07071 Palma de Mallorca
Spain.

H BREZIS

Liquid crystals and S^2-valued maps

This is a brief report about some recent results obtained in collaboration with J.N. Coron and E. Lieb (see [2] and [3]). The original motivation of our work comes from the theory of liquid crystals. Nematic liquid crystals consist of rod-like molecules whose orientation $\phi(x)$ is well defined at every point $x \in \Omega$ (the container) and varies smoothly except at a finite number of points. These points are called the defects of the liquid crystal and they are well observed experimentally (see e.g. [4]). Thus, we may view $\phi(x)$ as a map from Ω to S^2, the unit sphere of $\mathbb{R}^3$.

The mathematical modelling involves a deformation energy; the most general energy used by physicists (see [5], [6], [7]) is

$$\tilde{E}(\phi) = K_1 \int_\Omega (\operatorname{div} \phi)^2 + K_2 \int_\Omega (\phi \cdot \operatorname{curl} \phi)^2 + K_3 \int_\Omega |\phi \wedge \operatorname{curl} \phi|^2.$$

In the special case where $K_1 = K_2 = K_3 = 1$ (the so-called one constant approximation), $\tilde{E}(\phi)$ reduces (except for a boundary integral) to

$$E(\phi) = \int_\Omega |\nabla\phi|^2 \tag{1}$$

which we shall use as energy; it is an interesting open problem to extend our results to $\tilde{E}$.

A natural question, raised by J. Ericksen, is to study the least value of the deformation energy needed to produce singularities at a prescribed location. More precisely, let Ω be any domain in $\mathbb{R}^3$ and fix N points $a_1, a_2, \ldots, a_N$ in Ω (the observed position of the singularities) and consider the class of admissible maps

$$\mathcal{E} = \{\phi \in C^1(\Omega \setminus \bigcup_{i=1}^{N} \{a_i\}; S^2) \mid \int_\Omega |\nabla\phi|^2 < \infty \text{ and } \deg(\phi, a_i) = d_i\}.$$

Here $\deg(\phi, a_i)$ is the Brouwer degree of ϕ restricted to a small sphere centred at a_i and the d_i's are given integers (in $\mathbb{Z}$). So that the data of the problem are the domain Ω, the points (a_i) and the integers (d_i).

When $\Omega \neq \mathbb{R}^3$, the class E is always non-empty (this is not totally obvious, but follows from the construction sketched below). When $\Omega = \mathbb{R}^3$ the class E is non-empty iff $\Sigma d_i = 0$. Indeed the assumption $\int_{\mathbb{R}^3} |\nabla\phi|^2 < \infty$ implies that $\phi \to 0$ at infinity (in a weak sense) and therefore the total degree on a large sphere is zero.

Set

$$E = \inf_{\phi \in E} \int_\Omega |\nabla\phi|^2. \tag{2}$$

It tuns out - to our surprise and delight - that there is an explicit formula for E

THEOREM 1:

$$E = 8\pi L \tag{3}$$

where L is the length of a minimal connection.

The notion of a minimal connection will be explained below but we shall introduce first a mathematical extension of the above problem where points are replaced by "holes". Namely, consider N + 1 disjoint compact subsets of $\mathbb{R}^3$: $H_0, H_1, \ldots, H_N$ (in fact, one of them, say H_0, could possibly be just closed and not compact). Set

$$U = \mathbb{R}^3 \setminus \left(\bigcup_{i=0}^{N} H_i \right)$$

and consider the class of admissible maps

$$E = \{\phi \in C^1(U;S^2) \mid \int_U |\nabla\phi|^2 < \infty \text{ and } \deg(\phi,H_i) = d_i\}.$$

Here, $\deg(\phi,H_i)$ is the Brouwer degree of ϕ restricted to a smooth surface "surrounding" H_i (if H_i is the closure of a smooth domain and ϕ is smooth on $\bar{U}$, then $\deg(\phi,H_i)$ is simply the Brouwer degree of ϕ restricted to ∂H_i). Again, the d_i's are given integers (positive, negative or zero) and, as above, we must assume that

$$\sum_{i=0}^{N} d_i = 0 \tag{4}$$

in order to have $E \neq \emptyset$ (if H_0 is unbounded, then $\deg(\phi,H_0)$ is not well defined and we define it to be $-\sum_{i=1}^{N} \deg(\phi, H_i))$.

Using the same notation (2) as above, the conclusion of Theorem 1 still holds. Note that this setting is more general than the first one; just take $H_0 = \mathbb{R}^3 \setminus \Omega$ and $H_i = \{a_i\}$.

In order to state the definition of a minimal connection we have to start with the notion of a reduced distance D. Given two points x,y in $\mathbb{R}^3$ we set

$$\chi(x,y) = \begin{cases} 0 & \text{if for some } 0 \leq i \leq N,\ x \in H_i \text{ and } y \in H_i \\ |x-y| & \text{otherwise,} \end{cases}$$

and

$$D(x,y) = \operatorname*{Inf}_{C(x,y)} \sum_{i=1}^{p} \chi(x_{i-1}, x_i)$$

where the infimum is taken over all chains C(x,y) joining x to y, i.e. finite sequences $x_0, x_1, \ldots, x_p$ such that $x_0 = x$ and $x_p = y$.

Note that D(x,y) is only a semi-distance (since $D(x,y) = 0$ iff $x = y$ or $x, y \in H_i$ for some i); in other words the passage through the holes is "free". Given two sets A, B in $\mathbb{R}^3$, their reduced distance is defined by

$$D(A,B) = \operatorname{Inf} \{D(x,y) \mid x \in A \text{ and } y \in B\}.$$

The notion of a minimal connection is most easily explained first <u>in the case where all the degrees are ± 1</u>. Because of assumption (4), there are as many pluses as minuses. We relabel the holes (H_i) by distinguishing the positive holes $P_1, P_2, \ldots, P_k$ and the negative holes $N_1, N_2, \ldots N_k$. The definition of L is

$$L = \operatorname*{Min}_{\sigma} \sum_{i=1}^{k} D(P_i, N_{\sigma(i)}) \tag{5}$$

where the minimum is taken over the set of all permutations σ of the integers {1,2,...,k}.

In the <u>general case</u> where the d_i's are arbitrary integers one proceeds as above, except that:

(i) holes of degree zero do not appear in the (P), (N) list,

(ii) a hole H_i of positive (resp. negative) degree is repeated in the (P) list (resp. (N) list) according to its multiplicity $|d_i|$.

[Note that the holes of degree zero still influence L because they enter in the definition of D].

Sketch of the proof of Theorem 1 (for holes) There are two steps

Step 1 $E \leq 8\pi L$

Step 2 $E \geq 8\pi L$.

Step 1 The upper bound $E \leq 8\pi L$

Let σ be minimizing in (5). Realize $D(P_i, N_{\sigma(i)})$ as a union of line segments running between the holes (H_j) with a consistent sequence of orientations starting at P_i and ending at $N_{\sigma(i)}$. Then, take the union of all these line segments for $1 \leq i \leq k$; denote it (including multiplicity) by

$$\bigcup_{j \in J} [p_j, n_j].$$

All these line segments are either disjoint or coincide or intersect at their end points (but they have no self intersection because the connection is minimal).

On each segment $[p_j, n_j]$ one uses the basic construction described in

LEMMA 1: Given two points $p, n \in \mathbf{R}^3$, an integer $d > 0$ and $\varepsilon > 0$ there is a map $\phi_\varepsilon \in C^1(\mathbf{R}^3 \setminus \{p,n\}; S^2)$ such that

$$\deg(\phi_\varepsilon, p) = d, \quad \deg(\phi_\varepsilon, n) = -d, \tag{6}$$

$$\int |\nabla\phi_\varepsilon|^2 \leq 8\pi\, |d| \;\; |p-n| + \varepsilon \tag{7}$$

and

$$\phi_\varepsilon \text{ is constant outside an } \varepsilon\text{-neighbourhood of } [p,n]. \tag{8}$$

The construction of ϕ_ε is explicit (see [3]). The maps associated with all the segments $[p_j, n_j]$ glue well (because of (8) - in fact ϕ_ε is constant outside an ε-cone near p and n). In this way one obtains a map $\phi \in E$ such that

$$\int |\nabla\phi|^2 \leq 8\pi L + \varepsilon\, (\text{card } J).$$

Step 2 The lower bound $E \geq 8\pi L$

It is extremely convenient to associate with every map $\phi \in \mathcal{E}$ a vector field D (a kind of electric field) defined by its coordinates:

$$D = (\phi \cdot \phi_y \wedge \phi_z, \ \phi \cdot \phi_z \wedge \phi_x, \ \phi \cdot \phi_x \wedge \phi_y).$$

The vector field D has some remarkable properties. First, we have

$$|D| \leq \frac{1}{2} \ |\nabla\psi|^2. \tag{9}$$

Next, we have

$$\text{div } D = 0 \text{ in } \mathcal{D}'(U) \tag{10}$$

(note that $D \in L^1$ and thus (10) makes sense in $\mathcal{D}'$). Finally, if Σ is any smooth surface in U, then we have

$$\frac{1}{4\pi} \int_\Sigma D\cdot\nu = \deg (\phi|_\Sigma)$$

where ν is the normal to Σ; this follows from the fact that $D\cdot\nu = J_\phi$, the Jacobian determinant of ϕ restricted to Σ and also from the analytic formula for the degree (see e.g. [9])

$$\deg (\phi|_\Sigma) = \frac{1}{4\pi} \int_\Sigma J_\phi.$$

In particular, if we assume that each hole H_i has a smooth boundary and that ϕ is smooth on $\bar{U}$ we obtain

$$\int_{\partial H_i} D\cdot\nu = 4\pi \ d_i. \tag{11}$$

Let $\zeta : \mathbb{R}^3 \to \mathbb{R}$ be any function such that $\|\nabla\zeta\|_{L^\infty} \leq 1$ and ζ is constant on each H_i, $0 \leq i \leq N$. We have, from (9), (10), (11)

$$\int_U |\nabla\phi|^2 \geq 2 \int_U |D| \geq 2 \int_U D\cdot\nabla\zeta = 8\pi \sum_{i=0}^{N} d_i \ \zeta(H_i).$$

Note that a function $\zeta: \mathbb{R}^3 \to \mathbb{R}$ such that

$$|\zeta(x) - \zeta(y)| \leq D(x,y) \quad \forall x,y \in \mathbb{R}^3$$

satisfies $\|\nabla\zeta\|_{L^\infty} \leq 1$ and ζ is constant on each H_i, $0 \leq i \leq N$. Therefore, if we introduce the (P), (N) notation for the holes we find

$$\int_U |\nabla\phi|^2 \geq 8\pi \underset{\zeta\in K}{\text{Max}} \sum_{i=1}^{k} (\zeta(P_i) - \zeta(N_i))$$

where

$$K = \{\zeta:\mathbb{R}^3 \to \mathbb{R} \mid \ |\zeta(x) - \zeta(y)| \leq D(x,y) \quad \forall x,y \in \mathbb{R}^3\}.$$

We conclude easily with the following general Lemma:

LEMMA 2: Let M be a metric space with distance D(x,y) and let $P_1,P_2,\ldots,P_k$ and $N_1,N_2,\ldots,N_k$ be 2k points in M.

Then

$$\underset{\zeta\in K}{\text{Max}} \ \{\sum_{i=1}^{k} (\zeta(P_i) - \zeta(N_i))\} = L$$

where

$$K = \{\zeta:M \to \mathbb{R} \mid \ |\zeta(x) - \zeta(y)| \leq D(x,y) \quad \forall x,y \in M\}$$

and

$$L = \underset{\sigma}{\text{Min}} \sum_{i=1}^{k} D(P_i, N_{\sigma(i)}).$$

Lemma 2 may be derived (as in [4]) from a min-max equality of Kantorovich combined with Birkhoff's theorem on the extreme points of the set of doubly stochastic matrices; for an alternative direct proof see [1].

In most cases the infimum E (in (2)) is not achieved. But, it may be achieved for some exceptional geometries such as concentric balls (this is presumably the only case where E is achieved). Here is a complete result in a special case:

THEOREM 2: Assume $\Omega = \mathbb{R}^3$ and the holes are points. Then, E is not achieved. Moreover if (ϕ_n) denotes a minimizing sequence, then there exists a subsequence (ϕ_{n_k}) which converges to a constant a.e. and such that $|\nabla\phi_{n_k}|^2$

converges, in the sense of measures, to

$$8\pi \sum_{i=1}^{k} \delta_{[p_i,\, n_{\sigma(i)}]}$$

where σ is a minimizing permutation and δ_I is the one-dimensional Hausdorff measure uniformly distributed over the segment I.

For the proof of Theorem 2, see [3]. Let me just mention one ingredient which may be of interest for other problems:

LEMMA 3: Let $\mathcal{O}$ be a domain in $\mathbb{R}^2$ and let (ψ^n) be a sequence of maps from $\mathcal{O}$ to S^2 such that

$$\int |\nabla\psi^n|^2 \leq C.$$

We may always assume that

$$\psi^n \to \psi \quad \text{a.e. on } \mathcal{O}$$

$$\nabla\psi^n \rightharpoonup \nabla\psi \quad \text{weakly in } L^2 .$$

Set

$$f^n = \psi^n \cdot \psi^n_x \wedge \psi^n_y,$$

so that (f^n) is bounded in L^1 and a subsequence (f^{n_k}) converges weak * to some measure μ. Then

$$\mu = \psi \cdot \psi_x \wedge \psi_y + 4\pi \sum_{\text{finite}} n_j \delta_{c_j}$$

where the n_j are integers (in $\mathbb{Z}$) and the δ_{c_j} are Dirac masses at points c_j in $\mathcal{O}$.

This type of conclusion is related to the theory of concentration compactness developed by P.L. Lions (see [8]). Here, the correction term involves finitely many Dirac masses with integer weights because of geometric factors. A typical example is the following. Let ψ be a smooth map from $\mathbb{R}^2$ to S^2 which is a constant C far out. Let $\psi^n(x,y) = \psi(nx,ny)$, so that $\psi^n \to C$ a.e. and $\nabla\psi^n \rightharpoonup 0$ weakly in L^2. It is clear that $f^n \rightharpoonup \alpha\delta_0$ where $\alpha = \int \psi \cdot \psi_x \wedge \psi_y$, which is precisely 4π deg ψ (by stereographic projection ψ may be viewed as a map from S^2 to S^2, and thus deg ψ is well defined). The way Lemma 3 enters in the proof of Theorem 2 is by restricting (ϕ_n) to some appropriate (two dimensional) planes.

References

1. H. Brezis, Liquid crystals and energy estimates for S^2-valued maps (to appear in [7]).
2. H. Brezis - J.M. Coron - E. Lieb, Estimations d'energie pour des application de $\mathbb{R}^3$ à valeurs dans S^2, C.R.Acad.Sc.Paris 303 (1986), p. 207-120.
3. H. Brezis - J.M. Coron - E. Lieb, Harmonic maps with defects, Comm. Math. Phys. 107 (1986), p. 649-705.
4. W.F. Brinkman - P.E. Cladis, Defects in liquid crystals, Physics Today May 1982, p. 48-54.
5. P.G. De Gennes, The physics of liquid crystals, Clarendon Press, Oxford (1974).
6. J.L. Ericksen, Equilibrium theory of liquid crystals, in Advances in liquid crystals 2, G. Brown ed. Acad. Press New York (1976) p. 233-299.
7. J.L. Ericksen - D. Kinderlehrer ed., Proceedings I.M.A. Workshop on the theory and applications of liquid crystals, Springer (to appear).
8. P.L. Lions, The concentration compactness principle in the calculus of variations. The limit case, Riv. Mat. Iberoamericana 1 (1985) p. 45-121 and 145-201.
9. L. Nirenberg, Topics in nonlinear functional analysis, New York University Lecture Notes, New York (1974).

H. Brezis
Département de Mathématiques
Université Pierre Marie Curie
4, pl. Jussieu
75252 Paris Cedex 05
France.

J CARRILLO

On the uniqueness of the solution of a class of elliptic equations with nonlinear convection

Abstract

In this paper we study the existence and uniqueness of Kruskov solution of the problem

$$u \in H_0^1(\Omega),\ -\Delta u + \operatorname{div}(\beta(u)) + au \ni f,$$

where β is not necessarily continuous.

1. Introduction

1.1 Statement of the problem

Let Ω be a bounded, connected open set of $\mathbb{R}^N$, $N \geq 1$, with Lipschitz continuous boundary Γ, for $N \geq 2$.

Let a be a real valued function defined in Ω, such that:

$$a \in L^\infty(\Omega),\ a \geq 0 \text{ a.e. in } \Omega. \tag{1.1}$$

Let β be defined by:

$$\left.\begin{array}{l}\beta = (\beta_1, \beta_2, \ldots, \beta_N), \text{ where } \beta_i : \mathbb{R} \to P(\mathbb{R}) \text{ are multivalued} \\ \text{operators for } i = 1, \ldots, N.\end{array}\right\} \tag{1.2}$$

$$\beta = \alpha + \mu \operatorname{sig}^+ \tag{1.3}$$

where:

$$\alpha = (\alpha_1, \alpha_2, \ldots, \alpha_N), \text{ with } \alpha_i \in C^0(\mathbb{R}),\ \alpha_i(0) = 0 \text{ for} \tag{1.4}$$

$i = 1,2,\ldots,N.$

$$\operatorname{sig}^+(s) = \begin{cases} 1 & \text{if } s > 0 \\ [0,1] & \text{if } s = 0 \\ 0 & \text{if } s < 0 \end{cases} \tag{1.5}$$

and

$$\mu = (\mu_1, \mu_2, \ldots, \mu_N) \in \mathbf{R}^N \text{ is constant.} \tag{1.6}$$

Our goal is to study the problem (P):

$$(P) \begin{cases} \text{find a pair } (u,g) \in H_0^1(\Omega) \times L^\infty(\Omega) \text{ such that:} \\ \text{(i)} \quad g \in \text{sig}^+(u) \text{ a.e. in } \Omega \\ \text{(ii)} \quad -\Delta u - \text{div}(\alpha(u) + \mu g) + a\,u = f \text{ in } D'(\Omega). \end{cases}$$

where

$$f \in L^p(\Omega), \text{ for some p satisfying } p > 2 \text{ and } p > \frac{N}{2}. \tag{1.7}$$

We can also formulate (P) as:

$$(P) \begin{cases} \text{find a pair } (u,g) \in H_0^1(\Omega) \times L^\infty(\Omega) \text{ such that} \\ \text{(i)} \quad g \in \text{sig}^+(u) \text{ a.e. in } \Omega \\ \text{(ii)} \int_\Omega \{(\nabla u + \alpha(u) + \mu g).\nabla\xi + a.u.\xi\} = \int_\Omega f.\xi. \ \forall \xi \in H_0^1(\Omega). \end{cases}$$

Obviously, both formulations are equivalent.

Operators like this which appear in the formulation of (P), arise in many problems related to the filtrations of a fluid through a porous medium, the lubrication theory and others. (See [1], [2], [3]).

1.2 Preliminary results

We have:

THEOREM 1.1: Let us assume 1.1 to 1.7, then there exists a constant $C > 0$ independent of f, a, α, μ such that for every solution (u,g) of (P) we have:

$$-C\,\|f^-\|_{L^p(\Omega)} \leq u \leq C\,\|f^+\|_{L^p(\Omega)}, \text{ a.e. in } \Omega. \tag{1.8}$$

where $f^- = \min(f,0)$ and $f^+ = \max(f,0)$.

REMARK 1.1: Generally, if $\mu \neq 0$, the problem (P) has more than one solution. For instance, let (u_1, g_1) be

$$u_1(x) = \begin{cases} -\frac{x^2}{2} + \frac{3}{8}x & \text{if } 0 < x \leqq \frac{1}{2} \\ -\frac{x}{8} + \frac{1}{8} & \text{if } \frac{1}{2} \leqq x < 1 \end{cases}, \quad g_1(x) = 1 \text{ in } (0,1)$$

and let (u_2, g_2) be

$$u_2(x) = \begin{cases} -\frac{x^2}{2} + \frac{x}{4} & \text{if } 0 < x \leqq \frac{1}{2} \\ 0 & \text{if } \frac{1}{2} \leqq x < 1 \end{cases}, \quad g_2(x) = H(u_2(x)) \text{ in } (0,1)$$

then, for $i = 1,2$, (u_i, g_i) satisfies:

$$\begin{cases} -u_i'' + \frac{1}{4} g_i' = f = \begin{cases} 1 & \text{if } 0 < x \leqq \frac{1}{2} \\ 0 & \text{if } \frac{1}{2} < x < 1 \end{cases} \\ u_i(0) = u_i(1) = 0. \end{cases}$$

2. Kruskov solutions. Definition and existence

In this section, the assumptions are those of the one above. Moreover, we assume that

$$f \geqq 0 \text{ a.e. in } \Omega \tag{2.1}$$

REMARK 2.1: From (2.1) and theorem 1.1 we easily deduce that

$$u \geqq 0 \text{ a.e. in } \Omega. \tag{2.2}$$

First we shall prove some properties for the solutions of (P).

LEMMA 2.1: Let (u,g) be a solution of (P), then $\forall k \geqq 0$ and $\forall \xi \in H^1(\Omega)$, $\xi \geqq 0$ we have:

Proof see [5] theorem 1.

Also we have.

<u>THEOREM 1.2</u>: Let us assume 1.1 to 1.7, then there exists a solution (u,g) of (P). Moreover, all solutions (u,g) of (P) satisfy

$$u \in C^0(\bar{\Omega}).$$

<u>PROOF</u>. We approach the problem (P) by a problem (P_ε).

$$(P_\varepsilon)\begin{cases} \text{find } u_\varepsilon \in H_0^1(\Omega) \text{ such that } \forall\xi \in H_0^1(\Omega) \\ \int_\Omega \{(\nabla u_\varepsilon + \alpha(u_\varepsilon) + \mu\, H_\varepsilon(u_\varepsilon)).\ \nabla\xi + a\, u_\varepsilon\, \xi\} = \int_\Omega f\xi, \end{cases}$$

where H_ε is defined by

$$H_\varepsilon(s) = \begin{cases} 1 & \text{if } s > \varepsilon \\ \frac{s}{\varepsilon} & \text{if } \varepsilon \geqq s \geqq 0 \\ 0 & \text{if } 0 > s. \end{cases}$$

From theorem 1 of [4] we know that there exists a solution of (P_ε). Moreover, from theorem 1.1 we deduce that there exists a constant C such that $\forall\varepsilon$

$$\|u_\varepsilon\|_{L^\infty(\Omega)} \leqq C\, \|f\|_{L^p(\Omega)}.$$

Then it is well known (see [1], [3]) that there exists a subsequence of ε still denoted by ε such that

$$u_\varepsilon \to u \text{ in } H_0^1(\Omega) \text{ weakly}$$

$$H_\varepsilon(u_\varepsilon) \to g \text{ in } L^\infty(\Omega) \text{ weak-*}$$

where (u,g) is a solution of (P).

From theorem 8.30 of [5] we deduce that all solutions (u,g) of (P) satisfy: (see [3])

$$u \in C^0(\bar{\Omega}).$$

(i) $\lim_{\delta \to 0} \int_\Omega \nabla u . \nabla \min (\frac{(u-k)^+}{\delta}, \xi) \geq \int_{[u>k]} \nabla u . \nabla \xi$

(ii) $\lim_{\delta \to 0} \int_\Omega (\alpha(u) - \alpha(k)). \nabla \min (\frac{(u-k)^+}{\delta}, \xi)$

$= \int_{[u>k]} (\alpha(u) - \alpha(k)). \nabla \xi$

(iii) $\lim_{\delta \to 0} \int_\Omega (a\, u - f) \min (\frac{(u-k)^+}{\delta}, \xi) = \int_{[u>k]} (a\, u - f)\, \xi$,

where $[u > k] - \{x \subset \Omega / u(x) > k\}$.

PROOF. (i) and (iii) are obvious. To prove (ii) we define

$$G_i(r) = \int_0^r (\alpha_i(s+k) - \alpha_i(k))\, ds$$

for $i = 1,\dots,N$, and $G = (G_1,\dots,G_N)$. Then

$$\int_\Omega (\alpha(u) - \alpha(k)) \nabla \min (\frac{(u-k)^+}{\delta}, \xi)$$

$$= \int_{[u>k]} (\alpha(u) - \alpha(\min (u, \delta\xi + k))). \nabla \min (\frac{(u-k)}{\delta}, \xi)$$

$$+ \int_\Omega (\alpha(\min (u, \delta\xi + k)) - \alpha(k)). \nabla \min (\frac{(u-k)^+}{\delta}, \xi)$$

$$= \int_{[u>k]} (\alpha(u) - \alpha(\min(u, \delta\xi + k))). \nabla \xi$$

$$+ \frac{1}{\delta} \int_\Omega \operatorname{div} G (\min ((u-k)^+, \delta\xi)$$

$$= \int_{[u>k]} (\alpha(u) - \alpha(\min (u, \delta\xi + k))). \nabla \xi$$

since obviously $G(\min ((u-k)^+, \delta\xi)) \in (H_0^1(\Omega))^N$. Then by means of the Lebesgue theorem we get (ii).

LEMMA 2.2: Let (u,g) be a solution of (P), then $\forall k \geq 0$ $\forall \xi \in H^1(\Omega)$, $\xi \geq 0$ we have:

$$\int_{[u>k]} \{(\nabla u + \alpha(u) - \alpha(k)). \nabla \xi + (a\, u - f)\, \xi\} \leq 0.$$

PROOF. Obviously $\min\left(\frac{(u-k)^+}{\delta},\xi\right) \in H^1_0(\Omega)$ and

$$\int_\Omega g\,\mu.\ \nabla \min\left(\frac{(u-k)^+}{\delta},\xi\right) = 0$$

then from lemma 2.1 we deduce lemma 2.2.

Now we define:

DEFINITION 2.1: Let (u,g) be a solution of (P), then (u,g) is a Kruskov solution of (P) if and only if

$\forall\lambda \in [0,1]$, $\forall\xi \in H^1(\Omega)$, $\xi \geqq 0$ we have:

$$\int_\Omega \{(\nabla u + \alpha(u) + (g-\lambda)^+\mu).\nabla\xi + a\,u\,\xi\} \leqq \int_{[|\mu|g \geqq |\mu|\lambda]} f\xi.$$

THEOREM 2.1: There exists a Kruskov solution of (P).

PROOF. Let u_ε be a solution of the problem (P_ε) defined in the proof of theorem 1.2, then, from lemma 2.2 $\forall\xi \in H^1(\Omega)$, $\xi \geqq 0$, we have

$$0 \geqq \int_{[u_\varepsilon > \varepsilon\lambda]} \{(\nabla u_\varepsilon + \alpha(u_\varepsilon) - \alpha(\varepsilon\lambda) + (H_\varepsilon(u_\varepsilon) - H_\varepsilon(\varepsilon\lambda))\mu).\ \nabla\xi + (a\,u_\varepsilon - f)\xi\}$$

$$= \int_{[u_\varepsilon > \varepsilon\lambda\ ]} \{(\nabla u_\varepsilon + \alpha(u_\varepsilon) - \alpha(\varepsilon\lambda) + (H_\varepsilon(u_\varepsilon) - \lambda)\mu).\nabla\xi + (a\,u_\varepsilon - f)\xi\}\forall\lambda \in [0,1]$$

By taking the subsequence of ε, still denoted by ε, defined in the proof of theorem 1.2 and by letting $\varepsilon + 0$, since $f \geqq 0$ we get:

$$\int_\Omega\{(\nabla u + \alpha(u) + (g-\lambda)^+\mu).\ \nabla\xi + a\,u\,\xi\} \leqq \int_{[|\mu|g \geqq |\mu|\lambda]} f\xi$$

where (u,g) is a solution of (P) and therefore, a Kruskov solution of (P).

REMARK 2.1: When $\mu = 0$, i.e. β is continuous from lemma 2.2 we deduce that every solution of P is a Kruskov solution.

REMARK 2.2: When $\mu \neq 0$ we may have some solution which is not a Kruskov solution. For instance, in the example of the remark 1.1, the solution (u_2,g_2) is a Kruskov solution but (u_1,g_1) is not so.

3. Comparison - uniqueness of the Kruskov solution

The goal of this section is to compare a Kruskov solution with every other solution of (P).

First we shall prove a result concerning the solutions of (P):

LEMMA 3.1: Let (u,g) be a solution of (P), and let X_0 be the characteristic function of the set $\{x \in \Omega / u(x) = 0\}$, then we have:

(i) $\Delta u + \operatorname{div} \alpha(u) - au + f(1-X_0) \geqq 0$ in $D'(\Omega)$

(ii) $\mu . \nabla g + fX_0 \leqq 0$ in $D'(\Omega)$

(iii) $\int_{[|\mu|g \leq |\mu|\lambda]} (\lambda-g)\mu . \nabla\xi + f X_0 \xi \leqq 0 \quad \forall\lambda \in [0,1], \forall\xi \in D(\Omega), \xi \geqq 0.$

PROOF. We prove (i) and (ii) by applying lemma 2.1 with k = 0. If $\mu \neq 0$, from (ii), since $f \geqq 0$, we deduce that g is a nonincreasing function of $\mu.x$. Then, since Ω is Lipschitz, we can get a function F such that

$$\left.\begin{array}{lll} F \geqq 0 & \text{a.e. in } [g \leqq \lambda] \\ F = 0 & \text{a.e. in } [g > \lambda] \\ \mu . \nabla F = X_0 f & \text{a.e. } [g \leqq \lambda] \\ \mu . \nabla F = 0 & \text{a.e. [in } g > \lambda] \end{array}\right\} \quad (3.1)$$

Then, from (ii) we get:

$$\mu \nabla(g + F) \leqq 0 \text{ in } D'(\Omega) \quad (3.2)$$

and

$$[g + F \leqq \lambda] \Leftrightarrow [g \leqq \lambda] \quad (3.3)$$

and, therefore,

$$\int_{[|\mu|g \leq |\mu|\lambda]} (\lambda-g)\mu . \nabla\xi + f X_0 \xi = \int_\Omega (\lambda-g)^+ \mu . \nabla\xi + \mu . \nabla F \xi$$

$= \int_\Omega (\lambda-g-F)^+ \mu \nabla\xi \leqq 0$ since from (3.2) and (3.3) we get

$$\mu\nabla(\lambda-g-F)^+ \geqq 0.$$

If $\mu = 0$ then from (ii) we deduce that $f X_0 = 0$ and then (iii) is obvious. Also we prove

LEMMA 3.2: Let (u,g) be a Kruskov solution of (P), then we have:

$\forall\lambda \in [0,1]$, $\forall\xi \in H^1(\Omega), \xi \geq 0$

$$\lim_{\delta\to 0} \frac{1}{\delta} \int_{[u<\delta\xi]} \{|\nabla u|^2 + (\alpha(u) + (g-\lambda)^+\mu). \nabla u + (a\, u - f)\, u\}$$

$$+ \int_{[u=0]\cap[g\geq\lambda]} (f\xi - (g-\lambda)\mu\, \nabla\, \xi \geq 0$$

PROOF. Let $\xi \in H^1(\Omega)$, $\xi \geq 0$ and let $\zeta \in H^1_o(\Omega)$ be defined by

$$\zeta = \min\left(\frac{u}{\delta}, \xi\right),$$

and let $\lambda \in [0,1]$, then we have:

$$0 = \int_\Omega \{\nabla u.\ \nabla\zeta + (\alpha(u) + (g-\lambda)\ ).\ \nabla\zeta + (a\, u - f)\zeta\}.$$

Taking into account that (u,g) is a Kruskov solution and letting $\delta \to 0$ we get the result.

LEMMA 3.3: Let (u_1,g_1) be a Kruskov solution of (P) corresponding to a datum (f=) $f_1 \geq 0$ and let (u_2,g_2) be a solution of (P) corresponding to another datum (f=) $f_2 \geq f_1 \geq 0$, then $\forall\xi \in H^1(\Omega)$, $\xi \geq 0$ we have:

$$\int_{[u_1>u_2]} \{(\nabla(u_1-u_2) + (\alpha(u_1) - \alpha(u_2) + (g_1-g_2)\mu)\nabla\xi + \qquad (3.4)$$

$$+ (a(u_1-u_2) - f_1+f_2)\xi\}$$

$$+ \int_{[u_1=u_2=0]\cap[g_1\geq g_2]} \{(g_1-g_2)\mu\, \nabla\xi - (f_1-f_2)\xi\} \leq 0.$$

PROOF. Obviously it is sufficient to prove the inequality (3.4) for every nonnegative $\xi \in D(\bar{\Omega})$. Moreover, by means of a covering of $\bar{\Omega}$ and a corresponding C^∞-partition of unity, we can assume without loss of generality that $\xi \geq 0$ and $\xi \in D(\theta)$ where θ is an open subset of $\mathbb{R}^N$ such that $\theta \cap \partial\Omega$ is either a part of a Lipschitz graph, or empty.

Let $C \in D(\mathbb{R}^N)$, $C \geq 0$ $\operatorname{Supp}(C) \subset \beta(0,1)$, $\int_{\mathbb{R}^N} C dx = 1$ and let $C_\eta(x) = \frac{1}{\eta^N} C(\frac{x}{\eta})$, $\forall\eta > 0$.

It is easy to find a vector $\nu \in \mathbb{R}^N$, not depending on η, such that, for η

small enough, we can define a function $\zeta \in D(\mathbb{R}^{2N})$:

$$\zeta(x,y) = \xi(\frac{x+y}{2})\, C_{\eta,\nu}(\frac{x-y}{2}),$$

where $C_{\eta,\nu}(x) = C_\eta(x + \nu\eta)$, satisfying:

$$\forall x \in \Omega \quad \zeta(x,\cdot) \in D(\Omega). \tag{3.5}$$

Then, we consider the function $\psi \in H_0^1(\Omega \times \Omega)$ $\forall\delta > 0$:

$$\psi(x,y) = \min\left(\frac{(u_1(x)-u_2(y))^+}{\delta}, \zeta(x,y)\right).$$

Obviously, $\psi(x,\cdot) \in H_0^1(\Omega)$ for a.e. $x \in \Omega$ and $\psi(\cdot,y) \in H_0^1(\Omega)$ for a.e. $y \in \Omega$. Let x be a variable in $\mathbb{R}^N$ we set ∇_x for the operator $(\frac{\partial}{\partial x_1}, \ldots, \frac{\partial}{\partial x_N})$, then we have:

$$0 = \int_{\Omega\times\Omega}\{\nabla_x u_1(x)\nabla_x\psi(x,y) + (\alpha(u_1(x))-\alpha(u_2(y)) + (g_1(x)-g_2(y))\mu), \nabla_x\psi(x,y) + (a(x)u_1(x) - f_1(x))\psi(x,y)\}\,dx\,dy \tag{3.6}$$

and

$$0 = \int_{\Omega\times\Omega}\{\nabla_y u_2(y)\nabla_y\psi(x,y) + (\alpha(u_2(y)) - \alpha(u_1(x)) + (g_2(y) - g_1(x))\mu), \nabla_y\psi(x,y) + (a(y)u_2(y) - f_2(y))\psi(x,y)\}\,dx\,dy. \tag{3.7}$$

In (3.6) and (3.7) we introduce the change of variables: $t = \frac{x+y}{2}$ and $z = \frac{x-y}{2}$. Obviously the Jacobian of this change is a constant matrix. Then by substracting (3.7) from (3.6) we get:

$$\begin{aligned} 0 = \int_D \{&\tfrac{1}{2}\nabla_t(u_1(t+z)-u_2(t-z))\nabla_t\psi(t+z,t-z) + \\ &+ \tfrac{1}{2}\nabla_z(u_1(t+z)-u_2(t-z))\nabla_z\psi(t+z,t-z) + \\ &+ (\alpha(u_1(t+z))-\alpha(u_2(t-z))+(g_1(t+z)-g_2(t-z))\mu).\nabla_t\psi(t+z,t-z) + \\ &+ (a(t+z)u_1(t+z)-f_1(t+z)-a(t-z)u_2(t-z)+f_2(t-z))\psi(t+z,t-z)\}\,dt\,dz \end{aligned} \tag{3.8}$$

where $D = \{(t,z) \in \mathbb{R}^{2N}/(t+z,t-z) \in \Omega \times \Omega\}$.

By taking into account that $\psi(t+z,t-z) = \hat{\psi}(t,z) \in H_0^1(D)$ and that

$$\Delta_t(u_1(t+z)-u_2(t-z)) = \Delta_z(u_1(t+z)-u_2(t-z)) \text{ in } D'(D). \tag{3.9}$$

we get

$$0 = \int_D\{\nabla_t(u_1(t+z)-u_2(t-z))\nabla_t\psi(t+z,t-z) + \tag{3.10}$$

$$+ (\alpha(u_1(t+z))-\alpha(u_2(t-z))+(g_1(t+z)-g_2(t-z))\mu).\nabla_t\psi(t+z,t-z) +$$

$$+ (a(t+z)u_1(t+z)-f_1(t+z)-a(t-z)u_2(t-z)+f_2(t-z))\psi(t+z,t-z)\}\ dt\ dz.$$

and then

$$0 = \int_{D_\delta} \{\nabla_t(u_1(t+z)-u_2(t-z))\nabla_t\zeta(t+z,t-z) + \tag{3.11}$$

$$(\alpha(u_1(t+z))-\alpha(u_2(t-z))+(g_1(t+z)-g_2(t-z))\mu).\nabla_t\zeta(t+z,t-z)$$

$$+ (a(t+z)u_1(t+z)-f_1(t+z)-a(t-z)u_2(t\div z)+f_2(t-z))\zeta(t+z,t-z)\}\ dt\ dz.$$

$$+ \frac{1}{\delta} \int_{(D\setminus D_\delta)\cap[u_2>0]} |\nabla_t(u_1(t+z)-u_2(t-z))^+|^2\ dt\ dz$$

$$+ \frac{1}{\delta} \int_{(D\setminus D_\delta)\cap[u_2>0]} (\alpha(u_1(t+z))-\alpha(u_2(t-z))).\nabla_t(u_1(t+z)-u_2(t-z))^+ dt\ dz$$

$$+ \frac{1}{\delta} \int_{(D\setminus D_\delta)\cap[u_2>0]} (a(t+z)u_1(t+z)-f_1(t+z)-a(t-z)u_2(t-z)+f_2(t-z))$$

$$(u_1(t+z)-u_2(t-z))^+\ dt\ dz$$

$$+ \frac{1}{\delta} \int_{(D\setminus D_\delta)\cap[u_2=0]} \{|\nabla_t u_1(t+z)|^2+(\alpha(u_1(t+z)+(g_1(t+z)-g_2(t-z))\mu).$$

$$\nabla_t u_1(t+z) + (a_1(t+z)u_1(t+z)-f_1(t+z)+f_2(t-z))u_1(t+z)\}\ dt\ dz.$$

Obviously, when $\delta \to 0$, the first integral converges to:

$$I_1 = \int_{[u_1>u_2]} \{\nabla_t(u_1(t+z)-u_2(t-z)).\nabla_t\zeta(t+z,t-z) +$$

$$+ (\alpha(u_1(t+z)-\alpha(u_2(t-z)+(g_1(t+z)-g_2(t-z))\mu).\nabla_t\zeta(t+z,t-z)$$

$$+ (a(t+z)u_1(t+z)-f_1(t+z)-a(t-z)u_2(t-z)+f_2(t-z))\zeta(t+z,t-z)\} \, dt \, dz$$

The second integral in (3.11) is nonnegative. From the Lebesgue theorem the fourth integral converges to 0. If we denote by **c** the absolute value of the determinant of the change of variables we have:

$$\frac{1}{\delta}\int_{(D\setminus D_\delta)\cap[u_2>0]}(\alpha(u_1(t+z))-\alpha(u_2(t-z))).\nabla_t(u_1(t+z)-u_2(t-z))^+ dt \, dz$$

$$= \frac{c}{\delta}\int_{[u_1(x)-u_2(y)<\delta\zeta(x,y)]\cap[u_2(y)>0]}(\alpha(u_1(x))-\alpha(u_2(y))).(\nabla_x(u_1(x)-u_2(y))^+$$

$$+ \nabla_y(u_1(x)-u_2(y))^+)dx \, dy$$

As we have done in lemma 2.1(ii), it is easy to prove that this integral converges to 0 when $\delta \to 0$. Finally, denoting by J_δ the last integral in (3.11) we have:

$$J_\delta = \frac{c}{\delta}\int_{[u_2(y)=0]}\int_{[u_1(x)<\delta\zeta(x,y)]}\{|\nabla_x u_1(x)|^2+(\alpha(u_1(x))+(g_1(x)-g_2(y))\mu).$$

$$\nabla_x u_1(x) + (a_1(x)u_1(x)-f_1(x) + f_2(y))u_1(x)\} \, dx \, dy,$$

then, from lemma 3.2 we have:

$$\lim_{\delta\to 0} J_\delta \geq c\int_{[u_2(y)=0]}\int_{[u_1(x)=0]\cap[|\mu|g_1(x)\geq|\mu|g_2(y)]}\{(g_1(x)-g_2(y))$$

$$\mu\nabla_x\zeta(x,y)-f_1(x)\zeta(x,y)\} \, dx \, dy$$

and from lemma 3.1 we get:

$$\lim_{\delta\to 0} J_\delta \geq c\int_{[u_2(y)=0]}\int_{[u_1(x)=0]\cap[|\mu|g_1(x)\geq|\mu|g_2(y)]}\{(g_1(x)-g_2(y))\mu.$$

$$(\nabla_x\zeta(x,y) + \nabla_y\zeta(x,y)) - (f_1(x) - f_2(y))\zeta(x,y)\} \, dx \, dy$$

$$= \int_{[u_1=u_2=0]\cap[|\mu|g_1\geq|\mu|g_2]}\{(g_1(t+z)-g_2(t-z))\mu.\nabla_t\zeta(t+z,t-z)$$

$$- (f_1(t+z) - f_2(t-z))\zeta(t+z,t-z)\} \, dt \, dz = I_2$$

and then we get:

$$I_1 + I_2 \leqq 0. \tag{3.12}$$

Taking into account that I_1 and I_2 do not involve any derivate of $C_{\eta,\nu}$, and by letting $\eta \to 0$ we get (3.4), which achieves the proof.

Now we can prove:

THEOREM 3.1: Let (u_1,g_1) and (u_2,g_2) be defined as in lemma 3.3, with $f_2 \geqq f_1 \geqq 0$, and let us assume that at least one of the following conditions is true:

(i) $a > 0$ a.e. in Ω

(ii) $\exists i_0$, $1 \leqq i_0 \leqq N$ such that $\alpha_{i_0} + \mu_{i_0}$ sig$^+$ is either nonincreasing or nondecreasing,

then we have:

$$u_1 \leqq u_2 \text{ in } \Omega \text{ and } |\mu|g_1 \leqq |\mu|g_2 \text{ a.e. in } \Omega.$$

PROOF. First we suppose that (i) is true, then by choosing $\xi \equiv 1$ in (3.4) we get:

$$\int_{[u_1>u_2]} a(u_1 - u_2) \leqq 0$$

and then $u_1 \leqq u_2$ in Ω. Then, from (3.4) we get

$$\int_\Omega (g_1-g_2)^+\mu\nabla\xi \leqq 0 \quad \forall\xi \in H^1(\Omega) \quad \xi \geqq 0$$

$$\Rightarrow \int_\Omega (g_1-g_2)^+\mu \nabla \xi = 0 \quad \forall\xi \in H^1(\Omega)$$

$$\Rightarrow \quad |\mu|\, g_1 \leqq |\mu|g_2 \text{ a.e. in } \Omega.$$

Now we suppose that (ii) is true. Obviously, we have

$$a(u_1-u_2)^+ - f_1 + f_2 \geqq 0 \text{ a.e. in } \Omega,$$

then from (3.4) we deduce that $\forall\xi \in H^1(\Omega)$, $\xi \geqq 0$ we have:

$$0 \geqq \int_{u_1 \geqq u_2} \{\nabla(u_1-u_2)\nabla\xi+(\alpha(u_1)-\alpha(u_2)) + (g_1-g_2)^+\mu \nabla\xi\} \tag{3.13}$$

we choose $\xi \geqq 0$, $\xi \in D(\mathbb{R}^N)$ satisfying $\frac{\partial \xi}{\partial x_i} = 0$ when $i \neq i_0$ and satisfying:

$$\int_{u_1 \geqq u_2} (\alpha_{i_0}(u_1) - \alpha_{i_0}(u_2) + (g_1 - g_2)^+ \mu_{i_o}) \frac{\partial \xi}{\partial i_o} \geqq 0$$

and satisfying:

$$\frac{\partial^2 \xi}{\partial x_{i_0}^2} < 0,$$

then from 3.13 we get

$$0 \geqq \int_{u_1 \geqq u_2} -(u_1 - u_2) \frac{\partial^2 \xi}{\partial x_{i_o}} + \int_{u_1 \geqq u_2} (\alpha_{i_0}(u_1) - \alpha_{i_0}(u_2) + (g_1 - g_2)^+ \mu_{i_0}) \frac{\partial \xi}{\partial x_{i_0}}$$

then we deduce easily that $u_1 \leqq u_2$ in Ω. As in the first part of the proof, we easily deduce that $|\mu| g_1 \leqq |\mu| g_2$ which achieves the proof.

THEOREM 3.2: Under the assumptions of theorem 3.1, there exists only one Kruskov solution of (P). Moreover, any other solution is greater than or equal to the Kruskov solution.

PROOF: It is a direct consequence of theorem 3.1.

References

1. Brezis, H. Kinderlehrer, D. Stampacchia, G.: "Sur une nouvelle formulation du problème de l'écoulement a travers une digue". C.R. Acad. Sci. Paris. Série A-287 (1978), 711-714.
2. Carrillo, J.: "On the evolution Dam problem". To appear.
3. Carrillo, J. Chipot, M.: "On the Dam problem". Journal of Diff. Equat. Vol. 45 n°2, (1982), p. 234-271.
4. Carrillo, J. Chipot, M.: "On some nonlinear elliptic equations involving derivates of the nonlinearity". Proc. Roy. Soc. Edinburgh, 100A, (1985), p. 281-294.

5. Gilbarg, D. Trudinger, N.S.: "Elliptic partial differential equations of second order". Springer (1977).

The author was partially supported by the I.M.A. of the University of Minnesota and by C.A.Y.C.I.T. no. 3308/83.

J. Carrillo
Departamento de Matématica Aplicada
Universidad Complutense de Madrid
28040 - Madrid
Spain.

T CAZENAVE

Nonlinear Dirac equations: existence of stationary states

1. Introduction and the main result

We present here a joint work with L. Vazquez, where we prove the existence of stationary states for nonlinear Dirac equations of the form

$$i \sum_{\mu=0}^{3} \gamma^{\mu}\partial_{\mu}\Psi - m\Psi + F(a(\Psi))\Psi = 0 \qquad (1.1)$$

The notation is the following:
Ψ is defined on $\mathbf{R}^4$ with values in C^4, $\partial_{\mu} = \partial/\partial x_{\mu}$, m is a positive constant, $a(\psi) = (\gamma^{0}\psi,\psi)$, where (,) is the usual scalar product in C^4, and γ^{μ} are the 4 × 4 matrices of the Pauli-Dirac representation, given by

$$\gamma^{o} = \begin{pmatrix} I & 0 \\ 0 & -I \end{pmatrix}, \quad \gamma^{k} = \begin{pmatrix} 0 & \sigma k \\ -\sigma k & 0 \end{pmatrix}, \quad k = 1,2,3, \text{ where}$$

$$\sigma 1 = \begin{pmatrix} 0 & 1 \\ 1 & 0 \end{pmatrix}, \quad \sigma 2 = \begin{pmatrix} 0 & -i \\ i & 0 \end{pmatrix}, \quad \sigma 3 = \begin{pmatrix} 1 & 0 \\ 0 & -1 \end{pmatrix}$$

Finally, $F : \mathbf{R} \to \mathbf{R}$ models the nonlinear interaction.

We are interested in stationary states, or localized solutions of (1.1), that is solutions Ψ of (1.1) of the special form $\Psi(z) = e^{-i\omega t}\,\phi(x)$ where $z = (t,x)$ and $x = (x_1,x_2,x_3)$ in addition, since we want finite energy solutions, we want ϕ to be square integrable.

Clearly, the equation for ϕ: $\mathbf{R}^3 \to C^4$ is:

$$i \sum_{k=1}^{3} \gamma^{k}\partial_{k}\phi - m\phi + \omega\gamma^{o}\phi + F(a(\phi))\phi = 0. \qquad (1.2)$$

Equations (1.1) and (1.2) arise in particle physics. Nonlinear spinor fields giving rise to such equations have been considered first by D. Ivanenko [8], H. Weyl [25] and W. Heisenberg [7]. Later R. Finkelstein, C.F. Fronsdal and P. Kaus [5] considered the case of a spinor field with several types of fourth order self couplings. In 1970, M. Soler [17] was the first to investigate the stationary states of the nonlinear Dirac field with the scalar fourth order self coupling (corresponding to $F(x) = x$ in (1.1)) proposing them as a model of elementary extended fermions. Subsequently, the electromagnetic interaction was introduced ([18], [14], [15]) in order to construct a model of extended charged fermion, which describes with a reasonable accuracy the properties of the nucleons [13]. To improve the model, the pseudoscalar fields were introduced in order to represent the cloud of pions [16], [6]. A summary of the above models with the numerical computations and further developments are described by A.F. Rañada [11], [12]. Also, the case $F(x) = x$ was considered by J. Rafelski [10], K. Takahashi [20] and T. Van der Merwe [21].

Our main result is the following:

<u>THEOREM 1.1</u>: Assume that $F \in C^1(R_+,R_+)$ is such that $F(0) = 0$, $F'(x) > 0$ for $x > 0$, and $F(x) \to +\infty$ as $x \to +\infty$. Then for any $\omega \in (0,m)$ there exists a solution ϕ of (1.2) such that:

(i) $\phi \in C^1(R^3,C^4)$

(ii) $\phi \not\equiv 0$

(iii) $a(\phi(x)) \geqq 0$ for $x \in R^3$

(iv) ϕ and $\nabla\phi$ have an exponential fall-off as $|x| \to +\infty$.

Theorem 1.1 calls for seferal remarks:

- F is only assumed to be defined on $[0, +\infty)$ because of property (iii)
- Obviously, $\phi \equiv 0$ is a solution of (1.2) but (ii) ensures that we find a nontrivial solution.

- Most important is property (iv) which implies in particular that $\phi \in H^1(R^3,C^4)$. As we shall see in section 2, step 4, there are infinitely many solutions of (2.1), (i), (ii), (iv), those satisfying (iii) being exceptional.

- The hypothesis of theorem (1.1) is satisfied for example when $F(x) = x$, or more generally for $F(x) = x^p$ with $p \geqq 1$. This is in great contrast with the Klein-Gordon equation (cf. [2] and the references therein) since here we have no limitation on the growth of the nonlinearity.

- We do not know if the condition $\omega \in (0,m)$ is necessary. However, if $\omega > m$ or $\omega < -m$, it is proved by L. Vazquez [22] that there exists no solution of (1.2) satisfying the conclusions of theorem 1.1. Also, it is easily verified that when $\omega < 0$ there exists no solution of the type which we construct. Finally, it seems that no elementary transformation of a solution for $\omega > 0$ can provide a solution for $\omega < 0$. For example the charge conjugation changes ω to $-\omega$, but changes also F to -F.

- While we study here the stationary states of (1.1), little seems to be known about the Cauchy problem (indeed if we consider the variable $z = (t,x)$ with $t \in R$ and $x = (x_1,x_2,x_3) \in R^3$, (1.1) is an evolution equation). It is clear that the linear part of (1.1) generates a group of isometries in $H^m(R^3)$ for any $m \in Z$. Since, for example, the nonlinearity is Lipschitz continuous from bounded sets of $H^2(R^3)$ to $H^2(R^3)$ if $F \in C^2(R,R)$, the existence of local solutions of (1.1) in $H^2(R^3)$ is easily obtained. However, we do not know whether all the solutions are global in time or if there exist solutions which blow up in finite time. The main difficulty is the lack of conservation laws involving positive quantities.

- Equation (1.2) can be put into a variational form. Indeed, let

$$E(\phi) = (1/2) \int_\Omega \{(L\phi,\phi) - m(\phi,\gamma^0\phi) + \omega(\phi,\phi) + G(a(\phi))\}dx$$

where $\Omega = R^3$, $G' = F$, and L is given by

$$L\phi = i \sum_{k=1}^{3} \gamma^k \partial_k \phi .$$

Then the critical points of E are solutions of (1.2). However we shall not solve (1.2) by a variational method.

- Concerning the stability of the stationary states, some authors claim that they are unstable [3], [9], [24], while others find regions of stable behaviour by using numerical computations [1]. A very recent work of W.A. Strauss and L. Vazquez [19] indicates that, at least in the case $F(x) = x$, the stationary states should be stable.

2. A shooting method for a system of ODEs

Following M. Wakano [23] and M. Soler [17] we seek solutions of (1.2) that are separable in spherical coordinates (r,θ,ζ), of the form

$$\phi(x) = \begin{pmatrix} v(r) \\ 0 \\ iu(r)\cos(\theta) \\ iu(r)\sin(\theta)\exp(i\zeta) \end{pmatrix} \tag{2.1}$$

The equations for u and v are (compare [23], [17])

$$\begin{aligned} u'' + (2/r)u &= v(F(v^2 - u^2) - (m-\omega)) \\ v' &= u(F(v^2 - u^2) - (m+\omega)) \end{aligned} \tag{2.2}$$

What we want is a solution (u,v) of (2.2) which exists for all $r \geqq 0$, which satisfies for example $v(0) > 0$ (in order that $\phi \not\equiv 0$), and such that $(u,v) \to (0,0)$ as $r \to +\infty$. Both the numerical computations and the calculations below show that "most" of the solutions of (2.2) do not converge to (0,0) as $r \to +\infty$. Therefore we shall use a shooting method in order to exhibit a solution with the prescribed properties. The method is in a way similar to the one used by H. Berestycki, P.L. Lions and L.A. Peletier [2] for a second order equation but the technical problems involved are different. We shall now study a slightly more general problem than (2.2). Throughout this section we shall make the following assumptions and notation:

- m and ω are real numbers such that $0 < \omega < m$.
- α is a real number, $\alpha > 0$.
- $g \in C^1(R,R)$ is an increasing function satisfying

  $g(0) = 0$, there exists $\zeta > 0$ such that $g(\xi) > m + \omega$, and $g'(g^{-1}(m-\omega)) > 0$.
- For convenience we set $\ell = (g^{-1}(m-\omega))^{1/2}$
- We define the functions G and H by

$$G'(x) = g(x) \text{ for } x \in R,\ g(0) = 0 \tag{2.3}$$

$$H(u,v) = (1/2)[G(v^2-u^2) - m(v^2-u^2) + \omega(v^2+u^2)] \text{ for } (u,v) \in R^2. \tag{2.4}$$

We shall consider the following system

$$\begin{aligned} u' + (\alpha/r)u &= v(g(v^2 - u^2) - (m-\omega)) \\ v' &= u(g(v^2 - u^2) - (m + \omega)) \end{aligned} \tag{2.5}$$

Our main result of this section is the following:

<u>THEOREM 2.1</u>: There exists a solution $(u,v) \in C^1(R,R^2)$ of (2.5) such that $u(0) = 0$ and $0 < u(r) \leqq v(r) \leqq Cte \exp(-(1/2)(m-\omega)r)$ for $r > 0$.

<u>REMARK 2.2</u>: It is clear that Theorem 1.1 is a consequence of Theorem 2.1. Indeed, let F satisfy the hypothesis of Theorem 1.1. Let g be defined by $g(x) = F(x)$ for $x \geqq 0$ and $g(x) = -F(-x)$ for $x \leqq 0$. Then g satisfies the hypothesis of Theorem 1.2. Now ϕ defined by (2.1) is a solution of (1.2) where F is replaced by g defined above. Furthermore, since $(u,v) \in C^1(R,R^2)$ we have $\phi \in C^1(R^3,C^4)$. Since $u(r) > 0$ for $r > 0$ we also have $\phi \lneqq 0$. An explicit computation shows that $a(\phi) = v^2-u^2$. Because $u(r) \leqq v(r)$ for $r \geqq 0$ we have $a(\phi(x)) \geqq 0$ for $x \in R^3$. Finally, the exponential fall-off of u and v shows that ϕ satisfies all the conclusions of Theorem 1.1.

We shall only describe the main steps of the proof of Theorem 2.1. The complete proof can be found in T. Cazenave and L. Vazquez [4]. We distinguish six steps:

- <u>Step 1</u>: The associated conservative system
- <u>Step 2</u>: Local existence of solutions
- <u>Step 3</u>: The set F.
- <u>Step 4</u>: F is a nonempty open set
- <u>Step 5</u>: Estimates for solutions with initial datum in F.
- <u>Step 6</u>: Conclusion.

<u>STEP 1</u>: The associated conservative system.
We consider the following system (2.6) which is obtained from (2.5) by suppressing the term $(2/\alpha)u$.

$$\begin{aligned} u' &= v(g(v^2 - u^2) - (m-\omega)) \\ v' &= u(g(v^2 - u^2) - (m+\omega)) \end{aligned} \tag{2.6}$$

Then (2.6) is the Hamiltonian system associated with H given by (2.4). H achieves its minimum (which is negative) at the points $(0, \pm\ell)$ and its only other critical point is $(0,0)$. Since $H(u,v) \to +\infty$ as $|u| + |v| \to +\infty$, all the solutions of (2.6) are global and bounded. Let $(u_0,v_0) \in R^2$ and let (u,v) be the solution of (2.6) such that $(u(0),v(0)) = (u_0,v_0)$. Then:

- If $H(u_0,v_0) < 0$, (u,v) is periodic and v is bounded away from 0

- If $H(u_0,v_0) > 0$, (u,v) is periodic and both u and v take positive and negative values

- If $H(u_0,v_0) = 0$ and $(u_o,v_o) = (0,0)$, (u,v) converges exponentially to $(0,0)$.

STEP 2: Local existence of solutions

We now want to solve system (2.5). Since we want regular solutions, we shall assume that $u(0) = 0$ and $v(0) = x$. Then (2.5) can be put into the following form:

$$u(r) = (1/r^{\alpha})\int_0^r s^{\alpha} v(s)(g(v(s)^2 - u(s)^2) - (m-\omega))ds$$
$$v(r) = x + \int_0^r u(s)(g(v(s)^2 - u(s)^2) - (m+\omega))ds \qquad (2.7)$$

Since the second hand of (2.7) is a Lipschitz continuous function of (u,v), we may apply a classical contraction mapping argument. We find that, for any $x \in R$, there exists $R_x > 0$ and $(u_x,v_x) \in C^1([0,R_x],R^2)$, a unique solution of (2.5) such that $u_x(0) = 0$ and $v_x(0) = x$. In addition we can assume that either $R_x = +\infty$ or else $R_x < +\infty$ and $|u_x(r)| + |v_x(r)| \to +\infty$ as $r \to +\infty$.

Finally, an easy computation shows that:

$$(d/dr)H(u_x(r),v_x(r)) = (\alpha/r)u_x^2(r)(g(v_x(r)^2-u_x(r)^2)-(m+\omega)) \qquad (2.8)$$

for $r \in (0,R_x)$.

STEP 3: The set F.

We shall restrict our attention to the region $C = \{(u,v) \in R^2,\ u \geqq 0$ and $v \geqq 0\}$. For convenience we also define two subsets of C.

$$C^+ = C \cap \{(u,v) \in R^2,\ v^2-u^2 \geqq m+\omega\},\ C^- = C \cap \{(u,v) \in R^2,\ v^2-u^2 \leqq m+\omega\}.$$

Now let $x \geqq \ell$. From the first equation of (2.5), we see that $u'(0) \geqq 0$: Hence (u_x,v_x) will remain in C for some time. We set, for $x \geqq \ell$:

$$r_x = \text{Sup}\ \{r \geqq 0,\ (u_x(s),\ v_x(s)) \in C \text{ for } s \in [0,r]\}$$

Of course, either $r_x = R_x$ or $r_x < R_x$. In the latter case, we have $(u_x(r_x),\ v_x(r_x)) \in \partial C$ and since $(0,0)$ is an equilibrium point, $|u_x(r_x)| + |v_x(r_x)| > 0$. Therefore, either $u_x(r_x) = 0$ and $v_x(r_x) > 0$ or else $v_x(r_x) = 0$ and $u_x(r_x) > 0$. We are interested in the last situation and we set:

$$F = \{x \geq \ell,\ r_x < R_x \text{ and } u_x(r_x) = 0\} \tag{2.8}$$

Step 4: F is a nonempty open set.

The open character of F is an easy consequence of the continuous dependence of (u_x, v_x) on x.

Since $H(0,\ell) < 0$ and $H(0,x) \to +\infty$ as $x \to +\infty$, there exists $x_0 \in (\ell, +\infty)$ such that $H(0,x_0) = 0$. We set $x_1 = \text{Min}\{x_0, g^{-1}(m+\omega)\}$ and we claim that $F \supset (\ell, x_1)$. Indeed, consider $x \in (\ell, x_1)$. It follows from the definition of x_1 that $(0,x) \in C^-$ for some time. By (2.8) we have $H(u_x,v_x) < H(0,x)$ as long as (u_x,v_x) remain in C^-. Easy estimates on H show that $\{(u,v),\ H(u,v) \leq H(0,x)\}$ does not intersect C^+. It follows easily that $(u_x(r), v_x(r)) \in C^-$ for $0 \leq r \leq r_x$. We now consider two cases:

- $r_x = +\infty$: Since $H(u_x(r), v_x(r)) < H(0,x) < 0$ for $r > 0$, $(u_x(r), v_x(r))$ is bounded away from $(0,0)$. On the other hand and since $(u_x(r), v_x(r)) \in C^-$ for $0 \leq r < +\infty$, we have $(v_x)'(r) < 0$ for $r > 0$. Hence v_x goes to a limit as $r \to +\infty$. Passing to the limit in (2.5), we see that u_x also goes to a limit as $r \to +\infty$. Hence, the limit of (u_x,v_x) is necessarily $(0,\ell)$. An expansion of the solution around the point $(0,\ell)$ rules out this case.

- $r_x < +\infty$: The set $\{(u,v),\ H(u,v) \leq H(0,x)\}$ being bounded we have $r_x < R_x$. The same set being bounded away from the u-axis, we have $u_x(r_x) \neq 0$. Therefore we have $u_x(r_x) = 0$ and $x \in F$.

STEP 5: Estimates for solutions with initial datum in F.

LEMMA 2.3: Let $x \in F$. Then:

(i) $0 < u_x(r) \leq v_x(r)$ for $r \in (0, r_x)$

(ii) There exists τ_x such that $v_x(r) \leq v_x(\tau_x)$ for $r \in [0, r_x)$.

SKETCH OF PROOF: We set $K^+ = \{(u,v) \in C,\ u \leq v\}$ and $K^- = \{(u,v) \in C,\ u > v\}$. In order to prove (i) we have to show that (u_x,v_x) remain in K^+ for $r \in (0, r_x)$. We argue by contradiction and we assume that there exists $r \in [0, r_x)$ such that $(u_x(r), v_x(r)) \in K^-$. Since $(0,x) \in K^+$ and $(u_x(r), v_x(r_x)) \in K^+$, there exist $r_1 > r_2$ such that $u_x(r_1) = v_x(r_1)$, $u_x(r_2) = v_x(r_2)$, $(u_x)'(r_1) \geqq (v_x)'(r_1)$ and $(u_x)'(r_2) \leqq (v_x)'(r_2)$. From (2.5) we obtain $r_1 \geqq (\alpha/2\omega)$ and $r_2 \leqq (\alpha/2\omega)$, which is a contradiction. To prove (ii), we use a similar argument for C^+ and C^-.

The crux of the proof of Theorem 2.1 is the following

<u>LEMMA 2.4</u>: $\text{Sup}\ \{v_x(\tau_x),\ x \in F\} = M < +\infty$.

<u>SKETCH OF PROOF</u>: We argue by contradiction and we assume that there exist $x_n \in F$ such that, on replacing the subscript x_n by n, $v_n(\tau_n) \to +\infty$ as $n \to +\infty$. Since $u_n(r_n) = 0$, $(u_n)'(r_{n_2}) \leqq 0$ and $v_n(r_n) > 0$, we have by (2.5); $v_n(r_n) \leqq \ell$. Let $w > 0$ be such that $G(w^2) > m + \omega$ and let $\theta_n \in (\tau_n, r_n)$ be such that $v_n(\theta_n) = w$. By substituting the estimates of Lemma 2.3 into (2.5) we obtain that $\theta_n - \tau_n \to +\infty$ as $n \to +\infty$. Therefore, after θ_n, (u_n, v_n) satisfies 'almost' (2.6). However, since $v_n(\theta_n) > 0$, $u_n(\theta_n) > 0$ and $H(v_n(\theta_n), v_n(\theta_n)) > 0$, the corresponding solution of (2.6) is such that v vanishes before u. For n large enough this is a contradiction with the condition $x_n \in F$.

<u>STEP 6</u>: Conclusion.

From lemmas 2.3 and 2.4 we have $(\ell, M) \supset F$, and we can set $y = \text{Sup}\ (F)$. Let $(u,v) = (u_y, v_y)$ and $R = R_y$. Since $y > \ell$ we have $u > 0$ and $v > 0$ for $r > 0$ and small. Since F is open, $y \notin F$. Hence u cannot vanish before v does. On the other hand y belongs to the closure of F and therefore v cannot vanish before u. Finally, u and v cannot vanish simultaneously because (0,0) is an equilibrium point. Therefore we have $(u(r), v(r)) \in C$ for $r \in (0,R)$. Again since y belongs to the closure of F, it follows from lemmas 2.3 and 2.4 that (u,v) is uniformly bounded and that $0 \leqq u(r) \leqq v(r)$ for $r \in (0,R)$. This implies in particular that $R = +\infty$. An adaptation of the argument of step 4 shows that $(u(r), v(r)) \to (0,0)$ as $r \to +\infty$, the exponential decay being an easy consequence. This ends the proof of Theorem 2.1.

<u>References</u>

1. A. Alvarez, M. Soler: Energetic stability criterion for a nonlinear spinorial model, Phys. rev. Lett. 50 (1983), 1230-1233.

2. H. Berestycki, P.L. Lions, L.A. Peletier: An O.D.E. approach to the existence of positive solutions for semilinear problems in R^n. Indiana Univ. Math. J. 30 (1981), 141-157.
3. I.L. Bogolubski: On spinor soliton stability. Phys. Lett. A 73 (1979), 87-90.
4. T. Cazenave, L. Vazquez: Existence of localized solutions for a classical nonlinear Dirac field, Comm. Math. Phys. 105 (1986) 35-47.
5. R. Finkelstein, C.F. Fronsdal, P. Kaus: Nonlinear spinor fields, Phys. Rev. 103 (1956), 1571-1579.
6. L. Garcia, A.F. Rañada: A classical model of the nucleon. Prog. Theor. Phys. (Kyoto) 64 (1980), 671-693.
7. W. Heisenberg: Doubts and hopes in quantumelectrodynamics. Physica 19 (1953), 897-908.
8. D. Ivanenko: Sov. Phys. 13 (1938), 141-149.
9. P. Mathieu, T.F. Morris: Instability of stationary states for nonlinear spinor models with quartic self-interaction. Phys. Lett. B 126 (1983), 74.
10. J. Rafelski: Soliton solutions of a selfinteracting Dirac field in three space dimensions. Phys. Lett. B 66 (1977), 262-266.
11. A.F. Rañada: Classical nonlinear Dirac field models of extended particles: In: "Quantum theory, groups, fields and particles" Barut A.O. (ed) Amsterdam, Reidel 1982.
12. A.F. Rañada, M.F. Ranada: Nonlinear model of c-number confined Dirac quarks. Phys. Rev. D 24 (1984), 985-993.
13. A.F. Rañada, M.F. Ranada, M. Soler, L. Vazquez: Classical electrodynamics of a nonlinear Dirac field with anomalous magnetic moment. Phys. Rev. D 10 (1976), 517-525.
14. A.F. Rañada, M. Soler: Perturbation theory for an exactly soluble spinor model in interaction with its electromagnetic field Phys. Rev. D 8 (1973), 3430-3433.
15. A.F. Rañada, J. Uson, L. Vazquez: Born-Infeld effects in the electromagnetic mass of an extended Dirac particle. Phys. Rev. D 22 (1980), 2422-2424.
16. A.F. Rañada, L. Vazquez: Prog. Theor. Phys. (Kyoto) 56 (1976), 311-323.
17. M. Soler: Classical, stable, nonlinear spinor field with positive rest energy. Phys. Rev. D 1 (1970), 2766-2769.

18. M. Soler: Classical electrodynamics for a nonlinear spinor field: Perturbative and exact approaches. Phys. Rev. D 8 (1973), 3424-3429.
19. W.A. Strauss, L. Vazquez: On the stability under dilations of nonlinear spinor fields. Phys. Rev. D34 (1986), 641-643.
20. K. Takahashi: Soliton solutions of nonlinear Dirac equations. J. Math. Phys. 20 (1979), 1232-1238.
21. T. Van der Merwe: Nuovo Cimento 60 A (1980), 247.
22. L. Vazquez: Localized solutions of a nonlinear spinor field. J. Phys. A 10 (1977), 1361-1368.
23. M. Wakano: Intensely localized solutions of the classical Dirac-Maxwell field equations. Prog. Theor. Phys. (Kyoto) 35 (1966), 1117-1141.
24. J. Werle: Stability of particle-like solutions of nonlinear Klein-Gordon and Dirac equations. Acta Physica Polonica B 12 (1981), 601-616.
25. H. Weyl: A remark on the coupling of gravitation and electron. Phys. Rev. 77 (1950), 699-701.

T. Cazenave
Laboratoire d'Analyse Numérique
Tour 55-65, 5e étage,
Université Pierre et Marie Curie
4 place Jussieu
75230 Paris Cedex 05
France.

M CHIPOT

Some results on the compressible Reynolds equation

1. Introduction

Let Ω be a bounded domain in $\mathbb{R}^2$ with a smooth boundary Γ: Ω is the region of the plane where two solid bodies are in close contact. These two bodies are moving and the sum of their velocities is denoted by $V = (V_1, V_2)$. The pressure $p = p(X)$ which develops in a fluid layer confined between these two solids satisfies the so-called Reynolds lubrication equation:

$$\nabla.(h^3\rho \nabla p) = 6\mu V.\nabla(\rho h) \text{ in } \Omega \quad (1)$$

$$p = p_a \text{ on } \Gamma \quad (2)$$

Here ρ is the density of the fluid, $\mu > o$ its dynamic viscosity, $h = h(X)$ is the distance between the two bodies and $p_a \geqslant o$ is the given ambient pressure (we refer to [2] for a derivation of this equation using a simplified version of the Navier-Stokes equations).

We will assume that h is a Lipschitz continuous function such that:

$$o < h_1 \leqq h(X) \leqq h_2 \quad \text{a.e.} \quad X = (x,y) \in \Omega \quad (3)$$

$$|\nabla h(X)| \leqq H \quad \text{a.e.} \quad X = (x,y) \in \Omega \quad (4)$$

where h_1, h_2, H are positive constants.

Moreover, we will assume that the fluid is compressible that is to say that ρ depends on p and so we have:

$$\nabla.(h^3\rho(p)\nabla p) = 6\mu\, V.\nabla(\rho(p)h) \text{ in } \Omega \quad (5)$$

$$p = p_a \text{ on } \Gamma \quad (6)$$

When the fluid obeys the isothermal perfect gas relation we have (after rescaling)

$$\rho(p) = P \quad (7)$$

We would like to present here some new results for (5), (6) which go beyond the scope of (7). Our assumption on ρ will be

$$\rho \text{ is continuous, non decreasing, } \rho(o) = o \text{ and } \rho(s) > o \ \forall s > o. \quad (8)$$

This is reasonable for a density. Note that some of our techniques apply also to more general models. For instance, we refer the interested reader to [1] and [2] where the equation

$$\nabla.(h^3 p(1 + \frac{\lambda}{hp})\nabla p) = V.\nabla(hp)$$

is investigated. (λ is a non negative constant).

First we are going to recall how to get existence of a solution to (5), (6). Then we will show some techniques leading to uniqueness and in the last section we will do a rough study of the shape of the solution. For complete proofs and details the reader is referred to [3].

2. Existence

First note that we are looking for a positive solution to (5), (6).

Set

$$R(s) = \int_0^s \rho(t)dt$$

Clearly, under the assumption (8) this function is one-to-one from $[o,+\infty)$ onto itself. Let us denote by R^{-1} its inverse. If $p > o$ is a solution of (5), (6) (for instance in the classical sense when the data are smooth) then

$$u = R(p)$$

is a solution of

$$\nabla.(h^3\nabla u) = V.\nabla(6\mu h\rho\circ R^{-1}(u)) \text{ in } \Omega \quad (9)$$

$$u = \phi = R(p_a) \text{ on } \Gamma \quad (10)$$

Conversely, if $u > o$ is a solution of (9), (10) then $p = R^{-1}(u)$ is a solution to (5), (6). So setting

$$\beta(s) = 6\mu h\rho\circ R^{-1}(s)$$

we consider the problem

$$\nabla.(h^3\nabla u) = V.\nabla(h\beta(u)) \text{ in } \Omega \tag{11}$$

$$u = \phi \text{ on } \Gamma \tag{12}$$

(If we don't know a priori that u is non negative we can assume that β has been extended by 0 when s is negative).

Let us assume that β satisfies

$$\beta \text{ is a continuous function} \tag{13}$$

$$\exists c > o \text{ such that } |\beta(s)| \leq \frac{\sqrt{\nu}h_1^3}{2h_2|V|}|s| + c \qquad \forall s \in \mathbb{R} \tag{14}$$

where h_1, h_2 are the constants defined in (3), $|V|$ is the Euclidean norm of V and ν is the smallest eigenvalue of the problem

$$-\Delta W = \nu W \text{ in } \Omega$$

$$W = 0 \text{ on } \Gamma$$

<u>REMARK 1</u>: It is easy to check that the above assumptions hold for $s \geq o$ when ρ is given by

$$\rho(s) = Cs^\alpha \tag{15}$$

when C, α are two positive constants.

If $H^1(\Omega)$ and $H^1_o(\Omega)$ denote the usual Sobolev spaces on $L^2(\Omega)$, for $u \in H^1(\Omega)$ we will say that us is a weak solution to (11), (12) if

$$\int_\Omega (h^3\nabla u.\nabla\xi - h\beta(u)V.\nabla\xi)dx = 0 \qquad \forall\xi \in H^1_o(\Omega) \tag{16}$$

$$u-\phi \in H^1_o(\Omega) \tag{17}$$

<u>REMARK 2</u>: Note that when $u \in H^1(\Omega)$, $\beta(u) \in L^2(\Omega)$ and (16) makes sense (see (3), (13), (14)).

We can now prove:

THEOREM 1: Assume that $\phi \in H^1(\Omega)$ and that (3), (13), (14) hold. Then there exists a weak solution to (11), (12).

PROOF. The proof is a routine application of the Schauder fixed point theorem. For $v \in L^2(\Omega)$ one sets $u = T(v)$ the solution of the problem:

$$\int_\Omega (h^3 \nabla u . \nabla \xi - h\beta(v) V . \nabla \xi) dx = 0 \qquad \forall \xi \in H_0^1(\Omega)$$

$$u - \phi \in H_0^1(\Omega)$$

Then it is easy to prove that for R large enough, T maps the ball

$$B_R = \{v \in L^2(\Omega) | \quad |v|_2 \leqq R\}$$

into itself and is completely continuous. We then conclude by the Schauder fixed point theorem (see [3] for the details. In the definition of B_R, $|\ |_2$ denotes the usual norm in $L^2(\Omega)$).

3. Uniqueness

In this section we are going to assume that for some positive constant c we have:

$$\phi \geqq c > 0 \tag{18}$$

This is somehow relevant from a physical point of view. Then, under some assumptions on β and h, we are going to prove that u, the solution of (16), (17), remains bounded away from 0. (It is clear that this is not a trivial consequence of the maximum principles). As a corollary, for smooth data, u will be smooth. (Note that $R^{-1}(s)$ is not differentiable at 0). Moreover uniqueness will follow.

To prove that u is bounded away from 0 we construct a subsolution to (16), (17). The definition is the following: we will say that $v \in H^1(\Omega)$ is a subsolution to (16), (17) iff:

$$\int_\Omega (h^3 \nabla v . \nabla \xi - h\beta(v) V . \nabla \xi) dx \leqq 0 \quad \forall \xi \geqq 0, \quad \xi \in H_0^1(\Omega) \tag{19}$$

$$(v - \phi)^+ \in H_0^1(\Omega) \tag{20}$$

($(\ )^+$ denotes the positive part of a function).

For the sake of simplicity we are going to consider only the case where

$$\beta(s) = A\, s^\beta \qquad o < \beta < 1,\ o < A \tag{21}$$

On some interval (o,τ) around 0. One can of course strengthen this assumption and we refer to [3] for the details but the idea is the same. It is not difficult also to show that (21) holds in the case where ρ is given by (15) and so in particular in the isothermal case of (17).

Let us also assume that

$$h \in W^{2,\infty}(\Omega) \tag{22}$$

(see [5] for a definition of this space) and denote by k a smooth, bounded, positive function such that

$$V.\nabla k \leqq K < 0 \tag{23}$$

where K is a constant.

Then we have:

<u>PROPOSITION 1</u>: For $\varepsilon > o$ small enough the function $u_\varepsilon = (\varepsilon k/Ah)^{1/\beta}$ is a subsolution to (16), (17).

<u>PROOF</u>. Clearly $u_\varepsilon \in H^1(\Omega)$ and for ε small enough one has

$$u_\varepsilon = (\varepsilon k/Ah)^{1/\beta} \leqq c \leqq \phi \tag{24}$$

and thus (20) holds.

Next, we choose ε such that (24) holds together with

$$u_\varepsilon = (\varepsilon k/Ah)^{1/\beta} \leqq \tau$$

We have for $\xi \in H^1_o(\Omega)$ (see (21)):

$$\int_\Omega h^3\nabla u_\varepsilon.\nabla\xi - h\beta(u_\varepsilon)V.\nabla\,\xi\,dx = \int_\Omega h^3\nabla u_\varepsilon.\nabla\xi - \varepsilon k\ V.\nabla\xi\quad dx$$

$$= \varepsilon\ .\int_\Omega (-\varepsilon^{(1/\beta)-1}\ \nabla.(h^3\nabla(k/Ah)^{1/\beta}) + \nabla k.V).\xi\quad dx$$

(we have performed an integration by parts in the last integral).

Since $(1/\beta)-1 > 0$, for $\xi \geq 0$ the above integral is non positive for ε small enough (see (22) and (23)). This concludes the proof of the proposition.

For $\delta \geq 0$ let us denote by ω_δ the modulus of continuity of β on $(\delta, +\infty)$, i.e. set:

$$\omega_\delta(t) = \sup_{\substack{x,y \in (\delta,+\infty) \\ |x-y| \leq t}} |\beta(x) - \beta(y)| \tag{25}$$

and assume that

$$\int_0 \frac{dt}{\omega_\delta^2(t)} = +\infty \quad \forall\delta > 0 \tag{26}$$

REMARK 3: (26) holds for instance when β is Hölder continuous with exponent $\beta \geq 1/2$. This assumption can in fact be weaker – see Carrillo's lecture in this volume.

We have:

PROPOSITION 2: Let u be a solution of (16), (17) and v a subsolution of (16), (17). Assume that (26) holds and that for some $\delta > 0$ we have:

$$v \geq 2\delta \quad \text{a.e. in } \Omega \tag{27}$$

then

$$v \leq u \quad \text{a.e. in } \Omega \tag{28}$$

PROOF. Step 1. We first prove that for all $\zeta \geq 0$ smooth, one has

$$\int_{[v-u>o]} h^3\ \nabla(v-u).\nabla\zeta - h(\beta(v)-\beta(u))V.\nabla\zeta\quad dx \leq 0 \tag{29}$$

($[u-v>o]$ denotes the set $\{x \in\Omega|(v-u)\ (x) > 0\}$).

For δ satisfying (27) and for all $\varepsilon > 0$ there exists a number $\ell(\varepsilon) < \varepsilon$ such that (see (26)):

$$\int_{\ell(\varepsilon)}^{\varepsilon} dt/\omega_\delta^2(t) = 1$$

Define the function F_ε by

$$F_\varepsilon(t) = \begin{cases} 0 & \text{if } t \leqq \ell(\varepsilon) \\ \int_{\ell(\varepsilon)}^{t} dt/\omega_\delta^2(t) & \text{if } \ell(\varepsilon) \leqq t \leqq \varepsilon \\ 1 & \text{if } t \geqq \varepsilon \end{cases}$$

Then clearly $\xi = \zeta F_\varepsilon(v-u)$ is a positive function on $H_0^1(\Omega)$ and from the definition of u and v we get (dropping the measure dx):

$$\int_\Omega h^3\ \nabla(v-u)\nabla\xi - h(\beta(v) - \beta(u))V.\nabla\ \xi \leqq 0$$

which implies

$$I_\varepsilon = \int_\Omega [h^3\ \nabla(v-u)\nabla\zeta - h(\beta(v) - \beta(u))V.\nabla\zeta]F_\varepsilon(v-u)$$

$$\leqq -\int_\Omega [h^3|\nabla(v-u)|^2 - h(\beta(v) - \beta(u))V.\nabla(v-u)]\zeta.F_\varepsilon'(v-u)$$

By Young's Inequality we have

$$|h(\beta(v) - \beta(u))V.\nabla(u-v)|$$

$$\leqq \frac{1}{2}\ h^3|\nabla(v-u)|^2 + \frac{1}{2}\frac{|V|^2}{h}\ (\beta(v) - \beta(u))^2$$

so that (see (3))

$$I_\varepsilon \leqq \frac{1}{2}\frac{|V|^2}{h_1}\int_\Omega (\beta(v) - \beta(u)))^2\zeta\ F_\varepsilon'(v-u)$$

If $[\ell(\varepsilon) < v-u < \varepsilon]$ denotes the set $\{x \in \Omega | \ell(\varepsilon) < (v-u)\ (x) < \varepsilon\}$, we get

$$I_\varepsilon \leq \frac{1}{2} \frac{|V|^2}{h_1} \int_{[\ell(\varepsilon)<v-u<\varepsilon]} (\beta(v)-\beta(u))^2/\omega_\delta^2(v-u).\zeta \tag{30}$$

Now if we choose $\varepsilon \leq \delta$ we have on the set of integration

$$u > v-\varepsilon \geq 2\delta - \varepsilon > \delta$$

so that by (25), (30) becomes

$$I_\varepsilon \leq \frac{1}{2} \frac{|V|^2}{h_1} \int_{[\ell(\varepsilon)<v-u<\varepsilon]} \zeta$$

Letting ε go to zero (29) follows.

<u>Step 2</u>. If $V = (V_1,V_2)$ we choose $\zeta = M - \exp[s(V_2x-V_1y)]$ in (29). M is a constant chosen large enough in such a way that $\zeta \geq 0$ in Ω and s is a constant to be chosen later. Since $V.\nabla\zeta = 0$ we get:

$$\int_{[v-u>o]} h^3 \nabla(v-u).\nabla\zeta = \int_\Omega h^3 \nabla(v-u)^+.\nabla\zeta \leq 0.$$

Integrating by parts this leads to

$$\int_\Omega (v-u)^+.-\nabla.(h^3\nabla\zeta) \leq 0.$$

Now an easy computation shows that s can be selected large enough so that

$$-\nabla.(h^3\nabla\zeta) > 0.$$

This clearly leads to (28).

As a consequence of this proposition we have:

<u>THEOREM 2</u>: Assume that (3), (13), (14), (18), (21), (22), (26) hold, then there exists a unique solution to (16), (17). Moreover, this solution is strictly positive in Ω. If in addition the data h, ϕ, ρ, Ω are smooth then u is smooth and $p = R^{-1}(u)$ is the classical solution to (5), (6) with $P_a = R^{-1}(\phi)$.

<u>PROOF</u>. Let u, v be two solutions to (16), (17). Combining proposition 1 and proposition 2 we have for some δ small enough

$$u, v \geqq u_\varepsilon \geqq 2\delta > 0$$

But now applying again proposition 2 to u and v we get

$$u \geqq v \qquad v \geqq u$$

Hence v = u and (16), (17) has a unique solution which is strictly positive in Ω. By a bootstrap argument we obtain easily that u is smooth and the theorem follows.

REMARK 4: If instead of (26) we assume

$$\int_0 \frac{dt}{\omega_0^2(t)} = +\infty \tag{31}$$

it is clear that we can prove that the solution to (16), (17) is unique assuming only that $\phi \geqq 0$ on Γ. The proof follows that of proposition 2. However (31) does not hold when ρ is given by (15) with $\alpha < 1$. Note that in the case where ϕ can vanish on the boundary the problem becomes a free boundary one.

4. Shape of the solution

In this section we are going to suppose that ϕ, and thus p, is constant on Γ. Moreover, for simplicity we will assume that $V = (V_1,0)$ and so we will consider p to be the smooth solution of the problem:

$$\nabla.(h^3\rho(p)\nabla p) = \alpha \frac{\partial}{\partial x}(h\,\rho(p)) \text{ in } \Omega \tag{32}$$

$$p = p_0 > 0 \text{ on } \Gamma \tag{33}$$

where h and ρ are smooth and $\alpha = 6\mu\, V_1 > 0$.

Let us denote by

$$n = (n_x, n_y)$$

the outward normal to Γ and by $d\sigma$ the surface measure on Γ.

Then we have:

THEOREM 3: Assume that

$$\int_\Gamma h\, n_x\, d\sigma \leq 0 \tag{34}$$

and that

$$\exists\, x_o \in \Gamma \text{ such that } \frac{\partial h}{\partial x}(x_o) > 0 \tag{35}$$

Then p achieves a maximum strictly greater than p_o in Ω.

PROOF. If p does not achieve a maximum strictly greater than p_o in Ω then we have $p \leq p_o$ in Ω.

This implies $\frac{\partial p}{\partial n} \geq 0$ on Γ. Now around x_o we have by (32), (35)

$$\nabla.(h^3\rho(p)\nabla p) - \alpha h \frac{\partial p}{\partial x} = \alpha \frac{\partial h}{\partial x} p > 0 \tag{36}$$

Let us denote by N a neighbourhood of x_o in Ω where (36) holds. If $p < p_o$ in N then by the strong maximum principle we have

$$\frac{\partial p}{\partial n}(x_o) > 0 \tag{37}$$

Now integrating (32) over Ω and applying the divergence theorem we get

$$0 < \int_\Gamma h^3\ \rho(p_o)\frac{\partial p}{\partial n}\ dx = \int_\Gamma \alpha\ h\ \rho(p_o) n_x\ \ d\sigma = \alpha\rho(p_o)\int_\Gamma h\ n_x\ d\sigma \leq 0$$

(see (34)).

The above inequality being impossible, p achieves its maximum p_o in N. But, thanks to (36) and the strong maximum principle, this would imply that $p = p_o$ in N and a contradiction with (37). So we cannot have $p \leq p_o$ in Ω and this concludes the proof.

With the same proof as above we can prove:

THEOREM 4: Assume that

$$\int_\Gamma h\, n_x\, d\sigma = 0 \tag{38}$$

and that

$$\exists x_0 \in \Gamma \quad \text{such that } \frac{\partial h}{\partial x}(x_0) > 0 \tag{39}$$

$$\exists x_1 \in \Gamma \quad \text{such that } \frac{\partial h}{\partial x}(x_1) < 0 \tag{40}$$

then p the solution of (32), (33) achieves a maximum strictly greater than p_0 in Ω and a minimum strictly less than p_0 in Ω.

REMARK 4: (38) holds for instance when Ω and h are symmetric with respect to the y axis.

Using the maximum principle and (32) we can easily see that p can only achieve its maximum (resp. minimum) in the region

$$[\frac{\partial h}{\partial x} \leq 0] = \{X \in \Omega \mid \frac{\partial h}{\partial x}(X) \leq 0\}$$

$$(\text{Resp. } [\frac{\partial h}{\partial x} \geq 0] = \{X \in \Omega \mid \frac{\partial h}{\partial x}(X) \geq 0\})$$

When the dimension of Ω in the x-direction is small compared to its dimension in y, when h does not depend on y then finding the pressure p which develops between the two bodies can be reduced to solving a one-dimensional problem for which it is possible to describe completely the shape of p (see [3]). However the problem becomes more challenging in two dimensions.

In [3] we show that p can achieve several maxima or minima in Ω. It would be interesting to find conditions on h and Ω for such an extremum to be unique.

References

1. M. Chipot - M. Luskin - Existence and uniqueness of solutions to Reynolds lubrication equation (To appear in SIAM Journal of Analysis).
2. M. Chipot - M. Luskin - The compressible Reynolds lubrication equation (To appear in the Proceedings of a Workshop at the Institute of Minneapolis - Lecture Notes in Math.)
3. M. Chipot - On the Reynolds lubrication equation (I.M.A. preprint # 206 Minneapolis - to appear).

4. G. Cimatti - On certain non linear problems arising in the theory of lubrication. Appl. Math. Optim. II (1984) p. 227-245.
5. D. Gilbarg - N.S. Trudinger - Elliptic Partial Differential Equations of Second Order Springer Verlag - 2nd ed. 1985.

M. Chipot
Département de Mathématiques
Université de Metz
Ile du Saulcy
57045 Metz Cedex
France.

G DÍAZ

Uniqueness and interior L^∞– estimates for fully nonlinear equations in $\mathbb{R}^N$ without conditions at infinity

1. Introduction

Recently, H. Brezis [1] has proved that for every $f \in L^1_{loc}(\mathbb{R}^N)$ there exists a unique solution $u \in L^p_{loc}(\mathbb{R}^N)$ of the equation

$$- \Delta u + |u|^{p-1} u = f \text{ in } D' (\mathbb{R}^N),\ 1 < p < + \infty \tag{1}$$

hence no growth condition at infinity is required. This result is obtained by using a localization property (see Remark 3 below). The fact that (1) is in divergence form also plays an important role in proving it. Our main goal is to prove such a localization property for nondivergence equations in order to obtain several conclusions for fully nonlinear equations.

More precisely this contribution deals with the equation

$$F(x,u(x),\ D^2u(x)) = f(x),\ x \in \mathbb{R}^N \tag{2}$$

where $F \in C(\mathbb{R}^N \times \mathbb{R} \times S^{N \times N})$ verifies an assumption of accretivity. We recall that under the conditions

$$F(x,r,\hat{q}) \leqq F(x,r,q) \quad \forall x \in \mathbb{R}^N,\ r \in \mathbb{R},\ q-\hat{q} \geqq 0 \text{ in } S^{N \times N} \tag{3}$$

($S^{N \times N}$ denotes the real space of the symmetric $N \times N$ matrices) and

$$F(x,r,q) \leqq F(x,s,q),\quad \forall x \in \mathbb{R}^N,\ q \in S^{N \times N},\ r-s \in \mathbb{R}_+ \tag{4}$$

we may define, by the function - F, an accretive operator in $C_b(\mathbb{R}^N)$ over a class of functions satisfying some condition at infinity. Clearly (3) is consistent with the classical ellipticity (eventually degenerate) and (4) is an absorption property.

An important tool in our argument is the following accretivity assumption

$$0 \leqq F(x,r,q) - F(x,r,\hat{q}) \leqq \Theta(\text{trace }(q-\hat{q})), \quad \forall x \in \mathbb{R}^N,\ r \in \mathbb{R},$$

$$q-\hat{q} \in S_+^{N+N} \tag{5}$$

$$\gamma(r-s) \leqq -F(x,r,q) + F(x,s,q),\ \forall x \in \mathbb{R}^N,\ q \in S^{N\times N},\ r-s \in \mathbb{R}_+$$

where $\Theta, \gamma : \mathbb{R}_+ \to \mathbb{R}_+$ are two continuous functions such that $\Theta(0^+) = \gamma(0^+) = 0$.

From (5) it follows

$$F(x,r,q) - F(x,s,\hat{q}) \leqq \Theta(\text{trace }(q-\hat{q})) - \gamma(r-s),\ \forall x \in \mathbb{R}^N,\ r-s \in \mathbb{R}_+, \tag{6}$$

$$q-\hat{q} \in S_+^{N\times N},$$

then adequate relations between Θ and γ, involving a degeneracy condition at infinity, enable us to obtain in Section 2, the quoted interior L^∞-estimates. We use it in order to obtain comparison, uniqueness and accretivity properties for (2). We do this in Section 3. Finally, we give in Section 4 some examples and applications, that are the subject of forthcoming papers. We note that our results have the character of "a priori" properties for strong solutions of (2), i.e. they will be proved independently of the existence theory.

2. Interior estimates

The key result in the paper is the following

LEMMA 1: Let $\Theta, \gamma: \mathbb{R}_+ \to \mathbb{R}_+$ two continuous functions such that

$\Theta(0^+) = \gamma(0^+) = 0$ and verifying the condition at infinity

$$\text{(IC)}\qquad \int_\delta^{+\infty} \frac{ds}{\sqrt{\int_0^s \Theta^{-1}(\gamma(t))dt}} < +\infty, \text{ for some } \delta > 0,$$

$$\lim_{\delta \downarrow 0} \int_\delta^{+\infty} \frac{ds}{\sqrt{\int_0^s \Theta^{-1}(\gamma(t))dt}} = +\infty,$$

provided $\Theta^{-1}(\gamma) \in L^1(0,\infty)$. Then for each $d > 0$ there exists a function $\omega:]0, \frac{d}{\sqrt{N}}[\to \mathbb{R}_+$ such that

$$\Theta(\omega_d''(r)) = \gamma(\omega_d(r)),\ 0 < r < \frac{d}{\sqrt{N}}$$

$$0 < \omega_d(0) < \omega_d(r) < \omega_d\left(\frac{d}{\sqrt{N}}\right) = +\infty\ ,\ 0 < r < \frac{d}{\sqrt{N}} \tag{7}$$

$$\omega_d'(0^+) = 0$$

Moreover, for each $r > 0$ fixed, one satisfies

$$\lim_{d\to\infty} \omega_d(r) = 0. \tag{8}$$

PROOF. Due to the positivity of $\Theta^{-1}(\gamma)$ we may consider the decreasing function

$$I(h) = \int_h^{+\infty} \frac{ds}{\sqrt{\int_h^s \Theta^{-1}(\gamma(t))dt}}\ ,\ h > 0$$

By (IC) it is clear that $\lim_{h\to\infty} I(h) = 0$ and $\lim_{h\downarrow 0} I(h) = \infty$. Hence, there exists $h_d > 0$ verifying

$$I(h_d) = d\sqrt{\frac{2}{N}}\ ,\ d > 0. \tag{9}$$

So that we denote

$$I(h,h_d) \doteq \int_h^{+\infty} \frac{ds}{\sqrt{\int_{h_d}^s \Theta^{-1}(\gamma(t))dt}}\ ,\ h \geq h_d.$$

Since $I(\cdot,h_d):]h_d, +\infty[\to \mathbb{R}_+$ is a continuous decreasing function we may define ω_d given implicitly by

$$I(\omega_d(r),h_d) = \left(\frac{d}{\sqrt{N}} - r\right)\sqrt{2}\ ,\ 0 < r < \frac{d}{\sqrt{N}} \tag{10}$$

Then it is very easy to check that (7) and (8) hold. □

REMARK 1: Assumption (IC) is fulfilled for $\Theta(s) = as^n$, $\gamma(s) = bs^m$, $a,b > 0$ with $0 < n < m$. In this case one has

$$\omega_d\left(\frac{r}{\sqrt{N}}\right) \leq C(n,m,a,b,N)\,(d-r)^{\frac{2n}{n-m}}, \quad 0 < r < d \tag{11}$$

where

$$C(n,m,a,b,N) = \left[(m-n)\left(\frac{1}{2nN\,(n+m)}\right)^{\frac{1}{2}}\left(\frac{b}{a}\right)^{\frac{1}{2n}}\right]^{\frac{2n}{n-m}}. \quad \square$$

An interior L^∞-estimate can be derived from the above lemma on certain fully nonlinear equations.

<u>THEOREM 2</u>: Let $(x,r,q) \to F(x,r,q)$ a function on $\mathbb{R}^N \times \mathbb{R} \times S^{N \times N}$ such that

$$\begin{aligned} &x \to F(x,r,q) \text{ is measurable on } \mathbb{R}^N, \text{ for each } (r,q) \in \mathbb{R} \times S^{N \times N} \\ &(r,q) \to F(x,r,q) \text{ is continuous on } \mathbb{R} \times S^{N \times N}, \text{ a.e. } x \in \mathbb{R}^N \end{aligned} \tag{12}$$

and satisfying (3). Assume that there exist two continuous functions $\Theta, \gamma : \mathbb{R}_+ \to \mathbb{R}_+$ verifying

$$\begin{aligned} &F(x,r,q) - F(x,r,0) \leq \Theta(\text{trace } q), \text{ a.e. } x \in \mathbb{R}^N, \ r \in \mathbb{R}_+, \ q \in S_+^{N \times N} \\ &\gamma(r-s) \leq -F(x,r,q) + F(x,s,q), \text{ a.e. } x \in \mathbb{R}^N, \ r-s \in \mathbb{R}_+, \ q \in S^{N\times N} \end{aligned} \tag{13}$$

with $\Theta(0^+) = \gamma(0^+) = 0$ and γ being one-one in some interval $]0,c[$. Suppose the condition (IC). Let

$$R_d(x_o) = \{x \in \mathbb{R}^N : \sum_{i=1}^{N} |x_i - x_{io}| < d\}$$

and ω_d the function given in Lemma 2.1. Then one has

$$u(x) \leq \omega_d\left(\frac{\sum_{i=1}^{N} |x_i - x_{io}|}{\sqrt{N}}\right), \quad x \in R_d(x_o) \tag{14}$$

for any function $u \in W^{2,p}_{loc}(R_d(x_o))$ for some $p \geq N$, verifying

$$F(x,u(x), D^2u(x)) \geq F(x,o,o), \text{ a.e. } x \in R_d(x_o). \tag{15}$$

PROOF. We consider the function

$$u_{\tilde d}(x) = \omega_{\tilde d}\left(\frac{\sum_{i=1}^{N}|x_i - x_{io}|}{\sqrt{N}}\right),\ x \in R_{\tilde d}(x_o),\ \tilde d \leq d$$

Clearly by (7) one has

$$\Theta(\Delta u_{\tilde d}(x)) - \gamma(u_{\tilde d}(x)) = \Theta(\omega''_{\tilde d}(r)) - \gamma(\omega_{\tilde d}(r)) = 0,\ x \in R_{\tilde d}(x_o)$$

where

$$r = \frac{\sum_{i=1}^{N}|x_i - x_{io}|}{\sqrt{N}}\text{; then (13) implies}$$

$$F(x, u_{\tilde d}(x), D^2 u_{\tilde d}(x)) \leq F(x,0,0) + \Theta(\Delta u_{\tilde d}(x) - \gamma(u_{\tilde d}(x))$$

$$\leq F(x,u(x), D^2u(x)),\ \text{a.e. } x \in R_{\tilde d}(x_o)$$

On the other hand, for any fixed and arbitrary $\varepsilon > 0$ small enough we have

$$\omega_{d-\varepsilon}\left(\frac{d-\varepsilon}{\sqrt{N}}\right) = +\infty\text{ , then there exists } \delta_\varepsilon > 0 \text{ such that}$$

$$u(x) \leq \max_{x \in \bar R_{d-\varepsilon}(x_o)} u(x) < \omega_{d-\varepsilon}\left(\frac{d-\varepsilon-\delta_\varepsilon}{\sqrt{N}}\right),\ x \in \partial R_{d-\varepsilon-\delta_\varepsilon}(x_o)$$

(Without loss of generality we may assume $0 < \delta_\varepsilon < \varepsilon$). Now comparison results on $R_{d-\varepsilon-\delta_\varepsilon}(x_o)$ imply

$$u(x) \leq u_{d-\varepsilon}(x),\ \text{for all } x \in R_{d-\varepsilon-\delta_\varepsilon}(x_o)$$

and (14) follows by letting $\varepsilon \downarrow 0$. □

REMARK 2: From Theorem 2 one deduces that for each set $\Omega \subset R_d(x_o)$, such that $\bar\Omega \subset R_d(x_o)$, there exists a positive constant K such that

$$u(x) \leq K,\quad \forall x \in \bar\Omega$$

for any solution $u \in W^{2,p}_{loc}(R_d(x_o))$, for some $p \geq N$, where

$$K = K(\Theta^{-1}(\gamma), N, d_1(\bar{\Omega}, \bar{R}_d(x_o))$$

with $d_1(x,y) = \sum_{i=1}^{N} |x_i - y_i|$, $x,y \in \mathbb{R}^N$.

REMARK 3: Estimation (16) shows a uniform upper boundedness principle on the interior. It can be viewed as a localization property and it is peculiar to suitable nonlinear equations. A similar result for perturbation of the pseudo-Laplacian operator was obtained by J.L. Vazquez [4].

REMARK 4: By considering $R_{|x|/2}(x)$, for large $|x|$. The inequality (14) gives an estimation of the decay at infinity. In particular, if $\Theta(s) = as^n$, $\gamma(s) = bs^m$, $a,b > 0$, with $0 < n < m$ one has by (11)

$$u(x) \leq C(n,m,a,b,N)\,(2|x|)^{\frac{2n}{n-m}}, \quad |x| \geq 2d \tag{17}$$

for any $u \in W^{2,p}_{loc}(\mathbb{R}^N \setminus B_d(o))$, for some $p \geq N$, verifying (15) in $\mathbb{R}^N \setminus B_d(o)$. □

3. Comparison properties

THEOREM 3: Assume all hypotheses of Theorem 2. Then if $u \in W^{2,p}_{loc}(\mathbb{R}^N)$, for some $p \geq N$, verifies

$$F(x,u(x),D^2u(x)) \geq F(x,o,o), \text{ a.e. } x \in \mathbb{R}^N \tag{18}$$

one has $u(x) \leq 0$ for all $x \in \mathbb{R}^N$.

PROOF. By Theorem 2 one has

$$u(x) \leq \omega_d\left(\frac{\sum_{i=1}^{N}|x_i|}{\sqrt{N}}\right), \quad x \in R_d(o)$$

for any $d > 0$. Then the result follows from (8). □

A priori estimates can be now derived from Theorem 3.

COROLLARY 4: Let $(x,r,q) \to F(x,r,q)$ a function verifying (12). Assume two functions Θ, γ satisfying (5) and (IC), γ being increasing with $\lim_{r\to+\infty} \gamma(r) = +\infty$. Then if $u,v \in W^{2,p}_{loc}(\mathbb{R}^N)$, $f,g \in L^1_{loc}(\mathbb{R}^N)$, for some $p \geq N$, verify

$$\begin{aligned} &F(x,u(x),D^2u(x)) \geq f(x), \text{ a.e. } x \in \mathbb{R}^N \\ &F(x,v(x),D^2v(x)) \leq g(x), \text{ a.e. } x \in \mathbb{R}^N \end{aligned} \tag{19}$$

one has

$$\|(u-v)^+\|_{L^\infty(\mathbb{R}^N)} \leq \gamma^{-1}(\|(g-f)^+\|_{L^\infty(\mathbb{R}^N)}). \tag{20}$$

PROOF. Suppose $\|(g-f)^+\|_{L^\infty(\mathbb{R}^N)} < +\infty$. Then we define

$$\mathcal{F}(x,r,q) \doteq F(x,r + v(x), q + D^2v(x)) - F(x,v(x),D^2v(x)),$$

$$(x,r,q) \in \mathbb{R}^N \times \mathbb{R} \times S^{N\times N}.$$

Obviously $\mathcal{F}$ verifies (12) and (13) with $\mathcal{F}(x,o,o) \equiv 0$. On the other hand the function $w = u-v-\gamma^{-1}(\|(g-f)^+\|_{L^\infty(\mathbb{R}^N)}) \in W^{2,p}_{loc}(\mathbb{R}^N)$, $p \geq N$, verifies

$$\begin{aligned} \mathcal{F}(x,w(x),D^2w(x)) &\geq \mathcal{F}(x,(u-v)(x),D^2(u-v)(x)) + \|(g-f)^+\|_{L^\infty(\mathbb{R}^N)} \\ &\geq -(g-f)(x) + \|(g-f)^+\|_{L^\infty(\mathbb{R}^N)} \geq 0, \text{ a.e. } x \in \mathbb{R}^N, \end{aligned}$$

hence $w(x) \leq 0$ for all $x \in \mathbb{R}^N$. □

REMARK 5: By means of classical inequalities the function

$$F(x,r,q) = \left|\sum_{i=1}^N q_{ii}\right|^{n-1} \sum_{i=1}^N q_{ii} - |r|^{m-1}r, \ (r,q) \in \mathbb{R} \times S^{N\times N}$$

with $0 < n \leq 1$, $m \geq 1$, $n < m$ verifies the assumption of Corollary 4. □

REMARK 6: Under the hypothesis of Corollary 4 we may deduce some results:

(i) Uniqueness holds in the problem

$$F(x,u(x),D^2u(x)) = f(x), \text{a.e. } x \in \mathbb{R}^N,\ u \in W^{2,p}_{loc}(\mathbb{R}^N), \text{ for some } p \geq N, \tag{21}$$

and f measurable on $\mathbb{R}^N$, <u>without any condition at infinity</u>. Thus, the behaviour of solutions at infinity is fully determined for each equation. In particular, if $\Theta(s) = as^n$, $\gamma(s) = bs^m$, $a,b > 0$, $0 < n < m$, one has

$$|u(x)| \leq C(n,m,a,b,N)\ (2|x|^{\frac{2n}{n-m}}) + \left(\frac{1}{b} \max_{B_{|x|/2}} |F(\cdot,0,0) - f(\cdot)|\right)^{\frac{1}{m}}$$

for large $|x|$.

(ii) By symmetry one obtains

$$\|u-v\|_{L^\infty(\mathbb{R}^N)} \leq \gamma^{-1}(\|g-f\|_{L^\infty(\mathbb{R}^N)})$$

provided the equalities in (19). Consequently if ρ_u and ρ_f denote the moduli of continuity of u and f respectively we have

$$\rho_u(s) \leq \gamma^{-1}(\rho_f(s)),\ s \geq 0$$

for any solution (21).

(iii) If $\gamma(r) = br$, $b > 0$ we obtain

$$\|D_i u\|_{L^\infty(\mathbb{R}^N)} \leq \frac{1}{b} \|D_i f\|_{L^\infty(\mathbb{R}^N)},\ 1 \leq i \leq N$$

for any solution of (21).

(iv) Accretivity and T-accretivity in $L^\infty(\mathbb{R}^N)$ of the operator $A: D(A) \to L^\infty(\mathbb{R}^N)$, given by

$$D(A) = \{u \in W^{2,p}_{loc}(\mathbb{R}^N), \text{ for some } p \geq N: F(\cdot,u(\cdot),D^2u(\cdot)) \in L^\infty(\mathbb{R}^N)\}$$

$$u(\cdot) = -F(\cdot,u(\cdot),D^2u(\cdot)),\ u \in D(A)$$

also follows from Corollary 4, provided $F(\cdot,0,0) \in L^\infty(\mathbb{R}^N)$. □

The condition at infinity is, in some sense, sharp as it follows from

<u>THEOREM 5</u>: Let $(x,r,q) \to F(x,r,q)$ a function verifying (12). Assume that

there exist two continuous functions θ, $\Gamma: \mathbb{R}_+ \to \mathbb{R}_+$ such that

$$\begin{aligned}&\theta(\text{trace } q) \leqq F(x,r,q) - F(x,r,o), \text{ a.e. } x \in \mathbb{R}^N, r \in \mathbb{R}_+, \\ &q \in S_+^{N\times N} \\ &- F(x,r,o) + F(x,o,o) \leqq \Gamma(r), \text{ a.e. } x \in \mathbb{R}^N, r \in \mathbb{R}_+\end{aligned} \tag{22}$$

with $\theta(o^+) = \Gamma(o^+) = 0$ and θ being increasing. Under the condition

$$\int_\delta^{+\infty} \frac{ds}{\sqrt{\int_o^s \theta^{-1}(\Gamma(t))dt}} = +\infty \text{ for some } \delta > 0, \tag{23}$$

for each $M > 0$ there exists a function $u \in C^2(\mathbb{R}^N)$ such that

$$\begin{aligned}&F(x,u(x),D^2u(x)) \geqq F(x,o,o), \text{ a.e. } x \in \mathbb{R}^N \\ &u(x) \geqq M, \ \forall x \in \mathbb{R}^N.\end{aligned} \tag{24}$$

PROOF. Fixed $M > 0$ we consider

$$J(\ell) = \int_M^\ell \frac{ds}{\sqrt{\int_M^s \theta^{-1}(\Gamma(t))dt}}, \ M \leq \ell < +\infty.$$

It is clear that $J :]M, +\infty[\to \mathbb{R}_+$ is a continuous increasing function, then we may define the function ω given implicitly by

$$J(\omega(r)) = r\sqrt{2}, \ 0 < r.$$

It is very easy to check the properties

$$\theta(\omega''(r)) = \Gamma(\omega(r)), \ 0 < r$$

$$M = \omega(0) < \omega(r), \ 0 < r$$

$$\omega'(0) = 0.$$

So that, by means of the assumptions one proves that $u(x) = \omega(|x_1|)$,

$x = (x_1,\ldots,x_N) \in \mathbb{R}^N$ verifies

$$F(x,u(x),D^2u(x)) - F(x,0,0) \geq \theta(\Delta u(x)) - \Gamma(u(x)) = 0, \text{ a.e. } x \in \mathbb{R}^N$$

$$u(x) \geq M, \forall x \in \mathbb{R}^N. \quad \square$$

REMARK 7: From Theorem 5 one deduces that $m > 1$ is a necessary and sufficient condition in Theorem 3 relative to the semilinear case

$$\Delta u(x) - |u(x)|^{m-1}u(x) \geq 0 \text{ a.e. } x \in \mathbb{R}^N. \quad \square$$

4. Some other applications

(a) Semilinear equations

We may take as model the equation

$$\Delta u(x) - |u(x)|^{m-1}u(x) = f(x), \text{ a.e. } x \in \mathbb{R}^N, \ u \in W^{2,p}_{loc}(\mathbb{R}^N),$$

$$p \geq N, \ m > 1.$$

(i) Universal interior L^∞ and $W^{2,p}$ estimates can be obtained. In particular, for each set Ω such that $\bar{\Omega} \in R_d(x_o)$, there exist two positive constants K_1 and K_2 for which

$$||u||_{L^\infty(\Omega)} \leq K_1, \quad ||u||_{W^{2,p}(\Omega)} \leq K_2$$

for any solution u of (25), provided $f \in L^\infty_{loc}(\mathbb{R}^N)$.

(ii) Comparison and uniqueness can be also obtained in (25). Moreover if f is uniformly continuous in $\mathbb{R}^N$ we may obtain the existence by classical methods, finally if $f \in W^{1,p}_{loc}(\mathbb{R}^N)$ we have $u \in W^{3,p}_{loc}(\mathbb{R}^N)$ and consequently the solution is classical.

(b) Quasilinear equations

Now the model is the equation

$$\Delta u(x) - |Du(x)|^{\frac{2m}{m+1}} - |u(x)|^{m-1}u(x) = f(x), \text{ a.e. } x \in \mathbb{R}^N, \tag{26}$$

$$u \in W^{2,p}_{loc}(\mathbb{R}^N), \ p \geq N, \ m > 1.$$

Then all results of part (a) can be extended to (26). More general quasilinear equations are studied in G. Díaz [2].

(c) Fully nonlinear equations

In G. Díaz [2] we study similar properties for

$$F(x,u(x),Du(x),D^2u(x)) = f(x), \text{ a.e. } x \in \mathbb{R}^N, \ u \in W^{2,p}_{loc}(\mathbb{R}^N), \qquad (27)$$

$p \geq N$.

Some results can also be applied to other fully nonlinear equations. In G. Díaz and J.I. Díaz [3], we extend the above property to the Monge-Ampere equation

$$\det(D^2(u(x)) = |u(x)|^{\alpha-1}u(x) + f(x) \qquad (28)$$

assuming $\alpha > N$.

(d) Evolution equations

Let us consider $F \in C(\mathbb{R}_+ \times \mathbb{R}^N \times S^{N\times N})$ verifying

$$0 \leq F(t,x,p) - F(t,x,\hat{p}) \leq (\text{trace}(p-\hat{p}))^m, \ \forall t \in \mathbb{R}_+, \ x \in \mathbb{R}^N, \qquad (29)$$

$p-\hat{p} \geq 0$ in $S^{N\times N}$ with $0 < m < 1$.

By means of similar arguments we may obtain.

THEOREM 6: Suppose (29). Then if $u,v \in C(]0,\infty[;\ W^{2,p}_{loc}(\mathbb{R}^N))$, $f,g \in C(]0,\infty[;\ L^1_{loc}(\mathbb{R}^N))$, $p \geq N$, verify

$$\begin{aligned} &u_t(t,x) - F(t,u(t,x),D^2_x u(t,x) \leq f(t,x), \ t > 0, \text{ a.e. } x \in \mathbb{R}^N \\ &v_t(t,x) - F(t,v(t,x),D^2_x v(t,x) \geq g(t,x), \ t > 0, \text{ a.e. } x \in \mathbb{R}^N \end{aligned} \qquad (30)$$

one has

$$\|(u(t,\cdot)-v(t,\cdot))^+\|_{L^\infty(\mathbb{R}^N)} \qquad (31)$$

$$\leq |(1-m)t+1|^{\frac{1}{1-m}} \max\{\|(u(0,\cdot)-v(0,\cdot+)\|_{L^\infty(\mathbb{R}^N)}, \|f(t,\cdot)-g(t,\cdot))^+\|_{L^\infty(\mathbb{R}^N)}, t \geq 0.$$ □

REMARK 5: (i) Comparison and uniqueness of strong solutions of

$$u_t(t,x) - F(t,u(t,x),D_x^2u(t,x)) = f(t,x),\ t > 0,\ \text{a.e. } x \in \mathbb{R}^N; \quad (32)$$

$$u(0,x) = u_0(x),\ \text{a.e. } x \in \mathbb{R}^N$$

without any condition for large $|x|$, can be obtained. Other notions of solutions can be also used.

(ii) We may obtain uniform boundedness on the interior of any bounded set.

(iii) This evolution case, as well as other applications, will be studied in a forthcoming paper. □

References

1. Brezis, H.: "Semilinear equations in $\mathbb{R}^N$ without conditions at infinity". Applied Math. and Opt. 12 (1985) p. 271-282.
2. Díaz, G.: "Some nonlinear equations in $\mathbb{R}^N$ without conditions at infinity". To appear.
3. Díaz, G, and Díaz, J.L.: "Remarks on the Monge-Ampere equations. In preparation.
4. Vázquez, J.L.: "An a priori interior estimate for the solutions of nonlinear problems representing weak diffusion". Nonlinear Anal. Th. Meth. and Appl. 5 (1981) p. 95-103.

G. Díaz
Facultad de Matemáticas
Universidad Complutense de Madrid
28040 Madrid
Spain.

J ESQUINAS & J LÓPEZ-GÓMEZ

Optimal multiplicity in local bifurcation theory

1. Introduction

Let U, V be two real Banach spaces and

$$N : \mathbb{R} \times U \to V \tag{1.1}$$

a nonlinear operator such that

$$N(\varepsilon,0) = 0 \tag{1.2}$$

for ε in a neighbourhood of zero. We consider the nonlinear equation

$$N(\varepsilon,u) = 0 \tag{1.3}$$

and we look for nontrivial solutions to (1.3) bifurcating from $(\varepsilon,u) = (0,0)$. For this, we need to assume that $N(\varepsilon,u)$ can be written as

$$N(\varepsilon,u) = L(\varepsilon)u + F(\varepsilon,u), \tag{1.4}$$

where the following conditions are satisfied by $L(\varepsilon)$ and $F(\varepsilon,u)$:

HL1. — $L(\varepsilon) : U \to V$ is a linear continuous operator between U, V such that the mapping

$$\varepsilon \to L(\varepsilon),$$

from $\mathbb{R}$ to $L(U,V)$ is of class C^m, $m \geq 3$. Here we denote by $L(U,V)$ the space of the linear continuous operators between U and V.

HL2. — $L(0)$ is a Fredholm operator of index zero.

HF — $F(\varepsilon,u)$ is a C^2-mapping from a neighbourhood of $(0,0)$ in $\mathbb{R} \times U$ to V such that

$$F(\varepsilon,0) = 0, \quad D_uF(\varepsilon,0) = 0 \tag{1.5}$$

for ε sufficiently small.

By the implicit function theorem, a necessary condition for the origin to be a bifurcation point of (1.3) is

$$\dim N(L(0)) = n \geq 1. \tag{1.6}$$

Krasnoselskij in [7] gives a first general result related to the stated problem.

"If $U = V$, $L(\varepsilon) = I + (\varepsilon - \varepsilon_0)K$, K is a compact operator, and $\varepsilon_0 \neq 0$ is a characteristic value of K of odd multiplicity; that is, $1/\varepsilon_0$ is an eigenvalue of K of odd algebraic multiplicity, then $(0,0)$ is a bifurcation point of (1.3)".

Later, many authors have been looking for generalizations of this result.

Specifically, Magnus in [9] defines a generalized multiplicity of $L(\varepsilon)$ at zero, and proves that odd multiplicity entails bifurcation. In fact, he shows that, if $U = V$ are finite dimensional spaces, odd multiplicity implies a change in the sign of $\det L(\varepsilon)$ when ε crosses zero. Roughly speaking, this change in the sign of $\det L(\varepsilon)$ tell us that an odd number of eigenvalues of $L(\varepsilon)$ (counted with their algebraic multiplicities) are crossing the imaginary axis at $\varepsilon = 0$. So, the homotopy invariance of Brouwer's degree entails bifurcation.

Independently, Ize in [5] introduces another generalized multiplicity of $L(\varepsilon)$ at zero. As in the work of Magnus, odd multiplcity implies that the determinant of the linearized bifurcation equation changes its sign when it is gauged along the trivial solutions. So, he obtains bifurcation.

In our opinion, the Magnus concept of multiplicity, in the way it is explained in [9], is very complicated and rather difficult to apply.

The Ize concept of multiplicity is given after applying a Lyapunov-Schmidt reduction, so he needs to show the independence of his multiplicity concept on the choices of the projections used to reduce the original equation.

We refer to Chow, Hale [1] and Kielhöfer [6] for more extensive information.

All these notions of multiplicity are not sufficiently transparent since they do not show which intrinsic properties of $L(\varepsilon)$ entail an odd or an even multiplicity.

Moreover, in all cases there are "particular" counterexamples when the multiplicity is an even number. We say "particular" because they consist of concrete equations. In this direction, the best multiplicity would be that which enables us to give the following result

"Odd multiplicity entails bifurcation and, if the multiplicity is even, it is possible to find $F(\varepsilon,u)$ such that the only solutions to (1.3) in a neighbourhood of $(\varepsilon,u) = (0,0)$ are the trivial ones".

In this paper, under a suitable nondegeneracy condition, we define a concept of multiplicity, depending only on $L'(0), L''(0),\ldots,L^{(m)}(0)$, which is optimal in the above sense. Here, "prime" denotes derivation with respect to the parameter.

Our nondegeneracy condition is a natural extension of the condition of Crandall, Rabinowitz [2] and Westreich [10]. Our result generalizes the above ones by allowing $\dim N(L(0))$ to be even.

In the following section we shall define the necessary concepts which we shall use in the statement of our result. A simplified version of our result will appear in [3] and the complete version with the proofs will appear in [4]. In section three we shall give an example related to reaction-diffusion systems.

2. Concept of multiplicity and statement of results

In order to simplify the notation, we shall assume sufficient regularity of $L(\varepsilon)$ with respect to the parameter and denote

$$L_0 = L(0),\ L_j = (1/j!)L^{(j)}(0),\ j = 1,2,\ldots \tag{2.1}$$

Then, equation (1.3) can be written as

$$L_0 u + \varepsilon L_1 u + \ldots + \varepsilon^j L_j u + R_j(\varepsilon)u + F(\varepsilon,u) = 0, \tag{2.2}$$

where

$$R_j(\varepsilon) = O(\varepsilon^{j+1}),\ j = 1,2,\ldots \tag{2.3}$$

DEFINITION 1: We say that zero is a k-generic eigenvalue of $L(\varepsilon)$, $k \geqq 1$, if the following conditions are satisfied

$$\dim N(L_0) = n \geqq 1; \tag{2.4}$$

$$L_1(N(L_0)) \oplus \ldots \oplus L_k(N(L_0) \cap \ldots \cap N(L_{k-1})) \oplus R(L_0) = V. \tag{2.5}$$

REMARK 1: In Crandall, Rabinowitz [2] and Westreich [10], the condition (2.5) is substituted by

$$L_1(N(L_o)) \oplus R(L_o) = V. \tag{2.6}$$

Since L_o is a Fredholm operator of index zero, condition (2.6) entails

$$N(L_o) \cap N(L_1) = \text{span}\,[0],$$

hence (2.5) is more general than (2.6). In fact, k-genericity implies k+1-genericity and the genericity of Crandall and Rabinowitz is our 1-genericity.

DEFINITION 2: If zero is a k-generic eigenvalue of $L(\varepsilon)$, we shall call multiplicity of $L(\varepsilon)$ at zero the number

$$\chi = n_1 + 2n_2 + \ldots + kn_k, \tag{2.7}$$

where

$$n_j = \dim L_j(N(L_o) \cap \ldots \cap N(L_{j-1})),\ j = 1,\ldots,k. \tag{2.8}$$

REMARK 2: Observe that

$$\chi = \sum_{\substack{j=1 \\ j \text{ odd}}}^{k} n_j \ (\text{mod. } 2). \tag{2.9}$$

So, in particular, if zero is a k-generic eigenvalue of $L(\varepsilon)$, $k \geqq 2$, and it is not a (k-1)-generic eigenvalue of $L(\varepsilon)$, it is possible for $\dim N(L_o)$ to be even and χ odd.

REMARK 3: If we consider the generalized multiplicity of Magnus [9] and we assume zero to be a k-generic eigenvalue of $L(\varepsilon)$, it is not very difficult to show that the multiplicity defined by him and our multiplicity are the same.

It is important to remark that the Magnus multiplicity is much more difficult to calculate than ours. Moreover, in [9] no reference to generic situations appears.

Finally, as we shall show in the following section, our nondegeneracy condition is general enough to include many applications.

Now, with the above notations, we obtain the following result

THEOREM 1: The following conditions are equivalent:

C1. — χ is an odd number;

C2. — For all $F(\varepsilon,u)$ satisfying HF, the origin is a bifurcation point of the equation (2.2).

Let us observe that, by remark 3, C1 ⇒ C2 follows from the results in Magnus [9] after showing that his multiplicity coincides with ours. So, since the Magnus result is global, our theorem 1 admits a global version too.

However, the proof we use for C1 ⇒ C2 is strongly dependent on (2.5) and gives us the key to show the optimality of this multiplicity; that is, for proving C2 ⇒ C1.

The proof in case $k = 2$ can be seen in [3]. The proof in general case will appear in [4].

Finally, let us observe that, in the case χ is even, theorem 1 tells us that it is necessary to go to the full equation (2.2) in order to obtain necessary conditions for bifurcation. This is what López-Gómez does in [8].

3. Applications

Let us consider the problem

$$\sigma_1 u_1'' + p_{11}u_1 + p_{12}u_2 = au_1^2, \tag{3.1a}$$

$$\sigma_2 u_2'' + p_{21}u_1 + p_{22}u_2 = bu_1^2 + cu_1u_2 + du_2^2, \text{ on } (0,\pi) \tag{3.1b}$$

$$u_1'(0) = u_1'(\pi) = 0, \tag{3.2a}$$

$$u_2'(0) = u_2'(\pi) = 0. \tag{3.2b}$$

Let us observe that $(u_1,u_2) = (0,0)$ is a solution of (3.1), (3.2) for all values of the coefficients. In applications it is interesting to know the branching of nontrivial solutions from $(u_1,u_2) = (0,0)$ when the coefficients are varied. For this, it is usual to eliminate a variable in one of the equations and substitute it in the other equation. In (3.1) it is very easy to eliminate the variable u_2. After this elimination and under convenable assumptions on the coefficients, the problem is reduced to a problem of the form

$$u'''' + \alpha u'' + \beta u + a_{11}u^2 + a_{12}uu' + a_{22}u'u' + b_{12}uu'' + b_{22}u''u'' + [u,u',u'']_3 = 0, \tag{3.3}$$

$$u'(0) = u'(\pi) = u'''(0) = u'''(\pi) = 0, \tag{3.4}$$

for certainly coefficients α, β, a_{ij}, b_{ij}, where by $[u,u',u'']_i$ we denote the terms of order i in (u,u',u'').

The linear part about $u = 0$ of the right hand side of (3.3) is given by the operator

$$L(\alpha,\beta)u = u'''' + \alpha u'' + \beta u$$

for u verifying (3.4). If (α,β) is changed so that it crosses a value (α_0,β_0) in which $L(\alpha_0,\beta_0)$ is not invertible, then branching of solutions from $u = 0$ may occur. If $\dim N(L(\alpha_0,\beta_0)) = 1$, it is possible to apply the classical results to have bifurcation. If $\dim N(L(\alpha_0,\beta_0)) = 2$, it is possible to apply our results. We shall study an example in this direction.

For instance, let us consider the problem

$$L_0 u + \varepsilon L_1 u + \dots + \varepsilon^k L_k u + f(\varepsilon,u,u',u'') = 0, \tag{3.5}$$

$$u'(0) = u'(\pi) = u'''(0) = u'''(\pi) = 0, \tag{3.6}$$

where

$$L_0 u = u'''' + 5u'' + 4u, \tag{3.7a}$$

$$L_j u = a_j u'' + b_j u, \qquad j = 1,\dots,k \tag{3.7b}$$

and

$$f(\varepsilon,u,u',u'') = [u,u',u'']_2 + O(\varepsilon^{k+1})[u,u',u'']_1 + O(\varepsilon)[u,u',u'']_2. \tag{3.8}$$

We consider the above operators defined on

$$U = \{u \in C^4(0,\pi) : u'(0) = u'(\pi) = u'''(0) = u'''(\pi) = 0\}$$

with values in $V = C(0,\pi)$. Then

$$N(L_o) = \text{span}\,[\cos t, \cos 2t] \tag{3.9}$$

and, since L_o is a selfadjoint operator, the Fredholm theory assures us that

$$R(L_o) = \{v \in V : \int_0^\pi v(t)\cos t\,dt = \int_0^\pi v(t)\cos 2t\,dt = 0\}. \tag{3.10}$$

We have

$$L_j \cos t = (b_j - a_j)\cos t, \quad j = 1,\dots,k \tag{3.11a}$$

$$L_j \cos 2t = (b_j - 4a_j)\cos 2t, \quad j = 1,\dots,k \tag{3.11b}$$

and, by application of theorem 1, we obtain the following result related to (3.5), (3.6).

THEOREM 2: Suppose $a_1 \neq 0$. Then, any of the following conditions

(i) $b_i = a_i$, $i = 1,\dots,h-1$, $b_h \neq a_h$, h even, $2 \leq h \leq k$;

(ii) $b_i = 4a_i$, $i = 1,\dots,h-1$, $b_h \neq 4a_h$, h even, $2 \leq h \leq k$,

is sufficient to have bifurcation from $(\varepsilon,u) = (0,0)$.

PROOF. Suppose, for instance, (i) is satisfied. Then, by (3.11), we obtain

$L_1 \cos 2t = -3a_1 \cos 2t$,

$L_i \cos t = 0$, $i = 1,\dots,h-1$,

$L_h \cos t = (b_h - a_h)\cos t$.

So, we are able to apply theorem 1. □

Let us observe that the linear part of (3.5) about $u = 0$ does not give us any information if

$$b_1 \neq a_1 \text{ and } b_1 \neq 4a_1. \tag{3.12}$$

In fact, if (3.12) holds, then $\dim L_1(N(L_0)) = 2$ and theorem 1 forces us to go to the full equation (3.5) in order to study the existence of small solutions to (3.5), (3.6) about $(\varepsilon, u) = (0,0)$. Moreover, the terms $[u, u', u'']_3$ in (3.5) are "bad terms" for obtaining bifurcation (see [3]). However, the second order terms $[u, u', u'']_2$ are "good terms" for obtaining bifurcation. In fact, with the techniques in [8], it is possible to show that, under condition (3.12) and assuming the terms $[u, u', u'']_2$ in (3.8) are of the form

$$[u, u', u'']_2 = auu + buu' + cu'u' + duu'' + eu'u'' + fu''u'',$$

then there exist constants $\bar{a}, \bar{b}, \bar{c}, \bar{d}, \bar{e}, \bar{f}$ such that, if

$$a\bar{a} + b\bar{b} + c\bar{c} + d\bar{d} + e\bar{e} + f\bar{f} \neq 0,$$

then the origin is a bifurcation point of (3.5), (3.6).

References

1. S.N. Chow and J.K. Hale, Methods of Bifurcation Theory, Springer, New-York (1982).
2. M.G. Crandall and P.H. Rabinowitz, Bifurcation from Simple Eigenvalues, J. Funct. Anal. 8, 321-340 (1971).
3. J. Esquinas and J. López-Gómez, Optimal results in Local Bifurcation Theory, to appear in Bull Aust. Math. Soc.
4. J. Esquinas and J. López-Gómez, Optimal Multiplicity for Generalized Generic Eigenvalues in Local Bifurcation Theory, submitted to J.Diff.Eqns.
5. J. Ize, Bifurcation Theory for Fredholm Operators, Mem. Amer. Math. Soc. 7, no. 174 (1976).
6. H. Kielhöfer, Multiple Eigenvalue Bifurcation for Fredhom Operators, J. Reine Angew. Math. 358 (1985).

7. M.A. Krasnoselskij, Topological Methods in the Theory of Nonlinear Integral Equations, Oxford (1964).

8. J. López-Gómez, Multiparameter Local Bifurcation, Nonl. Anal. T.M.A. Vol. 10, No. 11, 1249-1259 (1986).

9. R.J. Magnus, A Generalization of Multiplicity and the Problem of Bifurcation, Proc. London Math. Soc. 32, 251-278 (1976).

10. D. Westreich, Bifurcation at Eigenvalues of Odd Multiplicity. Proc. Amer. Math. Soc. 41, 609-614 (1973).

J. Esquinas and J. López-Gómez
Departamento de Matematica Aplicada
Universidad Complutense de Madrid
28040 Madrid
Spain.

M J ESTEBAN

Variational approach to the existence of skyrmions

Introduction

In [S] T.H.R. Skyrme proposed a model for studying the stable configurations of a field of mesons. To do so he introduced a certain energy-functional and then, taking into account the problem's invariances, he defined a minimization problem which, at least in an approximative way, should answer the question of how to find the structure of the field around a meson. As we state below, his method provided also a way to study other physically interesting quantities, such as the interaction between two mesons or the baryon-baryon potential.

In order to outline what Skyrme's method is, let us consider the fields Φ which map $\mathbb{R}^3$ into the unit sphere of $\mathbb{R}^4$, S^3. Then define $|\nabla\Phi|^2$ as $\Sigma_{i=1,\dots,3} |\nabla\Phi^i|^2$. Skyrme's idea consisted of adding to the nonlinear σ-model ($\int_{\mathbb{R}^3} |\nabla\Phi|^2 dx$) another term to prevent the solitons collapsing at isolated points of $\mathbb{R}^3$. Indeed, as we will explain more clearly below, the classical nonlinear σ-model is not appropriate for dimensions $N \geq 3$. The term chosen was:

$$\int_{\mathbb{R}^3} |A(\Phi)|^2 dx, \text{ where } |A(\Phi)|^2 = \sum_{\alpha,\beta=1}^{3} \left| \frac{\partial\Phi}{\partial x_\alpha} \wedge \frac{\partial\Phi}{\partial x_\beta} \right|^2 \qquad (1)$$

where by $a \wedge b$ (resp. $a \wedge b \wedge c$) we denote the alternating exterior product of a, b (resp. a,b,c) $\in \mathbb{R}^4$ which is an element of $\Lambda^3(\mathbb{R}^4)$ (resp. $\Lambda^2(\mathbb{R}^4)$).

The term Skyrme chose to complete the energy functional was the simplest one fulfilling the following conditions: it should have terms of, at least, the fourth degree in the first derivatives and should not contain derivatives beyond the first order. Moreover he required the energy to be divergent as 1/m for the fields supported in a ball of radius m.

Skyrme observed also that when one considers a set of mesons with no external forces acting on them, there is a constant of motion which is:

$$N = \frac{1}{2\pi^2} \int_{\mathbb{R}^3} \det (\Phi, \nabla\Phi) dx, \qquad (2)$$

which represents the number of mesons present in the field. From the mathematical point of view, and at least when Φ is smooth, N is actually the degree of $\Phi\circ\tilde{E}$, $\tilde{E}$ being a stereographic projection from S^3 into $\mathbb{R}^3$, and in what follows we will denote it by $d(\Phi)$.

Before stating the complete mathematical problem to be considered, let us point out that Skyrme did not consider this problem in its full generality: he restrained himself to the consideration of functions satisfying some symmetry condition that we give below.

The energy finally considered by Skyrme has the following form:

$$E(\Phi) = \frac{\gamma}{4\pi^2} \int_{\mathbb{R}^3} (\kappa^2|\nabla\phi|^2 + |A(\phi)|^2)dx, \tag{3}$$

where γ and κ are two physical constants.

Once all the relevant functionals have been defined, we have to decide in which space we will pose the minimization problem. The natural set to work in would be the set of functions having finite energy E, i.e.,

$$Y = \{\phi : \mathbb{R}^3 \to S^3 \mid \nabla\phi \ , \ A(\phi) \in L^2(\mathbb{R}^3, \mathbb{R}^4)\}$$

but we do not know whether the functions in this set take only integer values, which is required to remain close to the physical meaning of it. On the other hand, we know that if Φ is at least of class C^1, then $d(\Phi)$ is in Z. Hence the set in which we will work is:

$$X = \{\phi \in Y \mid \phi \text{ satisfies condition } (P)\}, \tag{4}$$

where (P) is an approximation-condition defined by:

$$(P) \quad \begin{array}{l} \exists\{\phi_n\} \subset Y \cap C^1(\mathbb{R}^3,S^3) \text{ such that} \\ \nabla\phi_n \xrightarrow[n\to+\infty]{} \nabla\phi \ , \ A(\phi_n) \xrightarrow[n\to+\infty]{} A(\phi) \text{ in } L^2(\mathbb{R}^3, \mathbb{R}^4). \end{array}$$

We can now define the concrete mathematics we will look at as follows:

$$(I_k) \quad I_k = \inf \{E(\Phi)/\Phi \text{ is in } X_k\},\ k \in \mathbb{Z},$$

where $X_k = \{\Phi \in X/d(\Phi) = k\}$ and the problem is to determine whether I_k is achieved or not. The answer actually depends on the value of k. For $|k| \leq 1$ we solve the above problem completely showing that I_o, $I_{\pm 1}$ are achieved. For $|k| > 1$ we give a sufficient (but not necessary) condition for I_k to be achieved, but up to now we are unable to decide whether this condition is satisfied or not.

It is easy to conclude when $k = 0$. Indeed the functional E is positive and 0 is achieved by and only by the constant functions, which have degree 0. Moreover 0 is the global minimum of E in X.

When $k \neq 0$ we observe that $I_k = I_{-k}$, since the change of sign in the degree corresponds to a change of orientation in S^3 and it is obvious that this does not change the energy at all.

On the other hand, note that the problems (I_k) are invariant under the group of translations of $\mathbb{R}^3$ and this implies a lack of compacity for the problem. From this invariance it is not difficult to prove the following subadditivity inequality for the infima I_k:

$$I_k \leq I_\ell + I_{k-1} \text{ for all } k, \ell \text{ in } \mathbb{Z}. \tag{5}$$

This kind of inequality is typical of a large class of minimization problems and, as we will see below, the knowledge of whether the inequality is strict or not plays an important part in the achievability of I_k. The concentration-compactness method, due to P.L. Lions (see [L1,2]), is very useful for treating this kind of minimization problems invariant under different groups of transformations (translations, dilations, etc.) which make the problems not compact. This method can be applied to a large variety of problems, and gives necessary (and sometimes also sufficient) conditions for the corresponding infima to be achieved. Then in every particular case one has to find a method to verify that those conditions are satisfied. We will give here an important use of this method.

This paper is organized as follows: in Section 1 we state our general results about problem (I_k) and give an idea of how to prove them. Then in Section 2 we look at a more restrictive problem, with some symmetry conditions involved, which is the problem considered by Skyrme.

Let us emphasize that in this paper we briefly describe the results that we have obtained for the Skyrme's problem and the method to prove them, but that we do not give here any detailed proof. The interested reader may find all the details and some complementary results in [E1,2,3]. On the other hand, for the physical motivation of this work see [A, S, V, W]. The concentration-compactness method is extensively described in [L1] in the locally compact case and in [L2] in the limit case. Let us finally say that there are other mathematical problems which are very close to ours, such that the Yang-Mills' equations (see [T1,2; U1,2]) and that of harmonic maps in $\mathbb{R}^2$ (see [BC]), but that each of them has particular features which change the way of treating them completely.

Section 1

Let us start the description of our results by considering the case $k = 1$.

THEOREM 1: For all minimizing sequences $\{\Phi_n\}_n$ of I_1 there exists a sequence of points of $\mathbb{R}^3$, $\{y_n\}_n$, and a function Φ in X_1 such that for a subsequence of $\{\Phi_n(\cdot + y_n)\}_n$ denoted by $\{\underline{\Phi}_n\}$ we have:

$$\nabla\underline{\Phi}_n \to \nabla\Phi \quad \text{and} \quad A(\underline{\Phi}_n) \to A(\Phi) \text{ in } L^2(\mathbb{R}^3, \mathbb{R}^4) \tag{6}$$

and hence Φ is a minimum for I_1.

In the case $k > 1$ the situation is not so clear and this is due to the fact that we do not know much about the value of I_k with respect to k.

In [E1] we prove the following isoperimetric inequality:

$$\exists C > 0 \text{ s.t. } \forall\Phi \in X, \ |d(\Phi)|^{3/4} \leq C \int_{\mathbb{R}^3} \left| \frac{\partial\Phi}{\partial x_1} \wedge \frac{\partial\Phi}{\partial x_2} \wedge \frac{\partial\Phi}{\partial x_3} \right| dx \leq \tag{7}$$

$$\leq 3\, C \, \|\nabla\Phi\|_{L^2(\mathbb{R}^3)} \|A(\Phi)\|_{L^2(\mathbb{R}^3)}$$

from which we infer the existence of two positive constants, M, L, such that:

$$M\,|k| \leq I_k \leq L\,|k|, \quad L|k|^{3/4} \leq I_k.$$

But we do not have more precise information about the sequence $\{I_k\}$ (e.g., is it monotone?). As we see in the following, the knowledge of I_k's value would be very useful.

THEOREM 2: Let $k \in \mathbb{Z}$. Then if the following strict subadditivity inequality holds for all $\ell \in \mathbb{Z}$,

$$I_k < I_\ell + I_{k-1} \tag{8}$$

I_k is achieved. More explicitly, for any minimization sequence $\{\Phi_n\}$ the above condition is sufficient for the existence of a sequence of points of $\mathbb{R}^3$, $\{y_n\}$, such that the sequences $\{\nabla\Phi_n(\cdot + y_n)\}$ and $\{A(\Phi_n(\cdot + y_n)\}$ are relatively compact in $L^2(\mathbb{R}^3, \mathbb{R}^4)$.

REMARK. Notice that (8) is not a necessary condition and that minima of I_k may exist without (8) being satisfied. In that case those minima would be unstable in some sense. On the other hand, let us remark that the theorem still holds if we ask (8) to be satisfied only for the ℓ in $\mathbb{Z}$ such that $\sqrt{2}\,|k| \leqq |\ell| + |k-\ell|$.

IDEA OF THE PROOF OF THEOREM 1. Once we have chosen a minimizing sequence for I_1, $\{\Phi_n\}$, it is not difficult to infer the existence of a function Φ in X such that $\{\Phi_n\}$ (or a subsequence of it) converges weakly to it (for instance in $L^6(\mathbb{R}^3)$) and such that $E(\Phi) \leqq I_1$. It is obvious that if $d(\Phi) = 1$ the proof is finished.

In general let us define the functions $f_n = |\nabla\Phi_n|^2 + |A(\Phi_n)|^2 + \delta|\Phi_n|^6$ for some $\delta > 0$ to be chosen in a convenient way. Then for every n we consider the concentration function of f_n, P_n, which were defined by P. Levy in [Le] as follows:

$$\forall t \geqq 0,\quad P_n(t) = \sup_{y \in \mathbb{R}^3} \int_{B(y,t)} f_n(x)dx, \tag{9}$$

and then we apply the concentration-compactness method. Here I do not wish to explain in detail how this method works, basically it says that there are three possibilities for the functions P_n:

(a) (Vanishing) For any $y \in \mathbb{R}^3$ and for any $R > 0$, the energy spent by f_n in $B(y,R)$ is asymptotically null as n goes to $+\infty$.

(b) (Dichotomy) The support of f_n is "almost" cut into at least two pieces which slip infinitely from each other as n goes to $+\infty$ and when we say "almost" we mean that the energy spent by f_n away from these two sets goes to 0 when n goes to $+\infty$.

(c) (Compactness) There is a compact in $\mathbb{R}^3$ such that the energy spent by the functions f_n outside that compact (up to a translation) goes to 0 as n goes to $+\infty$.

The remainder of the proof consists of proving that the dichotomy and the vanishing cannot occur and that the compactness (in the above sense) implies the existence of a minimum of I_1.

The vanishing is avoided in two steps. First we show for all Φ in X there exists e in S^3 such that $\Phi - e$ is in $L^6(\mathbb{R}^3, \mathbb{R}^4)$. Then we prove that if the vanishing were to happen, then

$$\|\Phi_n - e_n\|_{L^p(\mathbb{R}^3)} \to 0 \text{ as n goes to } +\infty \text{ for all } p > 6. \tag{10}$$

Finally we prove that for all Φ in X, for all subset B of $\mathbb{R}^3$, the following inequality holds:

$$\int_B \left|\frac{\partial\Phi}{\partial x_1} \wedge \frac{\partial\Phi}{\partial x_2} \wedge \frac{\partial\Phi}{\partial x_3}\right| \, dx \leq \mathbb{E}(\Phi)^{3/4} \, |B|^{1/4} \tag{11}$$

and this together with (10) and the fact that for all n $\Phi_n(\mathbb{R}^3)$ contains S^3 implies that the vanishing cannot occur.

Let us now outline how the dichotomy is avoided. Apparently this is not a difficult case to avoid, since it has been done very often in [L1,2] using the strict inequality (8). The standard method of doing so is to split the functions Φ_n into two other functions, Φ_n^1, Φ_n^2, such that each of them reproduces the behaviour of Φ_n in the two sets into which Φ_n's support is cut. More explicitly, if Φ_n's support is asymptotically cut into $B(y_n,R)$ and $B(y_n,R_n)^c$, with $R_n \to +\infty$ as n goes to $+\infty$, and the energy spent by Φ_n outside these two sets goes to 0 as n goes to $+\infty$, then:

$$\Phi_n^1 \equiv \Phi_n \text{ in } B(y_n;R),\ \Phi_n^2 \equiv \Phi_n \text{ in } B(y_n,R_n)^c,$$

$$E(\Phi_n^1) + E(\Phi_n^2) \approx E(\Phi_n),\ d(\Phi_n^1) + d(\Phi_n^2) = 1, \qquad (12)$$

$$E(\Phi_n^1),\ E(\Phi_n^2) \geq \alpha \text{ for some positive constant } \alpha \text{ and for all } n$$

The remaining steps are then clear, but the difficulty arises here because the functions Φ_n are not scalar, but take their values in S^3. Hence the problem is not only to cut functions supported in all $\mathbb{R}^3$ to obtain functions supported, by instance, in a ball of $\mathbb{R}^3$, but to do so while we keep the images of the new functions in S^3 and this without spending much energy, i.e., keeping control of the energy spent in the cutting process. This is only a technical difficulty and we can handle it by writing the problem (I_k) in spherical coordinates of $\mathbb{R}^4$ and doing the cutting in the associated functions corresponding to Φ_n.

Now that the functions Φ_n have been cut in the above way, we use all the (partial) information we have on the value of I_k (minorations and majorations) to prove that (12) can never occur.

Once we have dealt with the vanishing and the dichotomy, we show that the compactness, in the sense of (c), allows us to end the proof. Indeed a lemma in [L2] (the concentration-compactness second lemma) can be applied to our problem to prove the following:

Either there is compactness in the sense that there are points of $\mathbb{R}^3$, y_n, such that the sequence $\{\Phi_n(\cdot + y_n)\}$ converges weakly to some function $\Phi \in X$ and $d(\Phi) = 1$, in which case the proof will be finished. Or there is a function $\Phi \in X$ and a finite subset of $\mathbb{R}^3$, $(\underline{x}_1,\ldots,\underline{x}_m$, such that the sequence of L^1-functions $\{\det(\Phi_n, \nabla\Phi_n)\}$ converge to $\det(\Phi,\nabla\Phi) + \Sigma c_i \delta(\underline{x}_i)$, where by $\delta(x)$ we denote the Dirac measure at $x \in \mathbb{R}^3$. Moreover in this second case there is also concentration of the L^1-functions

$$\frac{\partial\Phi_n}{\partial x_1} \wedge \frac{\partial\Phi_n}{\partial x_2} \wedge \frac{\partial\Phi_n}{\partial x_3}$$

at the points $\underline{x}_1,\ldots,\underline{x}_m$. But inequality (11) forbids this to happen; therefore this second possibility can never occur and this ends the proof. □

REMARK. The proof of Theorem 2 is done in quite a similar way, the only difference lying in the arguments we use to prove that the dichotomy cannot occur under the assumptions of the theorem.

REMARK. The estimates we have for I_k are sufficient to prove that the best constant in inequality (7) is achieved even if we do not know its precise value. Moreover we know that it is achieved in $X_1 \cup X_2 \cup X_3$.

Section 2 Here we do not consider the problems (I_k) in their full generality, but a family of somewhat more restrictive ones, where some symmetry conditions are considered. Note that some of these symmetry conditions were considered by Skyrme when he introduced the problem.

Assume now that we consider only the functions Φ in X which satisfy the following symmetry condition:

$$(S) \quad \exists\omega: \mathbb{R}^3 \to \mathbb{R} \text{ such that } \Phi(x) \equiv \left(\frac{x}{|x|} \sin \omega(x), \cos \omega(x)\right)$$

Then one may consider the following family of minimization problems:

$$(\underline{I}_k) \quad \underline{I}_k = \inf \{E(\Phi)/d(\Phi) = k, \Phi \text{ satisfies (S)}\}.$$

REMARK. If (Φ,ω) are such that (S) holds, then one can prove that Φ is in X if and only if $\omega(0)$ and $\omega(+\infty)$ are in 2π, where by $\omega(0)$ and $\omega(+\infty)$ we mean the following:

$$\omega(0) = \lim_{R\to 0} \frac{1}{|B_R|} \int_{B_R} \omega(x)dx, \quad \omega(\cdot) - \omega(+\infty) \in L^p(\mathbb{R}^3) \text{ for some } p > 1.$$

Moreover one shows that for the functions Φ satisfying (S), $d(\Phi) = k$ is equivalent to $\omega(+\infty) - \omega(0) = k\pi$.

This new class of problems is easier to deal with than that of Section 1 since now we have to study the minimization of a function which depends only on one variable function. This simplification allows us to prove a more complete result than above:

THEOREM 3 [E2]. There is a minimum for $\underline{I}_k$ for all k in $\mathbb{Z}$.

REMARK. The problems $(\underline{I}_k)$ are no more invariant under the group of translations of $\mathbb{R}^3$ and, therefore, the lack of compactness we had to deal with in the general case is not a problem here and the proof of this result is much easier than that of Theorems 1 and 2. This proof is completely done in [E2] and will not be given here.

REMARK If in (S) we assume that the function depends only on the modulus of x, $|x|$, then we can define a new family of problems, still less general than (I_k) that we denote by $(I_k{}^*)$, and these new problems are those which Skyrme considered (and solved for $k = \pm 1$), when he first introduced his model to study the stable configuration of a set of mesons in a field of weak energy in [S]. The corresponding result (existence of minima) for this kind of symmetry condition was given in [K] for k = 1 and in [E2] for all k in $\mathbb{Z}$.

It is obvious that the following inequalities hold for all k in $\mathbb{Z}$:

$$I_k{}^* \geqq \underline{I}_k \geqq I_k$$

and it would be interesting to know, for instance, whether $I_k{}^*$ is equal to I_k or not, i.e., whether the minimum of E in X_k is achieved by functions which have a particular symmetry or not. A partial answer to this question is given by the following:

PROPOSITION 4: For all $k \neq 0, \pm 1$, the following holds: $I_k{}^* > I_k$.

This proposition is proved as follows: we assume that for some $k \neq 0, \pm 1$, $I_k{}^* = I_k$; then we infer from it that for some m in $\{1,\ldots,k-1\}$ the Euler equation (O.D.E.) corresponding to the minimization problem $(I_m{}^*)$ possesses a solution with compact support, which is contradictory with the Cauchy-Lipschitz theorem.

REMARK. The question of whether I_1 is equal to $I_1{}^*$ remains open; however the knowledge of its answer would be very interesting, since the case k = 1 is the most interesting from the physical point of view.

References

[A] G.S. Adkins, C.R. Nappi, E. Witten, Static properties of nucleons in the Skryme model, Nuclear Phys. B 228 (1983), p. 552-566.

[BC] H. Brezis, J.M. Coron, Large solutions for Harmonic maps in $\mathbb{R}^2$. Comm. Math. Phys. 92 (1983), p. 203-215.

[E1] M.J. Esteban, An isoperimetric inequality in $\mathbb{R}^3$. To appear in Analyse non lineaire, Ann. IHP.

[E2] M.J. Esteban, Existence of symmetric solutions for the Skyrme's problcm. To appear in Ann. Mat. Pura ed Appl.

[E3] M.J. Esteban, A direct variational approach to Skyrme's model for meson fields. Comm. Math. Phys. 105 (1986), p. 571-591.

[K] L.B. Kapitanski, O.A. Ladyzenskaia. On the Coleman's principle concerning the stationary points of invariant functionals. Zapiski nauch sem. I.O.M.I 127 (1983), p. 84-102.

[Le] P. Levy, Théorie de l'addition des variables aléatoires. Gauthier-Villars Paris, 1954.

[L1] P.L. Lions, The concentration-compactness principle in the Calculus of variations. Part I : Ann. IHP, Anal. Non Lin. 1 (1984), p. 109-145. Part II : Ann, IHP, Anal. Non. Lin. 1 (1984), p. 223-283.

[L2] P.L. Lions, The concentration-compactness principle in the Calculus of variations. The limit case. Part I : Rev. Mat. Iberoamer. I, 1 (1985), p. 145-200. Part II : Rev. Mat. Iberoamer, I, 2 (1985), p. 45-121.

[S] T.H.R. Skyrme, A non-linear field theory. Proc. Roy. Soc. A260 (1961) p. 127-138.

[T1] C.H. Taubes, Min-Max theory for Yang-Mills-Higgs equations, Preprint.

[T2] C.H. Taubes, Monopoles and maps from S^2 to S^2; the topology of the configuration space. Preprint.

[U1] K.K. Uhlenbeck, Removable singularities in Yang-Mills fields. Comm. Math. Phys. 83 (1982), p. 11-29.

[U2] K.K. Uhlenbeck, Connections with L^p bounds on curvature. Comm. Math. Phys. 83 (1982), p. 31-42.

[V] G.S. Vinh Mau, M. Lacombe, B. Loiseau, W.N. Cottingham, P. Lisboa, The static baryon-baryon potential in the Skyrme model. Preprint.

[W] E. Witten, Baryons in the 1/N expansion, Nuclear Phys. B 160 (1979), p. 57-115.

M.J. Esteban
Laboratoire d'Analyse Numérique
Tour 55-65
Université Pierre et Marie Curie
4 place Jussieu
75252 Paris Cedex 05
France.

E FERNÁNDEZ–CARA & C MORENO

Exact regularization and critical point approximation

Abstract

This paper deals with the use of some iterative methods for the computation of the critical points of a class of non-differentiable functionals. The schemes rely upon partial and exact regularization. We focus our attention on those functionals which can be written as the difference of two convex functions. Primary interest will be devoted to applications stemming from the variational formulation of partial differential problems. More precisely, we study the convergence of the algorithm when it is used to solve an elliptic problem with discontinuous nonlinearities and two free boundary problems arising respectively in fluid dynamics and plasma physics.

1. Preliminaries: non-differentiable convex problems and exact regularization

The goal of this paper is to describe some iterative schemes for solving non-differentiable extremal problems or, more generally, to find critical points of a non-differentiable functional. In particular, we are interested in those functionals stemming from the variational formulation of a partial differential problem which can be written as the difference of two convex functions.

The methods rely upon exact regularization and nonconvex duality techniques and can be viewed as the nonconvex counterpart of augmented Lagrangian algorithms.

For the sake of clarity, we first recall some known facts concerning convex problems. Let H be a Hilbert space and $J:H \to (-\infty, +\infty]$ a proper, l.s.c. convex function and consider the following

PROBLEM (P_0): Find $u \in H$ such that $J(u) \leq J(v) \quad \forall v \in H$.

For any $\lambda > 0$, one obviously has

$$\operatorname*{Inf}_{v\in H} J(v) = \operatorname*{Inf}_{v,w\in H} \{J(v) + \frac{1}{2\lambda} |v-w|_H^2\} = \operatorname*{Inf}_{w\in H} J_\lambda(w),$$

where we have set by definition

$$J_\lambda(w) = \operatorname*{Inf}_{v\in H} \{J(v) + \frac{1}{2\lambda} |v-w|_H^2\}. \tag{1}$$

It is not difficult to see that J_λ is real-valued, convex and F-differentiable, with $J' : H \to H$ Lipschitz-continuous of constant $1/\lambda$. Also

$$\operatorname{Arg} \operatorname*{Inf}_H J = \operatorname{Arg} \operatorname*{Inf}_H J_\lambda.$$

As a consequence, (P_0) is completely equivalent to

PROBLEM (P_λ): Find $u \in H$ such that $J_\lambda(u) \leq J_\lambda(v) \quad \forall v \in H$.

We say that (P_λ) has been obtained from (P_0) by performing an exact regularization (inf-convolution). If we now apply the standard steepest descent method to (P_λ), we obtain the iterates

$$u_{k+1} = u_k - \rho_k J'_\lambda(u_k), \quad \rho_k = \operatorname{Arg} \operatorname*{Inf}_{\rho\geq 0} J_\lambda(u_k - \rho J'_\lambda(u_k)), \tag{2}$$

which, under some assumptions on J, converge to a solution of (P_0).

REMARKS

(a) It is well known that J'_λ coincides with $\frac{1}{\lambda}(\text{Id.}-(\text{Id.} + \lambda\partial J)^{-1})$, i.e. the Yosida approximation of the maximal monotone operator ∂J.

(b) Due to the fact that J'_λ is $(1/\lambda)$-Lipschitzian, one can easily deduce that the optimal stepsize parameters ρ_k must be $\geq \lambda$. Thus an immediate simplified version of (2) is

$$u_{k+1} = u_k - \lambda J'(u_k) = (\text{Id.} + \lambda\partial J)^{-1} u_k. \tag{3}$$

This is the proximal point algorithm, which seems to have been introduced in [1], further analysed by Rockafellar in [2]. Some important applications of (3) and some of its variants to the solution of variational inequalities can be found in [3] - [8] among other references.

2. The unconstrained problem under consideration

Let us now consider some more complicated problems arising in nonconvex optimization. Let V, H be Hilbert spaces (H is identified to its dual space H'), $f : V \to (-\infty, +\infty]$ and $g : H \to (-\infty, +\infty]$ proper, l.s.c. and convex functions and $B : V \to H$ a bounded linear operator. Set

$$J(v) = f(v) - g(Bv) \quad \forall v \in V. \tag{4}$$

Our problem is to minimize J or, more generally, to find a critical point of J. For a critical point we mean a solution $u \in V$ of

$$\partial f(u) - \partial(g \circ B)(u) \ni 0. \tag{5}$$

It is well known that $\partial(g \circ B)(v) \supset B^* \partial g(Bv)$ for all $v \in V$ and that this relation holds with equality under some specific (mild) conditions. Consequently, it suffices to solve the subdifferential inclusion

$$\partial f(u) - B^* \partial g(Bu) \ni 0 \tag{6}$$

to obtain a critical point.

Now, we can associate to J a (dual) function J*, with

$$J^*(q) = g^*(q) - f^*(B^*q) \quad \forall q \in H$$

(here f* and g* are the convex conjugate functions of f and g respectively), in such a way that the critical points of J and J* (resp. u and p) satisfy some specific extremality conditions:

$$u \in \partial f^*(B^*p), \quad p \in \partial g(Bu). \tag{7}$$

A crucial observation (already used in [8]) is that the second inclusion in (7) is equivalent to the equality

$$p = g'_\lambda(Bu + \lambda p),$$

where λ is <u>any positive real parameter</u>. This suggests the following scheme for solving (6):

ALG 1 ($\lambda > 0$ and $p_0 \in H$ are given):

$$\partial f(u_{k+1}) \ni B^*p_k, \quad p_{k+1} = g'_\lambda(Bu_{k+1} + \lambda p_k). \tag{8}$$

Concerning this algorithm, one has:

THEOREM 1 ([9]): Assume that one has:

$$B : V \to H \text{ is compact, } R(\partial f) \subset R(B^*), \tag{9}$$

$$\text{Inf } J > -\infty, \tag{10}$$

$$J \text{ is coercive, i.e. } J(v) \to +\infty \text{ as } v \in V, |v|_V \to +\infty. \tag{11}$$

Then every sequence $\{u_k\}$ generated by ALG 1 possesses subsequences weakly convergent in V whose limits are solutions of (6).

Let us make some remarks about ALG 1 and this convergence result. First, let us say that the proof of Theorem 1 relies on the following inequality:

$$J^*(p_{k+1}) \leq J^*(p_k) - \frac{\lambda}{2} |p_{k+1} - p_k|_H^2. \tag{12}$$

In particular, J^* is non-increasing along $\{p_k\}$ and this provides a useful tool for performing convergence acceleration methods.

Secondly, notice that (10) and (11) can be resp. replaced by

$$\text{Inf } J^* > -\infty, \tag{10*}$$

$$J^* \text{ is coercive (in H).} \tag{11*}$$

In some important applications (see below) the introduction of B leads to a significant simplification for the determination of g'_λ. Indeed, if (5) is a variational formulation of (e.g.) a semilinear elliptic problem, an appropriate choice for H is a space of square-integrable functions and (8) provides a pointwise definition formula for p_{k+1}.

3. An application : elliptic problems with discontinuous nonlinearities

As a first application, consider the following problem:

$$-\Delta u(x) \in \partial\Phi(x,u(x)) \text{ a.e. in } \Omega, \qquad u \in H^2(\Omega) \cap H_0^1(\Omega). \tag{13}$$

Here, $\Omega \subset \mathbb{R}^N$ is a bounded open set whose boundary $\partial\Omega$ is sufficiently smooth and $\Phi: \bar{\Omega} \times \mathbb{R} \to \mathbb{R}$ is continuous, convex with respect to its second variable and subquadratic, i.e.

$$\Phi(x,s) \leqq \frac{a}{2}|s|^2 + b \quad \forall(x,s) \in \bar{\Omega} \times \mathbb{R},\ a < \lambda_1. \tag{14}$$

($\partial\Phi(x,\bar{s})$ stands for the subdifferential of the function $s \to \Phi(x,s)$ at $\bar{s}$; λ_1 is the first eigenvalue of the operator $-\Delta$ in Ω with homogeneous Dirichlet conditions).

It is not difficult to see that (13) can be equivalently written as a problem of the kind (6), with

$$V = H_0^1(\Omega), \quad H = L^2(\Omega), \quad B \equiv H_0^1 \hookrightarrow L^2 \text{ (compact)},$$

$$f(v) \equiv \frac{1}{2}\int_\Omega |\nabla v|^2\, dx \quad \text{and} \quad g(q) \equiv \int_\Omega \Phi(x,q)\, dx.$$

In this case, ALG 1 leads to the following iterates:

$$-\Delta u_{k+1} = p_k \text{ in } \Omega, \quad u_{k+1} = 0 \text{ on } \partial\Omega, \tag{15}$$

$$p_{k+1}(x) = \Phi_\lambda'(x,u_{k+1}(x) + p_k(x)) \text{ a.e. in } \Omega. \tag{16}$$

Since (9) - (11) are satisfied (thanks to the fact that Φ is subquadratic) and using standard regularity, one easily deduces a strong convergence result in $H_0^1(\Omega)$ for a subsequence of $\{u_k\}$.

From a numerical viewpoint, (15) - (16) is well-suited: it needs at each step the solution of a linear elliptic problem and the computation of Φ_λ'. In practice, it will be necessary to choose an approximation on $H_0^1(\Omega)$ for the solution of the linear problems. Concerning the computation of p_{k+1}, it is clear that a good strategy consists of a previous tabulation of Φ_λ' together with appropriate interpolations.

Roughly speaking, (15) - (16) plays the role of the fixed point monotone scheme of Amann [10], [11], and Sattinger [12] in the discontinuous case. To convince ourselves, let us indicate that the former is order preserving, in the sense that the following result holds:

<u>THEOREM 2</u>: Assume that an upper solution u^+ and a lower solution u_- exist with $u^+ \geqq u_-$ a.e. Let u_k and p_k (resp. v_k and q_k) be defined by (15) - (16) with $p_0 = -\Delta u^+$ (resp. $q_0 = -\Delta u_-$). Then:

$$q_0 \leqq q_1 \leqq \dots \leqq q_k \leqq \dots \leqq p_k \leqq \dots \leqq p_1 \leqq p_0 \tag{17}$$

and

$$v_0 \leqq v_1 \leqq \dots \leqq v_k \leqq \dots \leqq u_k \leqq \dots \leqq u_1 \leqq u_0 \quad \text{a.e. in } \Omega, \tag{18}$$

$$\{u_k\} \text{ and } \{v_k\} \text{ converge strongly in } H_0^1(\Omega); \tag{19}$$

For $u = \lim u_k$, $v = \lim v_k$, one has $u \geqq v$ a.e.

An interesting consequence of this result is that, when $u = v$, this function, regarded as a stationary solution of the parabolic problem

$$\frac{\partial \omega}{\partial t} - \Delta\omega \in \partial\Phi(x,\omega) \text{ a.e. in } \Omega \times \mathbf{R}_+,$$

$$\omega = 0 \text{ on } \Omega \times \mathbf{R}_+,$$

is asymptotically stable, i.e. one has the following property:

$$|\omega(\cdot,0) - u|_L < \delta \rightarrow \lim_{t\to\infty} |\omega(\cdot,t) - u|_{L^\infty} = 0.$$

Thus, (17) - (19) are in fact stability results for (8) in the particular case of problem (13).

4. <u>We come back to the general unconstrained problem</u>

The negative aspect of ALG 1 is that the introduction of the positive parameter

λ can make the convergence <u>too slow</u>. Indeed, it is straightforward to verify that, in the setting of Section 3, increasing values of λ lead to "less decreasing" iterates u_k. For a better understanding, let us come back to the general formulation (6) and assume for simplicity that $V = H$ (B is the identity mapping) and f* is twice continuously differentiable. Rewrite (7) in the form

$$\partial g^*(p) - Df^*(p) \ni 0. \tag{20}$$

After partial linearization, an immediate scheme for (20) is:

$$\partial g^*(p_{k+1}) - \{Df^*(p_k) + D^2f^*(p_k)\cdot(p_{k+1} - p_k)\} \ni 0. \tag{21}$$

Now, <u>augmented Lagrangian techniques</u> suggest an approximation of $D^2f^*(p_k)$ of the form λId. with $\lambda > 0$. This changes (21) into

$$\partial g^*(p_{k+1}) - \{Df^*(p_k) + \lambda(p_{k+1} - p_k)\} \ni 9, \tag{22}$$

which can also be written:

$$u_{k+1} = Df^*(p_k), \quad \partial g^*(p_{k+1}) - \lambda p_k \ni u_{k+1} - \lambda p_k. \tag{23}$$

In other words, partial linearization of J* together with reasonable approximation of $D^2f^*(p_k)$ lead to the iterates (compare with ALG 1):

$$\partial f(u_{k+1}) \ni p_k, \quad p_{k+1} \in \partial g_{-}\,(u_{k+1} - \lambda p_k). \tag{24}$$

Here, $\partial g_{-\lambda}$ is the (set-valued) operator

$$\partial g_{-\lambda} = \frac{1}{-\lambda}\,(\mathrm{Id.}-(\mathrm{Id.}-\lambda\partial g)^{-1}). \tag{25}$$

It is quite tempting to conclude that a more appropriate algorithm corresponds to (8) with λ replaced by $-\lambda$.

Notice that, generally speaking, (24) is much more difficult to analyse ($\partial g_{-\lambda}(u_{k+1}-\lambda p_k)$ may be the empty set, since $D(\partial g_{-\lambda}) \neq H$ in general; furthermore, if it is not a singleton, how do we determine p_{k+1}?). To ensure well-definedness and convergence, one has to introduce new assumptions on the behaviour of g and this will depend on the particular problem under consideration. In this direction, a partial result is the following:

THEOREM 3: Assume that (9) - (11) are satisfied and

$$\left.\begin{array}{l} g \text{ is subquadratic at infinity, i.e.} \\ g(q) \leqq \frac{a}{2}|q|_H^2 + b \ \forall q \in H, \text{ for some } a,b > 0. \end{array}\right\} \quad (26)$$

Then, for any $p_0 \in H$ and $\lambda > 0$ sufficiently small, the iterates

$$\partial f(u_{k+1}) \ni Bp_k, \quad p_{k+1} \quad \partial g_-(Bu_{k+1} - \lambda p_k) \quad (27)$$

are well-defined and the conclusion of Theorem 1 holds for the sequence $\{u_k\}$.

5. Some remarks for a kind of constrained problem

In the general setting of Section 2, let $K \subset H$ be a closed convex set with nonempty interior. We consider the following extremal problem:

PROBLEM (P_K): Find $p \in K$ such that $g^*(p)-f^*(B^*p) \leqq g^*(q)-f^*(B^*q)$ for all $q \in \partial K$.

Using some results from non-differentiable optimization (see e.g. [13]), one sees that, under certain mild conditions, any solution p of (P_K) also satisfies:

$$\partial g^*(p) - B\partial f^*(B^*p) + N_{\partial K}(p) \ni 0, \quad p \in \partial K. \quad (28)$$

Here, for a given $q \in \partial K$, $N_{\partial K}(q)$ is the normal subcone to ∂K at q, i.e. $N_{\partial K}(q) = N_K(q) \cap N_{K^c}(q)$, where

$N_k(q)$ is the normal cone to K, that is the subdifferential of the indicator functional at q,

$N_{K^c}(q)$ is Clarke's generalized cone to $\bar{K}^c$, that is Clarke's generalized gradient of the function "distance from K^c" at q.

In this Section, we describe an iterative method for the solution of (28). We make the following regularity assumption on K:

$$N_{\partial K}(q) \text{ is a } \underline{\text{one-dimensional subspace}} \text{ of } H, \qquad N_{\partial K}(q) = \text{Span } \{\eta(q)\} \text{ for every } q \in \partial K. \tag{29}$$

Under this assumption, (28) can also be written in the form:

$$\partial g^*(p) - B\partial f^*(B^*p) + \beta\eta(p) \ni 0, \; p \in \partial K, \; \beta \in \mathbb{R}, \tag{30}$$

and, as suggested by the arguments of Sections 2 and 4, we set:

<u>ALG 2</u> ($\lambda \neq 0$ and p_0 are given):

$$u_{k+1} \in \partial f^*(B^*p_k) + \beta_{k+1}B^{-1}\{\eta(p_k)\}, \quad \beta_{k+1} \in \mathbb{R}, \tag{31a}$$

$$p_{k+1} \in \partial g \; (Bu_{k+1} + \lambda p_k), \; p_{k+1} \in \partial K. \tag{31b}$$

The meaning of (31a) - (31b) must be understood in the following sense: the component of u_{k+1} in $B^{-1}N_{\partial K}(p_k)$ (i.e. the real number β_{k+1}) must be chosen in such a way that the corresponding p_{k+1} belongs to ∂K.

6. <u>Other applications: plasma confinement and the equilibrium of vortex rings</u>

In this Section, we briefly present two applications of ALG 2. In both cases, it will be seen that (28) is the variational formulation of a free boundary problem of the semilinear elliptic kind.

6.1 <u>The equilibrium of a plasma confined in a toroidal or cylindrical cavity</u> (see [14] - [16])

Let $\Omega \subset \mathbb{R}^2$ be a bounded open set, with smooth boundary $\partial\Omega$ and let $I > 0$ be given. We consider the following problem:

To find $u \in H^2(\Omega)$ and $\beta > 0$ such that

$$\begin{aligned} &-\Delta u(x) \in \partial\Phi(x,u(x)) \text{ a.e. in } \Omega, \\ &u = \beta \text{ on } \partial\Omega, \\ &-\int_{\partial\Omega} \frac{\partial u}{\partial n}\, d\sigma = I. \end{aligned} \tag{32}$$

This problem has been studied by many authors (see, for example, [14] - [21]) and describes the equilibrium of a plasma confined in a toroidal cavity (for the co-responding axi-symmetrical three-dimensional situation, see references [14], [16]), the region filled by the plasma being $\Omega_p = \{x | x \in \Omega , u(x) > 0\}$.

Using the arguments in [20], one can write (32) as a problem of the kind (P_K) with

$$V = H_0^1(\Omega) \oplus \mathbb{R}, \quad H = L^2(\Omega), \quad B \equiv V \hookrightarrow H, \text{ f and g as in Section 3,}$$

$$K = \{q | q \in L^2(\Omega), \int_\Omega q \, dx \leq I\}.$$

It will be assumed that Φ is as in Section 3 (in (14) a is any positive number), also satisfying:

$$\Phi(x,s) = 0 \quad \forall x \in \bar{\Omega}, \forall s \leq 0, \tag{33}$$

$$\lim_{s \to \infty} \int_\Omega \partial\Phi(x,s-0) \, dx > I. \tag{34}$$

In this particular case, with $\lambda > 0$, ALG 2 gives the following iterates:

$$-\Delta w_{k+1} = p_k \text{ in } \Omega, \quad w_{k+1} = 0 \text{ on } \partial\Omega , \tag{35}$$

$$\left.\begin{aligned} & u_{k+1} = w_{k+1} + \beta_{k+1}, \quad \beta_{k+1} \in \mathbb{R}, \\ & p_{k+1}(x) = \Phi'_\lambda(x, u_{k+1}(x) + \lambda p_k(x)) \quad \text{a.e. in } \Omega, \\ & p_{k+1} \in \partial K. \end{aligned}\right\} \tag{36}$$

Thus, at each step, one must solve the linear elliptic problem (35) and the nonlinear equation in β

$$D_k(\beta) \equiv \int_\Omega \Phi'_\lambda(x, w_{k+1}(x) + \lambda p_k(x) + \beta) = I. \tag{37}$$

Regarding (35) - (36), one can prove that the equation (37) always possesses at least one solution. Also, there exist subsequences of $\{(u_k, \beta_k)\}$ converging strongly to a solution of (32) (for details, see [9]). From a computational viewpoint, (37) is fairly nice: It can be solved easily, for instance via the (natural) scheme

$$\beta^{\ell+1} = T_k(\beta^{\ell}), \quad \text{with } T_k(\beta) \equiv \beta - \frac{\lambda}{\text{meas}(\Omega)} \{D_k(\beta) - I\}. \tag{38}$$

Some numerical results can be found in [22] (see also [9]).

For smooth Φ, (36) with $\lambda = 0$ coincides (at least formally) with a fixed-point method analysed in [20] (see also the numerical results in [21]).

6.2 The equilibrium of a vortex ring or pair with vanishing flux parameter

Let $\Omega \subset \mathbb{R}^2$ be as in the previous Section, $\eta > 0$ a given constant and $\zeta \in C^1(\bar{\Omega})$, $\zeta > 0$ in Ω; denote by H the maximal monotone operator associated with the product of a given constant $\alpha > 0$ and Heaviside function, i.e.

$$H(s) = \begin{cases} 0 & \text{if } s < 0, \\ [0,\alpha] & \text{if } s = 0, \\ \alpha & \text{if } s > 0. \end{cases} \tag{39}$$

We consider the following problem:

To find $u \in H^2(\Omega)$ and $W > 0$ such that

$$\begin{aligned} &-\Delta u(x) \in H(u(x)-W\zeta(x)) \text{ a.e. in } \Omega, \\ &u = 0 \text{ on } \partial\Omega, \\ &\int_\Omega |\nabla u|^2\, dx = \eta. \end{aligned} \tag{40}$$

This problem serves to model the equilibrium of a vortex pair in an ideal fluid (Ω is assumed to be large enough) and, as (32), is related to the existence of a free boundary (the boundary of $\Omega_V = \{x | x \in \Omega, u(x)-W\zeta(x) > 0\}$, which is the region of vorticity of the fluid). Existence results for this and other related problems can be found in [23] - [28] among others (for the description and study of the analogous axi-symmetric three-dimensional model, see the pioneering paper of Fraenkel and Berger [23].

Assume that Ω is sufficiently large to satisfy the following property:

$$\text{If} \quad -\Delta u = \alpha \text{ in } \Omega,\ u = 0 \text{ on } \partial\Omega, \quad \text{then} \int_\Omega |\nabla u|^2\, dx > \eta \,. \tag{41}$$

Then, (40) possesses at least one solution also satisfying (28) with V, H and B as in Section 3,

$$f(v) \equiv \begin{cases} 0 & \text{if } v + \zeta = 0, \\ +\infty & \text{otherwise,} \end{cases} \qquad g(q) \equiv \alpha \int_\Omega q_+ dx,$$

$$K = \{q \mid q \in L^2(\Omega),\ \int_\Omega (-\Delta)^{-1} q\cdot q\, dx \leqq \eta\}.$$

Now, $\partial K = \{q \mid q \in L^2(\Omega),\ \int (-\Delta)^{-1} q\cdot q\, dx = \eta\}$ and, for any $q \in \partial K$ one has $N_{\partial K}(q) = \text{Span}\ \{(-\Delta)^{-1}q\}$. For this problem, ALG 2 reads:

$$-\Delta w_{k+1} = p_k \text{ in } \Omega,\ \ w_{k+1} = 0 \text{ on } \partial\Omega\ , \tag{42}$$

$$\left.\begin{aligned} &u_{k+1} = -\zeta + \mu_{k+1} w_{k+1}, \quad \mu_{k+1} \in \mathbb{R}, \\ &p_{k+1}(x) = \text{Proj}_{|0,\alpha|}\ \{p_k(x) + \frac{1}{\lambda} u_{k+1}(x)\} \ \text{ a.e.}, \\ &p_{k+1} \in \partial K. \end{aligned}\right\} \tag{43}$$

For $\lambda > 0$ sufficiently small, one can derive for (43) results of the same kind as for (35) - (36). In particular, the sequence $\{(u_k,\ 1/\mu_k)\}$ contains subsequences strongly convergent in $H_0^1(\Omega) \times \mathbb{R}$ to a solution (u,W) of (40) (for details, see [9]).

Again (43) is related to a fixed-point method, introduced by Berestycki [27] and further used in [29] to obtain numerical results.

References

1. B. Martinet.- Comptes Rendus Acad. Sc. Paris, Série A. 274 (1972), p. 163-165.
2. R.T. Rockafellar.- SIAM J. Control Optim. 14 (1976), p. 877-88.
3. R. Glowinski.- Numerical Methods for Nonlinear Variational Problems, 2nd ed. Springer-Verlag, New York 1984.

4. M. Fortin.- Appl. Math. Optim. 2 (1976), p. 236-250.
5. M. Fortin & R. Glowinski.- Résolution Numérique des Problemes aux Limites par des Méthodes de Lagrangien Augmenté, North-Holland, Amsterdam 1983.
6. D. Gabay & B. Mercier.- Comp. and Math. with Appl., 2 (1976), 1, p. 17-40.
7. P.L. Lions & B. Mercier.- SIAM J. Numer. Anal., 16 (6) (1978), p. 964-979.
8. A. Bermúdez & C. Moreno.- Comp. and Math. with Appl., 7 (1981), p. 43-58.
9. E. Fernández-Cara & C. Moreno.- Critical point approximation through exact regularization, to appear.
10. H. Amann.- Proc. Royal Soc. Edinburgh, 81A (1978), p. 37-47.
11. H. Amann.- Indiana U. Math. J. 21 (1971), p. 125-146.
12. D. Sattinger.- Indiana U. Math. J. 21 (1972), p. 979-1000.
13. J.B. Hiriart-Urruty.- Math. Op. Res., Vol. 4, 1 (1979), p. 79-97.
14. R. Temam.- Arch. Rat. Mech. Anal., 60 (1975), p. 51-73.
15. R. Temam.- Comm. P.D.E. 2 (1977), p. 563-585.
16. C. Mercier.- In EURATOM-CEA, Comm. of the European Commun., Luxenbourg 1974 (Rep. EUR 5127/1).
17. J.M. Puel.-Comptes Rendus Acad. Sc. Paris, Série A, 284 (1977), p.861-863.
18. A. Damlamiam.- Comptes Rendus Acad. Sc. Paris, Série A, 286 (1978), p. 153-155.
19. D.G. Schaeffer.- Comm. P.D.E. 2 (1977), p. 587-600.
20. H. Berestycki & H. Brezis.- Nonlinear Anal. T.M.A. 4 (1980), p. 415-436.
21. M. Sermange.- Thesis, University Paris XI, 1982.
22. J. Pons.- Memoria de Licenciatura, Universidad Autónoma, Madrid 1984.
23. L.E. Fraenkel & M.S. Berger.- Acta Math. 132 (1974), p. 13-51.
24. A. Ambrosetti & G. Mancini.- In Recent Contributions to Nonlinear Partial Differential Equations, H. Berestycki & H. Brézis eds., Pitman, 1981.
25. A. Friedman & B. Turckington.- J. Functional Anal., 37 (1980), p. 136-163.
26. J. Norbury.- Comm. Pure Appl. Math. 28 (1975), p. 679-700.
27. H. Berestycki.- Thesis, University Paris VI, 1980.
28. H. Berestycki & P.L. Lions. - To appear.

29. H. Berestycki, E. Fernández-Cara & R. Glowinski.- R.A.I.R.O. Anal. Numér., Vol. 18, 1 (1984), p. 7-85.

E. Fernández-Cara
Facultad de Matemáticas
Universidad de Sevilla
Sevilla
Spain.

C. Moreno
División de Matemáticas
Universidad Antónoma de Madrid
28049 Madrid
Spain.

J HERNÁNDEZ

Continuation and comparison methods for some nonlinear elliptic systems

1. Introduction

Reaction-diffusion systems have been widely studied during the last years (cf., e.g., [17]), and a variety of methods (sub and supersolutions, degree theory, global bifurcation; etc) have been used in order to study existence and multiplicity of solutions for some classes of semilinear elliptic systems arising in applications. In particular, positive solutions are especially interesting because they are often the only physically meaningful solutions.

We consider in this paper a class of reaction-diffusion systems arising in the study of activator-inhibitor interaction in morphogenesis, namely

$$\begin{aligned} -\Delta u &= \lambda u - f(u) - v && \text{in } \Omega \\ -\Delta v &= \delta u - \gamma v && \text{in } \Omega \qquad (1.1) \\ u &= v = 0 && \text{on } \partial\Omega , \end{aligned}$$

where Ω is a bounded domain, δ and γ are positive physical constants, λ is a real parameter and f is a nonlinear function such that $f(0) = 0$ satisfying some additional assumptions (as $f(u) = u^3$). Our aim is to find solutions of (1.1) with both components u and v different from zero and we are particularly concerned with positive solutions. Problem (1.1) for f odd was considered by Rothe [16] by using a decoupling technique and critical point theory. Later, the case of f not necessarily odd was studied by Lazer-McKenna [14]: they proved, by using degree theory, that there are exactly three solutions for some range of the parameter λ, without giving information about the sign of these solutions. Finally, de Figueiredo-Mitidieri [10] improved those results by showing (using sub and supersolutions and a Maximum Principle for (1.1)) the existence of a positive and a negative solution for, roughly speaking, the same values of λ.

Here we deal with a variant of the problem treated in the preceding papers [16], [14], [10], namely the case of asymptotically linear f. In this case we are able to give an almost complete description of the bifurcation diagram for positive solutions. First, we use sub and supersolutions, as in [10], to prove existence and uniqueness for positive solutions. Then, by using a local inversion theorem by Crandall-Rabinowitz [7], [15] and some continuation arguments, we obtain an alternative existence proof. Moreover, we show that positive solutions form actually a smooth curve. This follows from the fact that every nontrivial positive solution is non-degenerate, i.e. the linearized problem along this solution is invertible. An extended version of this paper including other kinds of nonlinear terms will appear elsewhere [13]. For related results concerning, in particular, a system describing predator-prey interactions see [5], [4], [8], [9].

Finally, the author would like to thank J. Blat and E. Mitidieri for some useful conversations.

2. Main results and proofs

We consider the elliptic system

$$
\begin{aligned}
-\Delta u &= \lambda u - f(u) - v && \text{in } \Omega \\
-\Delta v &= \delta u - \gamma v && \text{in } \Omega \qquad (2.1)\\
u &= v = 0 && \text{on } \partial\Omega,
\end{aligned}
$$

where Ω is a smooth bounded open subset of R^N, δ, $\gamma > 0$ are given real numbers, λ is a real parameter, and $f: \mathbb{R} \to \mathbb{R}$ satisfies

f is C^2, increasing, $f(0) = 0$, (2.2)

$f(u)/u$ is strictly increasing for $u > 0$ and strictly decreasing for $u < 0$, (2.3)

$$\lim_{|u| \to \infty} \frac{f(u)}{u} = f'(\infty) < +\infty . \qquad (2.4)$$

REMARK 2.1: It follows immediately from (2.2) (2.3) that $f'(0) < f'(\infty)$.

The study of (2.1) can be reduced to the consideration of a single equation by using a decoupling technique. More precisely; for $u \in L^2(\Omega)$

given, the linear problem

$$-\Delta v + \gamma v = \delta u \text{ in } \Omega$$
$$v = 0 \text{ on } \partial\Omega$$

has a unique solution $v = Bu \in H^2(\Omega)$. (We employ the usual notations $C^{k,\alpha}(\bar{\Omega})$, $0 < \alpha < 1$, for the spaces of Hölder continuous functions, and $W^{m,p}(\Omega)$, with $H^m(\Omega) = W^{m,2}(\Omega)$ for Sobolev spaces). Moreover, by the Maximum Principle, if $u \in C(\bar{\Omega})$, $u \geq 0$, then $Bu \equiv 0$ or $Bu > 0$ on Ω. Now, system (2.1) is equivalent to the nonlinear equation

$$-\Delta u + Bu + f(u) = \lambda u \text{ in } \Omega \tag{2.5}$$
$$u = 0 \text{ on } \partial\Omega,$$

where the (nonlocal) operator $B : L^2(\Omega) \to L^2(\Omega)$ is defined as above.

If we define $T = -\Delta + B$ with domain $D(T) = H^2(\Omega) \cap H_0^1(\Omega)$, then T is a positive, symmetric closed operator. We need some spectral theory for T (cf. [10], [14], [16]).

<u>LEMMA 2.1</u> (Comparison Lemma, [14]): Let $\rho \in C(\bar{\Omega})$. Then there exists an increasing sequence γ_n, $n = 1,2,\ldots,\gamma_n$ counted according to its multiplicity, such that $\gamma_n \xrightarrow[n\to\infty]{} +\infty$ and the problem

$$-\Delta w + Bw + \rho(x)\, w = \gamma w \text{ in } \Omega \tag{2.6}$$
$$w = 0 \quad \text{on } \partial\Omega$$

has a nontrivial solution if and only if $\gamma = \gamma_k$ for some k. Moreover, if $\rho(x) \leq \bar{\rho}(x)$, with $\bar{\rho} \in C(\bar{\Omega})$, then for every n, $\gamma_n \leq \bar{\gamma}_n$, where $\bar{\gamma}_n$ are the corresponding eigenvalues of (2.6) with ρ replaced by $\bar{\rho}$.

In the following we denote by $\hat{\lambda}_k(\rho)$ the eigenvalues of (2.6) for $\rho \in C(\bar{\Omega})$. For $\rho \equiv 0$ we put $\hat{\lambda}_k \equiv \hat{\lambda}_k(0)$. In the same vein, for $B \equiv 0$, we employ the usual notation $\lambda_k(\rho)$ for the eigenvalues, now $\lambda_k \equiv \lambda_k(0)$.

Suppose that ϕ_k is the eigenfunction for the eigenvalue λ_k, normalized by $\|\phi_k\|_{L^\infty} = 1$, i.e. that we have

$$-\Delta\phi_k = \lambda_k\phi_k \text{ in } \Omega$$
$$\phi_k = 0 \quad \text{on } \partial\Omega. \tag{2.7}$$

It is well-known that ϕ_1 can be chosen in such a way that $\phi_1 > 0$ on Ω. Hence

$$T\phi_k = \hat{\lambda}_k\phi_k$$

where

$$\hat{\lambda}_k = \lambda_k + \frac{\delta}{\gamma+\lambda_k} \tag{2.8}$$

On the other hand, (ϕ_k) is a complete orthonormal system in $L^2(\Omega)$, and this implies that $\hat{\lambda}_k$ are the only eigenvalues of T and that the spectrum $\sigma(T)$ contains these eigenvalues only.

An interesting remark is that the sequence $\hat{\lambda}_k$ is not necessarily increasing. A sufficient condition is

$$\gamma + \lambda_1 > \sqrt{\delta} \tag{2.9}$$

and a necessary and sufficient condition is

$$(\gamma + \lambda_1)\ (\gamma + \lambda_2) > \delta. \tag{2.10}$$

(Obviously, (2.9) implies (2.10)). We assume (2.9) for all that which follows. In particular, $\hat{\lambda}_1$ is simple with eigenfunction ϕ_1.

LEMMA 2.2 [10]: For every μ in the resolvent of T, the resolvent operator $T_\mu = (T - \mu I)^{-1}$ is compact.

The next Lemma is the main tool for our results.

LEMMA 2.3 (Maximum Principle, [10]): If (2.9) is satisfied, then for every μ such that $2\sqrt{\delta} - \gamma \leq \mu < \hat{\lambda}_1$, T_μ is positive, i.e. $u \geq 0$ implies $T_\mu u \geq 0$.

REMARK 2.2: If $u \in C(\bar{\Omega})$, $u \geq 0$, then $T_\mu u > 0$ in Ω and $\frac{\partial(T_\mu u)}{\partial n} < 0$ on $\partial\Omega$. This follows immediately from the representation formula (1.5) in [10].

A first result concerning problem (2.5) is the following.

LEMMA 2.4: If $u \geq 0$ is a nontrivial solution of (2.5), then we have $\hat{\lambda}_1 + f'(0) < \lambda$

PROOF: If $u \geq 0$ is a nontrivial solution, then

$$-\Delta u + Bu + h(x)u = \lambda u \text{ in } \Omega$$

$$u = 0 \quad \text{on } \partial\Omega,$$

where the function

$$h(x) = \begin{cases} \dfrac{f(u(x))}{u(x)} & \text{if } u(x) > 0 \\ f'(0) & \text{if } u(x) = 0 \end{cases}$$

is continuous and bounded, with $f'(0) \leq h(x)$, and the conclusion follows from the comparison results (Lemma 2.1).

LEMMA 2.5: If $u \geq 0$ is a nontrivial solution of (2.5) with f satisfying (2.2)-(2.4) and

$$f'(\infty) - f'(0) < \hat{\lambda}_1 + \gamma - 2\sqrt{\delta}\,, \tag{2.11}$$

then $u > 0$ in Ω.

PROOF: Problem (2.5) can be rewritten equivalently as

$$-\Delta u + Bu - \bar{\mu}u = \lambda u - f(u) - \bar{\mu}u \text{ in } \Omega$$

$$u = 0 \qquad \text{on } \partial\Omega.$$

Suppose that there is a $\bar{\mu}$ such that (i) the Maximum Principle holds for $-\Delta + B - \bar{\mu}I$ and (ii) $\lambda u - f(u) - \bar{\mu}u$ is increasing for $u \geq 0$. Then the conclusion will follow by Remark 2.2. But (i) is satisfied if $2\sqrt{\delta} - \gamma \leq \mu < \hat{\lambda}_1$ and (ii) will be satisfied if $\lambda - f'(u) - \bar{\mu} \geq 0$ for every $u \geq 0$ and every λ in the corresponding interval, i.e. if $\bar{\mu} + f'(\infty) \leq \hat{\lambda}_1 + f'(0)$. It suffices

$$2\sqrt{\delta} - \gamma \leqq \bar{\mu} \leqq \hat{\lambda}_1 + f'(0) - f'(\infty)$$

which is precisely (2.11).

Now we use the method of sub and supersolutions (see, for example, [1]) to obtain existence and uniqueness for positive solutions of (2.5).

<u>THEOREM 2.1</u>: Assume that f satisfies (2.2)-(2.4), (2.9) and (2.11). Then for every λ in the interval $\hat{\lambda}_1 + f'(0) < \lambda < \hat{\lambda}_1 + f'(\infty)$, there is a unique nontrivial positive solution of (2.5).

<u>PROOF</u>: First we remark that Lemma 2.2 and the proof of Lemma 2.5 show that the method of sub and supersolutions can be applied if (2.11) holds.

Now we need a subsolution u_o and a supersolution u^o such that $0 < u_o \leqq u^o$. As in [10], it is easy to show that $u_o = c\phi_1$ with $c > 0$ sufficiently small is a subsolution for $\hat{\lambda}_1 + f'(0) < \lambda$.

To find a supersolution u^o, we first point out that from the classical results in [6] and the explicit formula (2.8) it follows that $\hat{\lambda}_k$ depends continuously and monotonically on the domain Ω. If $\lambda < \hat{\lambda}_1 + f'(\infty)$, there exist $\bar{\lambda}$ with $\lambda - f'(\infty) < \bar{\lambda} < \hat{\lambda}_1$ and $\tilde{\Omega} \supset \Omega$ such that $\bar{\lambda}$ is the first eigenvalue for $-\Delta + B$ in $\tilde{\Omega}$. If we denote by ψ the corresponding eigenfunction satisfying $\psi > 0$ on $\tilde{\Omega}$ and $||\psi||_\infty = 1$, it is not difficult to see that $u^o = C\psi$, with $C > 0$ sufficiently large, is a supersolution verifying $0 < u_o < u^o$, and existence is proved. The fact that we obtain classical solutions is an easy consequence of well-known regularity results.

Concerning uniqueness, suppose that $u, v \geqq 0$ are nontrivial solutions of (2.5). By Lemma 2.5, $u,v > 0$ in Ω. Assume they are ordered, e.g. $0 < u \leqq v$. Multiplying (2.5) by v, integrating over Ω with Green's formula, and interchanging u and v, we obtain

$$\int_\Omega \nabla u.\nabla v + Bu.v + f(u)v = \lambda \int_\Omega uv = \int_\Omega \nabla u.\nabla v + Bv.u + f(v)u.$$

Since B is symmetric we get

$$\int_\Omega \left(\frac{f(u)}{u} - \frac{f(v)}{v}\right)u\ v = 0$$

and this, together with (2.3) gives $u \equiv v$. If u and v are not ordered, an additional argument involving sub and supersolutions gives $u \equiv v$.

REMARK 2.3: The existence of a unique nontrivial negative solution for the same interval of λ's can be proved in exactly the same way.

REMARK 2.4: An analogous existence proof was given in [10] for f behaving like, for example, u^3. Uniqueness is not proved there, but it follows necessarily from the results by Lazer-McKenna [14] giving exactly three solutions.

An alternative proof of the preceding results can be given by using local inversion and continuation arguments, as in [15] or [7]; refer, also, to [1], [2], [5], [11], [12] for similar results.

First, we define the map $F : \mathbb{R} \times C_0^{2,\alpha}(\bar{\Omega}) \to C^{\alpha}(\bar{\Omega})$

$$F(\lambda,u) = -\Delta u + Bu + f(u) - \lambda u.$$

It is clear that F is C^2 and $F(\lambda,0) = 0$. It follows immediately that

$$F_u(\mu,v)w = -\Delta w + Bw + f'(v)w - \mu w$$

$$F_{\lambda u}(\mu,v)w = -w.$$

Hence

$$F_u(\hat{\lambda}_1 + f'(0), 0)w = -\Delta w + Bw - \hat{\lambda}_1 w$$

and

$$\ker F_u(\hat{\lambda}_1 + f'(0), 0) = \text{span}(\phi_1).$$

As a complementary subspace we take

$$Z = R(F_u(\hat{\lambda}_1 + f'(0), 0)) = \{z \in C^{\alpha}(\bar{\Omega}) : \int_{\Omega} z\,\phi_1 = 0\}.$$

It follows that $F_{\lambda u}(\hat{\lambda}_1 + f'(0), 0)\phi_1 = -\phi_1 \notin Z$ and the local inversion theorem by Crandall-Rabinowitz [7] can be applied. Reasoning as in [15], we obtain the existence of a "small" branch of nontrivial positive solutions in a right-neighbourhood of $\hat{\lambda}_1 + f'(0)$. This branch is parametrized by λ and can be continued to the right by using the Implicit Function Theorem. This yields a C^2 curve of positive solutions. The main auxiliary result is the following.

LEMMA 2.6: If $v > 0$ is a solution of (2.5) for $\hat{\lambda}_1 + f'(0) < \lambda < \hat{\lambda}_1 + f'(\infty)$ with f satisfying (2.2)-(2.4), (2.9), (2.11) and

$$f'(\infty) - f'(0) \leqq \hat{\lambda}_2 - \hat{\lambda}_1, \tag{2.12}$$

then v is non-degenerate, i.e. $F_u(\lambda,v)$ is invertible.

PROOF: If $F(\lambda,v) = 0$, $v > 0$, $\hat{\lambda}_1 + f'(0) < \lambda$, we have

$$- \Delta v + Bv + \frac{f(v)}{v} v = \lambda v \text{ in } \Omega \tag{2.13}$$

$$v = 0 \quad \text{on } \partial\Omega.$$

We claim that $F_u(\lambda,v)$ is an isomorphism. If not, there is a $w \not\equiv 0$ such that

$$- \Delta w + Bw + f'(v)w = \lambda w \text{ in } \Omega \tag{2.14}$$

$$w = 0 \quad \text{on } \partial\Omega.$$

From (2.13), (2.14), and a comparison argument involving Lemma 2.1 and (2.3) we infer

$$\lambda \geqq \hat{\lambda}_1(f'(v)) > \hat{\lambda}_1(f(v)/v)$$

and we shall get a contradiction if $\lambda = \hat{\lambda}_1(f(v)/v)$. Indeed, it is clear from (2.13) that $\lambda = \hat{\lambda}_j(f(v)/v)$ for some $j \geqq 1$. If $j \geqq 2$, by a similar comparison argument $\lambda > \hat{\lambda}_2(f'(0)) = \hat{\lambda}_2 + f'(0)$ and this is impossible if $\hat{\lambda}_1 + f'(\infty) \leqq \hat{\lambda}_2 + f'(0)$ or, equivalently, if (2.12) holds.

REMARK 2.5: If $B \equiv 0$, the fact that $\lambda = \hat{\lambda}_1(f(v)/v)$ is an easy consequence of (2.13), since it is the only eigenvalue having a positive eigenfunction, and (2.12) is by no means necessary. But, apparently, the same result for the operator $- \Delta + B + \rho(x)I$, with non-constant $\rho(x)$, is not available, and (2.12) is needed. In view of Theorem 2.1 it can be conjectured that (2.12) is only a technical condition which is not really necessary for the proof of

Lemma 2.6.

REMARK 2.6: It follows from (2.8) and (2.10) that

$$\hat{\lambda}_2 - \hat{\lambda}_1 = (\lambda_2 - \lambda_1)\left(1 - \frac{\delta}{(\gamma + \lambda_2)(\gamma + \lambda_1)}\right) < \lambda_2 - \lambda_1$$

and $\hat{\lambda}_2 - \hat{\lambda}_1$ tends to $\lambda_2 - \lambda_1$ (monotonically) when $\gamma \to +\infty$ (for $\delta > 0$ fixed). This means that if

$$f'(\infty) - f'(0) < \lambda_2 - \lambda_1 \tag{2.15}$$

holds, then (2.12) will be satisfied for γ large enough, a condition which is clearly compatible with (2.8) (or (2.9)).

We return to the existence proof. Starting from $\lambda = \hat{\lambda}_1 + f'(0)$, we obtain a branch of nontrivial positive solutions which can be continued since the Implicit Function Theorem is applicable by Lemma 2.6. We sketch the proof that these solutions are still positive. It follows from the continuity of the map $\lambda \to u(\lambda)$ that if $u(\lambda)$ is not in the cone of positive functions for every λ there should be a minimal $\bar{\lambda} < \hat{\lambda}_1 + f'(\infty)$ such that $u(\bar{\lambda})$ is on the boundary of this cone, and this implies $u \equiv 0$ by Lemma 2.5. Therefore, $\bar{\lambda}$ is a bifurcation point for positive solutions, a contradiction (cf. [1], [3]). We have proved the existence of a unique positive solution for a maximal interval $(\hat{\lambda}_1 + f'(0), \lambda^+)$. We claim that $\lambda^+ = \hat{\lambda}_1 + f'(\infty)$. If not, it is possible to show, reasoning as in [15] or [11], that $\|u(\lambda)\|_\alpha$ goes to $+\infty$ when $\lambda \uparrow \lambda^+$, and λ^+ would be an asymptotic bifurcation point for positive solutions, again a contradiction ([1], [3]).

Therefore we have proved the following result.

THEOREM 2.2: Assume that f satisfies (2.2)-(2.4), (2.11) and (2.12). If $u(\lambda)$ is the unique nontrivial positive solution of (2.5), the map $\lambda \to u(\lambda)$ from $[\hat{\lambda}_1 + f'(0), \hat{\lambda}_1 + f'(\infty))$ into $C^\alpha(\bar{\Omega})$ is C^2. Moreover

$$\lim_{\lambda \uparrow \hat{\lambda}_1 + f'(\infty)} \|u(\lambda)\|_\alpha = +\infty$$

REMARK 2.7: The same technique can be used to prove that the positive solutions whose existence is shown in [10] actually form a smooth curve.

REMARK 2.8: Similar continuation arguments can be used if (2.3) is replaced by a concavity assumption. Refer to [11], [12] for the case $B \equiv 0$, and [1] - [3] for related results. However, sub and supersolutions are not available in this case, and uniqueness should be proved, under suitable assumptions, in a different way. See [13] for details.

References

1. H. Amann, Fixed point equations and nonlinear eigenvalue problems in ordered Banach spaces. SIAM Review 18 (1976), 620-709.
2. H. Amann, Nonlinear eigenvalue problems having precisely two solutions. Math. Z. 150 (1976), 27-37.
3. H. Amann and T. Laetsch, Positive solutions of convex nonlinear eigenvalue problems. Indiana Univ. Math. J. 25 (1976), 259-270.
4. J. Blat and K.J. Brown, Bifurcation of steady-state solutions in predator-prey and competition systems. Proc. Roy. Soc. Edinburgh 97A (1984), 21-34.
5. E. Conway, R. Gardner and J. Smoller, Stability and bifurcation of steady-state solutions for predator-prey equations. Adv. in Appl. Math. 3 (1982), 288-334.
6. R. Courant and D. Hilbert, Methods of Mathematical Physics, New York, Interscience; 1953.
7. M. Crandall and P.H. Rabinowitz, Bifurcation, perturbation of simple eigenvalues and linearized stability. Arch. Rat. Mech. Anal. 52 (1973), 161-180.
8. E.N. Dancer, On positive solutions of some pairs of differential equations. I. Trans. Amer. Math. Soc. 284 (1984), 729-743.
9. E.N. Dancer, On positive solutions of some pairs of differential equations II. J. Diff. Eqns. 60 (1985), 236-258.
10. D.G. de Figueiredo and E. Mitidieri, A Maximum Principle for a elliptic system and applications to semilinear problems. M.R.C. Technical Report 2653, Madison, 1984.
11. J. Hernández, Bifurcación y soluciones positivas para ciertos problemas de tipo unilateral. Thesis. Madrid, Universidad Autonoma, 1977.

12. J. Hernández, Qualitative methods for nonlinear diffusion equations. Lecture Notes, CIME Course 1985. To appear.
13. J. Hernández. To appear.
14. A.C. Lazer and P.J. McKenna. On steady-state solutions of a system of reaction-diffusion equations from biology. Nonlinear Anal. 6 (1982), 523-530.
15. P.H. Rabinowitz, A note on a nonlinear eigenvalue problem for a class of differential equations. J. Diff. Eqns. 9 (1971), 536-548.
16. F. Rothe, Global existence of branches of stationary solutions for a system of reaction-diffusion equations from biology. Nonlinear Anal. 5 (1981), 487-498.
17. J. Smoller, Shock Waves and Reaction-Diffusion Equations. Berlin, Springer, 1983.

This work was partially sponsored by project no. 3308/83 of CAICYT, Spain.

J. Hernández
Departamento de Matemáticas
Universidad Autónoma
28049 Madrid
Spain.

J M LASRY

EDP et anticipations rationnelles (à partir du modèle de J Scheinkman et L Weiss)

1. Equations aux derivées partielles en économie

Le premier exemple d'E.D.P. en économie est du à Slutsky. Cet exemple n'a pas eu de suite jusqu' au développement dans les annés 70 de la théorie de Blake et Schole, c'est-a-dire, l'application aux mathématiques financières des équations paraboliques non lineaires de type Hamilton-Jacobi-Bellman (voir par exemple [1]).

Puis, très recemment (depuis 1980 environ) Jean-Charles Rochet a montré qu'un grand nombre de problèmes en théorie des jeux et en économie comportant une information asymétrique des agents conduisaient à des E.D.P. non lineaires [4], [5], [6].

Dans un domaine voisin, la généralisation du modèle de J. Scheinkman et L. Weiss [7] sur les fluctuations macro-économiques, nous a conduit à des équations differentielles et aux derivées partielles fonctionelles non lineaires qui vont constituer l'objet de ce texte. Aussi bien dans le cas des équations proposées par J.C. Rochet, que dans celui des équations de cet exposé, un grand nombre de resultats mathématiques ont été déjà obtenus, mais il reste encore plus de questions ouvertes et de nature mathématique originale. Bref, l'application des E.D.P. à l'économie théorique et appliquée, et à la théorie des jeux, est un domaine nouveau en plein developpement.

2. Le modèle de J. Scheinkman et L. Weiss

Nous n'exposerons pas ici en detail les considérations économiques sur lesquelles s'appuie le modèle proposé par J. Scheinkman et L. Weiss (le lecteur peut les trouver détaillées dans l'article [7]).

Contentons nous d'indiquer qu'il s'agit d'un modèle destiné à expliquer les fluctuations macroéconomique à partir de la demande d'épargne dans une économie dont le marché financier est incomplet: dans ce modèle les agents économiques ne peuvent pas s'assurer contre les aléas de leur productivité ni emprunter de façon illimitée. Il s'agit d'économie théorique, et ce modèle se situe dans le cadre géneral de la théorie des anticipations rationnelles.

Par contre, nous allons exposer le probléme d'optimisation résolu par J. Scheinkman et L. Weiss, ce qui permettre au lecteur de constater que sa structure est d'un type nouveau du point de vue de la théorie de l'optimisation.

Un préliminaire technique

Avant même de formuler le problème il nous faut donner un résultat d'existence pour un système differentiel dont le rôle ne pourra être compris qu'après coup (i.e. après le théorème 2).

Théorème 1 (J. Scheinkman et L. Weiss [7]): Suit λ et ρ deux constantes strictement positives. Il existe au moins une paire (A,B) telle que:

(1) $A \in C^1(]0,1])$, $B \in C^1([0,1])$, $A > 0$, $B > 0$

(2) $A'(z) = A(z)\,(\lambda\, B(z) - (\lambda + \rho)\, A(z)) \quad \forall z \in\,]0,1]$

(3) $B'(z) = A(1-z)\,((\lambda+\rho)B(z) - \lambda A(z)) \quad \forall z \in [0,1[$

(4) $A(z) \to +\infty$ quand $z \to 0+$

(5) $(\lambda+\rho)\, B(1) = \lambda A(1)$.

On constate que le système (2) (3) n'est pas une équation différentielle ordinaire, ni même une équation différentielle avec retard, à cause du terme A(1-z) dans l'équation (3), L'uncité de la solution (A,B) n'est ni prcuvée, ni conjecturée.

La démonstration du théorème 1 proposée par J. Sheinkman et L. Weiss consiste à montrer:

(i) que le système (2) + (3) + (*), où (*) est la condition:

(*) $A(1/2) = \alpha$, $B(1/2) = \beta$

possède une unique solution définie sur un intervalle maximal J(avec $1/2 \in J$)

(ii) que l'on peut choisir le couple de paramètres (α,β) de telle sorte que l'intervalle J contient]0,1] et que les conditions (4) et (5) sont vérifiées.

La démonstration du premier point (i) est une adaptation assez facile de la démonstration du théorème Cauchy-Lipschitz. Par contre la deuxième partie demande une analyse qualitative subtile du pseudo-plan de phase du pseudo-système dynamique (2) + (3) (le préfixe pseudo est là pour rappeler que si z est le temps, le mouvement du point (A(z), B(z)) donné par les équations différentielles (2) + (3) dépend de A(1-z)).

Anticipations rationnelles

Les solutions (A,B) du système (1 à 5) vont servir à définir deux problèmes d'optimisations. C'est la résolution de ces problèmes qui va permettre de comprendre après coup le rôle du système(1 à 5).

Nous justiferons alors le titre de ce paragraphe.

Donc dans la suite on désigne par (A,B) des fonctions qui satisfont les conditions (1) et (4).

On se donne aussi un processus de Markov en temps continu $\alpha_t = \alpha_t(w)$ à valeurs dans $\{0,1\}$ avec une probabilité de transition $0 \to 1$ ou $1 \to 0$ proportionnelle à λdt, c'est-a-dire:

$$\text{prob } \{\alpha_{t+h}(\omega) = 1 \text{ sachant } \alpha_t(\omega) = 0\} = \lambda h + o(h)$$

et de même en échangeant 1 et 0.

On suppose en outre que α verifie

(6) $\alpha_0(\omega) = 0$ presque surement.

REMARQUE: La donnée de α suppose que l'on se donne un quadruplet $(\Omega, P, F, (F_t)_{t\geq 0})$ où F est une tribu sur Ω, P une probabilité sur F, où $(F_t)_{t\geq 0}$ la plus petite famille croissante de tribus telles que F_t rend mesurable l'application $\omega \to \alpha_t(\omega)$ et où F est la réunion des F_t, $t \geq 0$.

A partir du processus α on définit le processus $z_t(\omega)$ par l'équation différentielle stochastique suivante

(7) $z_0(\omega) = z_0$ p.s.

(8) $\dot{z}_t = -1/A(z_t)$ si $\alpha_t = 0$

(9) $\dot{z}_t = 1/A(1-z_t)$ si $\alpha_t = 1$

Le processus z ainsi défini est progressivement mesurable par rapport à α

(c'-est-a-dire: z_t est mesurable par rapport à F_t, pour tout $t \geq 0$).

On appelle contrôles admissibles les couples (C,ℓ) de processus aléatoires progressivement mesurables par rapport à α qui vérifient

(10) $0 \leq C_t(w) \leq 1$ et $\ell_t(w) \geq 0$, p.s., $\forall t \geq 0$.

Pour tout couple (C,ℓ) admissible (C pour consommation, ℓ pour travail), on pose

(11) $J(C,\ell) = E \int_0^{+\infty} e^{-\rho t}(\log C_t - \ell_t)dt$

où E désigne l'espérance mathématique (on remarque que la condition (10) assure que $J(C,\ell)$ est bien défini par (11) avec $J(C,\ell) \geq 0$ et eventuellement $J(C,\ell) = -\infty$).

Pour tout couple (C,ℓ) admissible on définit aussi deux processus aléatoires progressivement mesurables R et S par les équations differentielles stochastiques:

(12) $R_0 = 1-z_o$

(13) $\left.\begin{array}{l} \dot{R}_t = -C_t/B(1-z_t) \text{ si } \alpha_t = 1 \\ \dot{R}_t = (\ell_t - C_t)/B(z_t) \text{ si } \alpha_t = 0 \end{array}\right\}$

(14) $S_o = z_o$

(15) $\left.\begin{array}{l} \dot{S}_t = -C_t/B(1-z_t) \text{ si } \alpha_t = 0 \\ \dot{S}_t = (\ell_t - C_t)/B(z_t) \text{ si } \alpha_t = 1 \end{array}\right\}$

Nous pouvons maintenant définir deux problèmes d'optimisation stochastique P et Q:

(P) Maximiser $J(C,\ell)$ sur les couples admissibles (C,ℓ) tels que le processus R associé est ≥ 0.

(Q) Maximiser $J(C,\ell)$ sur les couples admissibles (C,ℓ) tels que le processus S associé est ≥ 0.

On a alors le:

THÉORÈME 2 (J. SCHEINKMAN ET L. WEISS [7]): Les problèmes P et Q ont chacun une unique solution. Soit R (resp. S) le processus associé à la solution optimale de P (resp. de Q). On a:

(16) $R = 1 - z$ et $S = z$

Plus précisément: c'est seulement pour la conclusion du théorème 2, pour démontrer les égalités (16) que l'on utilise le système (1 à 5). On voit que le rôle joué par le système (1 à 5) par rapport à la problèmatique d'optimisation exposée ci-dessus est d'un type nouveau. Par exemple: il ne s'agit pas de programmation dynamique, les fonctions A et B ne sont pas des fonctions de Bellman ou d'Isaacs.

Pour éclairer (un peu) la problèmatique (1 à 16) voici schématiquement son interpolation économique. Pour éviter tout malentendu rappelons tout d'abord qu'il s'agit d'économie théorique, c'est-a-dire d'une reflexion conceptuelle.

La démarche intellectuelle est la même qu'en physique théorique: on ne recherche un réalisme immédiat (penser aux expériences mentales d'Einstein: "Je vais à la vitesse de la lumière et je me regarde dans une glace; que vois-je? ").

Donc J. Scheinkman et L. Weiss considèrent une économie avec un unique bien b_1 consommable non stockable, et un bien b_2 non consommable non dégradable dont le stock ne varie pas; ils supposent qu'il y a deux types d'agents économiques τ_1, τ_2; quand les agents de type τ_1 sont productifs, les agents de type τ_2 ne le sont pas et vis versa (la productivité des agents τ_1 est decrite par le processus de Markov α mentionné plus haut); chaque agent économique de type τ_1 (resp. τ_2) va s'efforcer de posséder une quantité R_t (resp. S_t) de bien b_2 pour payer sa consommation lorsqu'il est improductif; les contraintes $R \geqq 0$ et $S \geqq 0$ correspondent à l'absence de crédit; la quantité z_t est la quantité de b_2 possédée en moyenne par les agents de type τ_2 à l'instant t; $1/B\,(1-z_t)$ si $\alpha_t = 1$, $1/B(z_t)$ si $\alpha_t = 0$ est le prix sur le marché annoncé par la théorie (d'une unité de bien b_1, en unité de bien b_2); enfin l'équation (16) exprime la cohérence de la théorie au sens des anticipations rationnelles: les agents ayant calculé leur optimum individuel en tenant compte des prix annoncees par la théorie (résolution de P et Q) ils concourent par leur comportement à la réalisation de l'équilibre annoncé (équation (16)). On encore: la "théorie (A,B)" se comporte comme un équilibre de Nash par rapport aux agents économiques; prise comme référence dans les calculs d'optimisation ("égoïstes") de chaque agent, la théorie engendre des comportements individuels dont l'effet est de réaliser ce que la théorie annonce.

Cette cohérence de la théorie avec elle même est nécessaire à sa credibilité: c'est même cette cohérence qui fonde sa crédibilité auprès des agents (cf. [1]). Tel est du moins le paradigme proposé par la théorie des anticipations rationnelles de Muth et R. Lucas.

Il est apparu que les problèmes mathématiques sur lesquels debouchent les modèles basés sur la théorie des anticipations sont trés souvent non triviaux et nous pensons qu'ils vont provoquer des développements mathématiques nouveaux. Le modèle de J. Scheinkman et L. Weiss (ci-dessus) et ses extensions (voir plus loin) nous parait à cet égard un cas particulier d'un phénomène général.

REMARQUE: On peut toute fois faire plusieurs rapprochement entre le système (1 à 5) et les équations d'Hamilton-Jacobi-Bellman (HJB). Un rapprochement méthodologique consiste à remarquer que comme pour les équations HJB la théorie se coupe naturellement en deux: d'une part un théorème d'existence (ici le théorème 1) pour un système parachuté (ici la système (1 à 5)) ou plus exactement, un système construit à partir d'heureuses considérations heuristiques (cf. [7]); d'autre part un théorème de vérification (ici le théorème 2) qui montre que les solutions de ce système permettent de resoudre le problème d'optimisation que l'on a en vue(Voir aussi plus loin un autre rapprochement avec les équations HJB).

3. Equations aux dérivées partielles

Du point de vue mathématique la difficulté est du côté du théorème 1: le théorème 2 est (relativement) facile. En élargissant le modèle économique on obtient tout une série d'autres systèmes différentiels fonctionnels. Outre l'intérêt pour l'économie théorique, ces systèmes, présentent des difficultés mathématiques interessantes.

Une première direction se trouve déjà dans l'article de J. Scheinkman et L. Weiss. Elle consiste à remplacer le logarithme par une fonction concave u de $]0, +\infty[$ dans $\mathbb{R}$, de classe C^1, telle que u' est une bijection decroissante de $]0, +\infty[$ sur lui-même.

Les equations (2) et (3) sont alors remplacé par

$$(17) \quad A'(z) = B(1-z)[g(A(z)/B(1-z))]^{-1} \quad (\lambda B(z) - (\lambda+\rho)A(z))$$

$$(18) \quad B'(z) = B(z)[g(A(1-z)/B(z))]^{-1} \; ((\lambda+\rho)B(z) - \lambda A(z))$$

où g est la bijection inverse de u'. Dans la cas particulières u = log les équations (17) et (18) redonnent les équations (1) et (2).

Une autre direction consiste à remplacer le processus α à valeurs dans $\{0,1\}$ par un processus Markovien à valeurs dans un ensemble fini $\Pi = \{\alpha_1, \ldots, \alpha_{2N}\}$ avec des probabilités de transition λ_{ij} dt pour $i \to j$. On suppose pour des raisons provenant de l'interprétation économique que l'on a

$$(19) \quad 0 < \alpha_i < 1 \text{ et } 1-\alpha_i = \alpha_{2N+1-i}$$

$$(20) \quad \lambda_{ij} = \lambda_{km} \text{ pour } k = 2N+1-i \text{ et } m = 2N + 1 - j.$$

Le système (1 à 5) est alors remplacé par le système (21 à 25) suivant

$$(21) \quad A_i \in E \quad \forall i = 1,\ldots,2N$$

où E est l'espace des fonctions de classe C^1 strictement positives et strictement décroissantes

$$(22) \quad A_i'(z)\phi_i(z) = \sum_{i,j=1}^{2N} \lambda_{ij}(A_j(z) - A_i(z)) - \mu A(z)$$

$$(23) \quad \phi_i(z) = 1/A_i(z) \text{ si } \alpha_i A_i(z) \le \alpha_j A_j(z)$$

$$(24) \quad \phi_i(z) = -1/A_j(1-z) \text{ si } \alpha_i A_i(z) \ge \alpha_j A_j(z)$$

$$(25) \quad \alpha_i A_i(0) \le \alpha_j A_j(1)$$

(avec j = 2N + 1 - i dans (23) (24) (25)).

On a alors le

<u>THÉORÈME 3</u>: Il existe une constante c (dépendant des α_i) telle que le système (21 à 25) a au moins une solution si les λ_{ij} sont tous $\ge$ c.

REMARQUE: On peut donner une CNS sur les (λ_{ij}), Voir [3].

Enfin l'étape naturelle suivante consiste à remplacer le processus de Markov α à valeurs dans l'ensemble fini Π, par un brownien (où plus géneralement une diffusion, eventuellement avec sauts) à valeurs dans $[\eta, 1-\eta]$ avec $0 < \eta < 1/2$. Le système (21 à 25) est alors remplacé par l'équation aux dérivees partielles fonctionnelle non lineaire suivante (dans le cas d'un brownien standart avec reflexion aux deux bouts de l'intervalle $[\eta, 1-\eta]$):

(26) $A(\cdot,a) \in E \quad \forall a \in [\eta,1-\eta]$

(27) $-\dfrac{\partial^2 A}{\partial a^2} + \dfrac{\partial A}{\partial z}\phi + \mu A = 0 \quad \forall(z,a) \in [0,1] \times [\eta,1-\eta]$

(28) $\phi(z,a) = 1/A(z)$ si $aA(z,a) \leq (1-a)A(1-z, 1-a)$

(29) $\phi(z,a) = -1/A(1-z)$ si $aA(z,a) \geq (1-a)A(1-z,1-a)$

(30) $aA(0,a) \geq (1-a)A(1,1-a)$

Les résultats concernant le système (26 à 30) sont pour l'instant parcellaires.

4. Comparaison avec une équation d'Hamilton-Jacobi-Bellman

Des considerations provenant de l'interprétation économique, conduisent à comparer le modèle de J. Scheinkman et L. Weiss avec le problème de gestion de stock:

(31) $\underset{c(\cdot)>0,\ \ell(\cdot)>0}{\text{Minimiser}} \quad E\int_0^{+\infty} e^{-\mu t}\,(\ell(t) - \log C(t))dt$

sous la contrainte d'état $S(t) \geq 0$, où S est defini par l'équation différentielle stochastique:

(32) $S(0) = S_0,\ \dot{S}(t) = \alpha_t \ell(t) - C(t) \quad \forall\, t \geq 0$

(où α est le processus markovien à valeurs dans $\{0,1\}$ considéré au début du paragraphe 2).

Pour traiter ce problème par la programmation dynamique il faut (cf. [2])

introduire deux fonction de Bellman u et v: $u(S_0)$ est l'infimum dans (31) sous la contrainte (32) et $S \geq 0$; $v(S_0)$ est l'infimum dans (31) sous la contrainte $R \geq 0$ où R est defini par

(33) $R(0) = S_0,\ \dot{R}(t) = (1-\alpha_t)\ \ell(t) - C(t), \quad \forall\ t \geq 0$

On montre alors (cf. [2]) que (u,v) est l'unique couple de fonctions concaves qui verifié le système (34 à 36) suivant:

(34) $v'(S) \geq -1$

(35) $v'(S) > -1 \Rightarrow \log(-v') + 1 - \lambda u + (\lambda+\rho)v = 0$

(36) $\log(-u') - 1 + \lambda v - (\lambda+\rho)u = 0.$

(On peut calculer une solution explicite de ce problème.)

Si l'on pose $A = -u'$ et $B = -v'$ on obtient, en derivant l'équation (36)

(37) $\dfrac{A'(s)}{A(s)} - \lambda B(s) + (\lambda+\rho)A(s) = 0$

soit la même équation que l'équation (2) du modèle de J. Scheinkman et L. Weiss.

Plus généralement (cf. [2]) on peut definir un problème de gestion de stock stochastique pour lequel la dérivée $A = \partial u/\partial z$ de la fonction de Bellman u(z,a) vérifié dans une partie de Ω_0 de domaine $[0,1] \times [\eta, 1-\eta]$ l'équation

(38) $\dfrac{-\partial^2 A}{\partial a^2} + \dfrac{1}{A}\ \dfrac{\partial A}{\partial z} + \rho A = 0$

soit la même équation que (27) (28). Mais d'une part Ω_0 ne coincide pas avec le sous domaine associé à (28), d'autrepart dans la reste du domaine équations différent profondément et c'est justement là qu'apparait la terme non standart 1-z (cf. (29)) qui fait la spécificité du système (26 à 30)

Cette coincidence et cette différence s'interpretent toutes deux dans le modèle économique (cf. [2]).

Bibliographie

1. D. Lewis, Convention - a philosophical study, Harvard University Press.
2. S. De Laguiche, Comparaison de certaines économies de stockage avec le modèle de J. Scheinkman et L. Weiss, en préparation.
3. J.M. Lasry et J. Scheinkman, Influence des transferts dans une économie avec crédits limités, en preparation.
4. J.C. Rochet, Thèse, Université Paris-Dauphine, 1985.
5. J.C. Rochet, Bilateral monopoly with imperfect information, CORE discussion paper 833.
6. J.C. Rochet et M. Quinzii, Multidimensional signalling, Ecole Polytechnique, Laboratoire d'Econometrie, preprint A2771082.
7. J. Scheinkman et L. Weiss, Borrowing constraints and aggregate fluctuations, JET, 1986.

J.M. Lasry
CEREMADE
Université Paris-Dauphine,
Place de Lattre de Tassigny,
F-75775 Paris Cedex 16,
France.

Y MEYER

Les ondelettes

Nous nous proposons de faire le point sur les ondelettes. Nous dirons d'abord ce que sont les ondelettes, à quoi elles servent, quelles sont les diverses familles d'ondelettes dont on dispose aujourd'hui et enfin quelles sont les limitations inhérentes à l'usage de telle ou telle famille.

Les ondelettes sont soit de nouvelles "fonctions élémentaires" permettant de donner des représentations simples, efficaces et robustes des fonctions générales sous la forme de séries d'ondelettes; soit des fonctions déjà connues mais que l'on peut (sous des conditions encore mystérieuses) utiliser comme constituants de base dans les algorithmes en séries (qui relèvent alors de la théorie des "frames" et non de bases orthomornées d'ondelettes

Les séries d'ondelettes se rattachent aux séries de Fourier mais présentent par rapport à ces dernières un triple avantage

(1) on s'affranchit de la contrainte de périodicite inhérente a l'utilisation de séries de Fourier et on peut décomposer des fonctions tout à fait arbitraires (d'une ou plusieurs variables réelles) en séries d'ondelettes.

(2) les séries d'ondelettes constituent une analyse de la fonction vis à vis d'une échelle dimensionnelle. Une série d'ondelettes est une série double parceque l'ondelette analysante joue le rôle d'un microscope dont on règle d'abord le grossissement (le plus généralement l'échelle est celle des puissances 2^j, $j = 0, \pm 1, \pm 2, \ldots$) puis la position (on le centre successivement en chacun des points, $0, 2^{-j}, 2.2^{-j}, 3.2^{-j}, \ldots$ et de même, $-2^{-j}, -2.2^{-j}, -3.2^{-j}, \ldots$ (ce qui est naturel si l'on veut voir des détails de dimension 2^{-j})

(3) cette analyse dimensionnelle fournit également une analyse en fréquences, du moins si la transformée de Fourier $\hat{\psi}$ de l'ondelette analysante ψ est portée par un intervalle de fréquences assez bien délimité.

La robustesse des décompositions en séries d'ondelettes signifie que les séries écrites continuent à converger même après perturbation de l'ordre des termes ou après multiplications des termes par des facteurs arbitraires compris entre 0 et 1. Cette robustesse signifie également que les algorithmes écrits convergent pour d'autres normes que la norme de référence (à savoir la norme L^2) du moins lorsque la fonction f que l'on analyse appartient à l'espace normé correspondant.

Nous verrons que ces exigences de robustesse ne sont pas toujours satisfaites et dépendent en fait de l'ondelette analysante $\psi(x)$. Cette ondelette analysante sera toujours choisie la plus régulière possible, la mieux localisée possible et la plus oscillante possible; cette dernière exigence se traduit par des conditions de la forme $\int_{-\infty}^{\infty} x^k \psi(x)dx = 0$ pour $k = 0,1,\ldots,m-1$ et l'on cherchera à choisir m le plus grand possible. Les autres ondelettes sont alors (en général) les fonctions $\psi_{(j,k)}(x) = 2^{j/2}\psi(2^j x-k)$ où $j = 0, \pm 1, \pm 2,\ldots$ et où $k = 0, \pm 1, \pm 2,\ldots$ de sorte que les séries d'ondelettes s'écrivent

$$(*) \qquad \sum_{-\infty}^{\infty} \sum_{-\infty}^{\infty} \alpha(j,k)\psi_{(j,k)}(x).$$

Les problèmes algorithmiques concernent le calcul des coefficients $\alpha(j,k)$ tels que que la série (*) représente une fonction donnée f(x). Selon le choix de l'ondelette analysante $\psi(x)$, les algorithmes que l'on devra utiliser influeront la robustesse des séries (*).

Nous allons maintenant décrire les diverses familles $\psi_{(j,k)}$ dont on dispose aujourd'hui.

1. Les ondelettes de Guy David/P.G. Lemarié/Y. Meyer

Elles sont définies à l'aide d'un entier $q \geq 1$ et l'échelle $(1 + \frac{1}{q})^j$, $j = 0, \pm 1, \pm 2,\ldots$ généralise l'échelle 2^j que nous avons mentionnée jusqu'ici. Les plages de frequnces définies par les ondelettes D.L.M. sont de la forme $[(1 + \frac{1}{q})^j a, (1 + \frac{1}{q})^j b]$ où a est légèrement inférieur à $q\pi$ et b légèrement superiéur à $(q + 1)\pi$. Ces plages se chevauchent légèrement et la rapport entre la plus haute et la plus basse fréquence dépasse légèrement $1 + 1/q$. Cela signifie que la "note émise" par une ondelette D.L.M. est assez précise (dans une échelle logarithmique) si q est assez grand.

Il est temps de décrire l'ondelette analysante ψ de la famille D.L.M. Cette fonction ψ est indéfiniment dérivable, à décroissance rapide à l'infini ainsi que toutes ses dérivées (elle appartient donc à la classe $\mathcal{S}(\mathbb{R})$ de Schwartz). Elle est finalement définie à l'aide de sa transformée de Fourier $\hat{\psi}$. Pour simplifier les notations, on préfère écrire $\psi(x) = \eta(x-1/2)$ où $\eta(x)$

est une fonction paire, à valeurs réelles et dont la transformée de Fourier $\hat{\eta}(\xi)$ va maintenant être explicitée. On a $\hat{\eta}(\xi) = 0$ si $\xi \geq b$ et si $0 \leq \xi \leq a$. Définissons a' et b' par $\frac{1}{2}(a + a') = q\pi$, $\frac{1}{2}(b + b') = (q + 1)\pi$ et supposons $a' \leq b'$ et $b' = (1 + \frac{1}{q})\, a$ avec $b = (1 + \frac{1}{q})a'$. Pour alléger les notations posons $\theta = 1 + 1/q$. Finalement $\hat{\eta}(\xi)$ vérifie $\hat{\eta}^2(\xi) + \hat{\eta}^2(\xi') = 1$ si $a \leq \xi \leq \xi' \leq a'$ et $\frac{1}{2}(\xi + \xi') = q\pi$ et de même $\eta^2(\xi) + \hat{\eta}^2(\xi') = 1$ si $b' \leq \xi \leq \xi' \leq b$ et $\frac{1}{2}(\xi + \xi') = (q+1)\pi$. La dernière relation est $\hat{\eta}^2(\xi) + \hat{\eta}^2(\theta\xi) = 1$ si $a \leq \xi \leq a'$. Nous venons de décrire la "mère" de la famille D.L.M. Les autres ondelettes de la famille D.L.M. sont obtenues à l'aide de la mère ψ en faisant, dans cet ordre, les deux opérations suivantes

translations entières: on remplace $\psi(x)$ par (1.1)

$\psi(x-k)$ où $k = 0, \pm 1, \pm 2, \ldots$

changements d'échelle dans le rapport $\theta = 1 + 1/q$: (1.2)

on remplace $\psi(x-k)$ par $\theta^{j/2}\ \psi(\theta^j x-k)$ où $j = 0, \pm 1, \pm 2, \ldots$

Finalement on pose $\psi_{(j,k)}(x) = \theta^{j/2}\ \psi(\theta^j x-k)$.

Cette construction très précise implique que <u>les fonctions</u> $\psi_{(j,k)}$ <u>forment une base hilbertienne de</u> $L^2(\mathbb{R})$. On a donc pour toute fonction f de $L^2(\mathbb{R})$,

$$f(x) = \sum_{-\infty}^{\infty} \sum_{-\infty}^{\infty} \alpha(j,k)\ \psi_{j,k}(x) \quad \text{où} \tag{1.3}$$

$$\alpha(j,k) = \int_{-\infty}^{\infty} f(u)\psi_{j,k}(u)du.$$

Donnons l'analogue bi-dimensionnel de (1.3). Il faut introduire le "père" ϕ qui est une fonction appartenant également à la classe de Schwartz, paire et réelle et dont la transformée de Fourier $\hat{\phi}(\xi) \geq 0$ vérifie

$$\hat{\phi}(\xi) = 1 \text{ si } |\xi| \leq a, \quad \hat{\phi}(\xi) = 0 \text{ si } |\xi| \geq a'$$

et $\hat{\phi}^2(\xi) + \hat{\eta}^2(\xi) = 1$ si $a \leq |\xi| \leq a'$.

Alors la base orthonormée de $L^2(\mathbb{R}^2; dxdy)$ est composeé des fonctions $\theta^j \phi(\theta^j x-k)\, \psi(\theta^j y-\ell)$, $\theta^j\psi(\theta^j x-k)\phi(\theta^j y-\ell)$ et $\theta^j\psi(\theta^j x-k)\psi(\theta^j y-\ell)$ où $j = 0, \pm1, \pm2,\ldots$ $k = 0, \pm1, \pm2,\ldots$ et $\ell = 0, \pm1, \pm2,\ldots$

La construction précédente se généralise immédiatement à plusieurs dimensions.

La robustesse de (1.3) signifie que (1.3) converge vers f pour, à peu près, toutes les échelles de régularité que l'on utilise en analyse et ceci inconditionnellement, c'est à dire même si l'on a perturbé l'ordre des termes. Commençons par l'échelle des espaces de Sobolev $H^s(\mathbb{R})$ où s varie de $-\infty$ à $+\infty$. L'appartenance de f à H^s est caractérisée par la condition $\sum_{-\infty}^{\infty} \sum_{-\infty}^{\infty} |\alpha(j,k)|^2 (1 + \theta^j)^{2s} < +\infty$. Si cette condition est satisfaite, la série écrite est équivalente au carré de la norme H^s de f et (1.3) converge en norme H^s.

Continuons notre voyage dans l'analyse fonctionnelle en examinant ce qui se passe si f appartient à l'espace de Lebesgue $L^p = L^p(\mathbb{R};dx)$ où $1 \le p \le +\infty$. On commence par les cas extrèmes $p = 1$ où $p = +\infty$.

Grâce au choix de ψ, on a $\int_{-\infty}^{\infty} x^k\psi(x)dx = 0$ pour $k = 0,1,\ldots$ et, en particulier, l' intégrale de chaque ondelette $\psi_{(j,k)}$ est nulle. Mais supposons que f(x) soit une fonction (aussi régulière et bien localisée que l'on veut) d'intégrale égale à 1. Alors (1.3) ne peut converger en norme L^1 sinon on pourrait intégrer terme à terme pour obtenir $1 = 0$.

Il est facile de voir que si f(x) appartient à L^∞, la série (1.3) ne convergera pas vers f(x) pour cette norme (tout simplement parceque f(x) peut très bien être discontinue).

En revanche les normes $L^p(\mathbb{R})$, $1 < p < +\infty$, sont adaptées aux séries d'ondelettes (1.3). En fait l'appartenance de f à L^p est une propriété que l'on lit sur les <u>modules</u> $|\alpha(j',k)|$ des coefficients d'ondelettes. On pose

$$g(x) = \Big(\sum_{(j,k) \in M(x)} \sum |\alpha(j,k)|^2\, \theta^j\Big)^{1/2} \text{ où } M(x) \qquad (1.4)$$

est l'ensemble de tous les couples (j,k) tels que $\theta^{-j}k \le x < \theta^{-j}(k+1)$. Alors $\|g\|_p$ et $\|f\|_p$ sont deux normes équivalentes sur $L^p(\mathbb{R})$ pour $1 < p < +\infty$. La condition $g(x) \in L^1(\mathbb{R})$ caractérisé l'espace de Hardy $H^1(\mathbb{R})$ tel qu'il fut défini par E.M. Stein et G. Weiss.

Terminons notre voyage en examinant le cas des espaces de Hölder $C^r(\mathbb{R})$ où $r > 0$. Rappelons que si $0 < r < 1$, alors $f \in C^r$ signifie l'existence d'une constante $C \geq 0$ telle que pour tout x et tout x' réels on ait $|f(x') - f(x)| \leq C|x' - x|^r$. Par exemple la fonction $\sqrt{x}$ appartient à $C^{1/2}$. Si $r = 1$, on abandonne l'espace C^1 usuel que l'on remplace tout de suite par la <u>classe de Zygmund</u> Λ_* composée des fonctions f(x) vérifiant

$$|f(x + h) + f(x - h) - 2f(x)| \leq C|h| \tag{1.5}$$

pour tout x réel et tout h réel.

Finalement si $r > 1$, on écrit $f \in C^r$ si et seulement si f est continûment dérivable et si $f' \in C^{r-1}$. Avec ces définitions précisées (et en remplaçant systématiquement C^1 par Λ_*, C^2 par les primitives des fonctions de Λ_*,...) on a la résultat suivant

$$f \in C^r \iff |\alpha(j,k)| \leq C\theta^{-(r+1/2)j} \tag{1.6}$$

pour une certaine constant C (équivalente à la norme de f dans C^r), tout j et tout k.

En dimension n l'exposant 1/2 doit être remplacé par n/2.

2. Les ondelettes de G. Battle et P.G. Lemarié

Les ondelettes que nous allons maintenant décrire ont été décourtes par P.G. Lemairé, le 30 Mai 1986, à Tunis (Tunisie) puis indépendamment pas G. Battle à Ithaca, le 24 Août 1986. Nous les appelerons les ondelettes B.L.

Les ondelettes B.L. ont une qualité que ne possèdent pas les ondelettes D.L.M. Elles ont une décroissance exponentielle par rapport à la variable x : $|\psi(x)| \leq C \exp(-\varepsilon|x|)$ pour un certain $\varepsilon > 0$ (que l'on pourra calculer) et une constante C. Les ondelettes D.L.M. ne peuvent decroître exponentiellement à l'infini. Des essais sur ordinateur montrent que la localisation des ondelettes D.L.M. est assez mauvaise (du point de vue numérique).

Les ondelettes B.L. ont une seconde qualité : ce sont des fonctions splines, de classe C^{m-1}, obtenues en raccordant des polynômes de degré m aux extrémités d'intervalles explicites.

Là encore les ondelettes B.L. sont (en dimension 1) de la forme

$2^{j/2}\ \psi(2^j x-k)$ où ψ, la mère des ondelettes B.L. est décrite par l'énoncé ci-dessous.

On part d'un entier $m \geq 1$. Il existe une (et une seule) fonction $\psi(t)$, définie sur la droite reélle et ayant les propriétés suivantes

$\psi(t) = \eta(t - 1/2)$ où $\eta(t)$ est une fonction paire, (2.1)

à valeurs reélles

$\eta(t)$ coincide sur chaque intervalle $[\frac{j}{2}\ ,\ \frac{j+1}{2}]$ (2.2)

avec un polynôme $P_j(t)$, de degré égal à m

ces polynômes sont ajustés les uns aux autres de sorte (2.3)

que $\psi(t)$ ainsi que toutes ses dérivées, jusqu'à l'ordre

m-1, soient continues sur la droite reélle

on a $\psi(t) = O(e^{-\varepsilon|t|})$, pour un certain $\varepsilon > 0$ quand $|t|$ (2.4)

augmente indéfiniment

on a $\int_{-\infty}^{\infty} t^k\ \psi(t)dt = 0$ pour $k = 0,1,\ldots,m-1$ (2.5)

les fonctions $2^{j/2}\ \psi(2^j t-k)$, $j = 0, \pm 1, \pm 2,\ldots$ (2.6)

$k = 0, \pm 1, \pm 2,\ldots$ forment une base orthonormée de

$L^2(\mathbb{R})$ (ou base hilbertienne).

Pour obtenir une base orthonormée en dimension supérieure, on construit le "père" ϕ de la famille B.L. C'est encore une fonction de classe C^{m-1}, obtenue en raccordant des polynômes de degré m sur chaque intervalle $[j - \frac{1}{2}\ ,\ j + \frac{1}{2}]$ et ayant une localisation exponentielle. La construction de la base B.L. de $L^2(\mathbb{R}^2;\ dxdy)$ s'obtient en copiant celle de la base D.L.M.

Les ondelettes B.L. rendent les mêmes services que les ondelettes D.L.M., à ceci pres que la régularité ou l'irrégularité des fonctions ou des distributions que l'on décompose en séries d'ondelettes sont limitées par m-1.

Les ondelettes B.L. sont donc mieux localisées en variable d'espace (ou de temps) que les ondelettes D.L.M. En revanche ceci se paye par la localisation en fréquences. Une fonction qui a une localisation exponentielle a pour transformée de Fourier la restriction à l'axe réel d'une fonction holomorphe dans une bande horizontale et ceci interdit toute localisation aussi précise que celle de la transformée de Fourier des ondelettes D.L.M.

3. Les ondelettes de Tchamitchian

Elles constituent en un sens un progrès supplémentaire par rapport aux ondelettes B.L. La fonction ψ de départ est de classe C^{m-1}, à support compact, coincide avec des polynômes de degré m sur les intervalles $[\frac{j}{2}, \frac{j+1}{2}]$ et a un nombre arbitrairement élevé de moments nuls. La construction de Tchamitchian consiste à construire un objet $\tilde{\psi}$ (qui peut être une fonction régulière ou une distribution sauvage suivant le choix précis de ψ mais) qui dans tous les cas sert à calculer les coefficients d'ondelette $\alpha(j,k)$, $j = 0, \pm 1, \pm 2, \ldots, k = 0, \pm 1, \pm 2, \ldots$ dans la décomposition

$$f(t) = \sum_{-\infty}^{\infty} \sum_{-\infty}^{\infty} \alpha(j,k)\, \psi_{(j,k)}(t) \tag{3.1}$$

où $\psi_{(j,k)}(t) = 2^{j/2}\, \psi(2^j t-k)$

par

$$\alpha(j,k) = \int_{-\infty}^{\infty} f(t)\, \tilde{\psi}_{(j,k)}(t)dt \tag{3.2}$$

où $\tilde{\psi}_{(j,k)}(t) = 2^{j/2}\, \tilde{\psi}(2^j t-k)$.

On ne dispose donc pas, à l'heure actuelle, d'un critère simple portant sur la "mère" $\psi(t)$ des ondelettes de Tchamitchian et assurant, par exemple que (3.1) fonctionne pour $f \in L^2(\mathbb{R})$, que l'on ait alors $(\Sigma\Sigma|\alpha(j,k)|^2)^{1/2} \leq C\,\|f\|_2$ et que la série (3.1) converge dans $L^2(\mathbb{R})$ vers f.

Ce que Tchamitchian sait faire est de produire des choix très particuliers de ψ où $\tilde{\psi}$ soit, elle aussi de classe C^{m-1} et ait un certain nombre (arbitraire par ailleurs) de moments nuls.

Donnons brièvement les règles de construction de ψ et de $\tilde{\psi}$.

On a $\tilde{\psi}(\xi) = e^{-i\xi/2}\, g(\xi)/\xi^m$ où $g(\xi) = \sum_{-M}^{M} \gamma_k\, e^{ik\xi/2}$

et où M et les coefficients γ_k sont choisis de sorte que l'on ait $0 \le g(\xi) \le 1$ et $g(\xi) + g(\xi + 2\pi) = 1$ avec, en outre, $g(0) = g'(0) = \ldots = g^{(m)}(0)$. A l'aide de cette fonction $g(\xi)$, on forme $G(\xi) = \prod_1^\infty (1 - g(2^{-j}\xi))$. Toute la difficulté est de choisir $g(\xi)$ de sorte que le produit $\xi^m G(\xi)$ soit, au moins, de carré intégrable et que l'on puisse donc définir $\tilde{\psi} \in L^2(\mathbb{R})$ par sa transformée de Fourier $\mathcal{F}(\tilde{\psi})(\xi) = e^{-i\xi/2}\, \xi^m G(\xi)$. Tchamitchian montre que l'on peut faire un tel choix mais la fonction $\tilde{\psi}$ sera nécessairement très compliquée. Le support de $\tilde{\psi}$ est compact, comme on le montre facilement.

Il semble donc que ce que l'on a pu gagner en choisissant l'ondelette analysante ψ à support compact soit aussitôt perdu par le fait que les fonctions $2^{j/2}\,\psi(2^j x - k)$, $j = 0, \pm1, \pm2, \ldots$, $k = 0, \pm1, \pm2, \ldots$ ne forment plus une base hilbertienne de $L^2(\mathbb{R})$. Dans certains cas les fonctions $\psi_{(j,k)}$ forment une <u>base inconditionnelle</u> de $L^2(\mathbb{R})$, c'est à dire, l'image d'une base hilbertienne par un isomorphisme (non nécessairement isométrique). La base duale est alors précisément la base $\tilde{\psi}_{(j,k)}$ déjà mentionnée, base dont on ne peut, en général, prédire la régularité. Mais il y a, rappelons le, des choix particuliers de ψ tels que $\tilde{\psi}$ soit une fonction continue, à support compact, ayant un certain nombre de dérivées continues. Il y a aussi des choix de ψ tels que $\tilde{\psi}$ n'appartienne pas à L^p pour $p > 2$ donné ([4]).

Nous ne savons pas résoudre le problème suivant: existe-t-il pour tout entier $m \ge 1$ une fonction $\psi(x)$, à support compact, de classe C^{m-1}, telle que la suite $\psi_{(j,k)}(x) = 2^{j/2}\,\psi(2^j x - k)$ $j = 0, \pm1, \pm2, \ldots, k = 0, \pm1, \pm2, \ldots$ soit une base orthonormée de $L^2(\mathbb{R})$? Si l'on ne demande aucune régularité, le système de Haar convient.

4. Les "frames" de Daubechies et Morlet

J. Morlet est un spécialiste du traitement numérique des signaux sismiques que l'on recueille au cours des campagnes de prospection pétrolière. Il a inventé des algorithmes de décomposition avec reste qui, après itération, donneront des algorithmes exacts ressemblant à ceux de Tchamitchian mais sans unicité des décompositions écrites.

Donnons d'abord la formulation abstraite. Soit H un espace de Hilbert où le produit scalaire est noté $\langle x|y\rangle$ et soit $(e_j)_{j\in J}$ une suite de vecteurs de H. On dit que $(e_j)_{j\in J}$ est un frame s'il existe deux constantes $C_2 > C_1 > 0$ telles que, pour tout vecteur $x \in H$, on ait la double inégalité

$$C_1 \|x\|^2 \leqq \sum_{j\in J} |\langle x|e_j\rangle|^2 \leqq C_2 \|x\|^2. \tag{4.1}$$

Quitte à multiplier tous les vecteurs e_j par une même constante $\delta > 0$, on peut évidemment supposer $C_2 < 2$ ce gue nous ferons désormais.

On considère alors l'opérateur linéaire continu $T : H \to H$ défini par

$$\mathcal{T}(x) = \sum_{j\in J} \langle x|e_j\rangle \; e_j. \tag{4.2}$$

et l'on a, au sens des opérateurs auto-adjoints,

$$C_1 1 \leqq \mathcal{T} \leqq C_2 1 \tag{4.3}$$

Il en résulte que $\|\mathcal{T}-1\| \leqq r$ où $r = \inf(|C_1-1|, |C_2-1|) < 1$. On a donc, pour tout $x \in H$,

$$x = \sum_{j\in J} \langle x|e_j\rangle \; e_j + x_1 \tag{4.4}$$

$$x_1 = \sum_{j\in J} \langle x_1|e_j\rangle \; e_j + x_2$$

$$\cdots\cdots\cdots\cdots\cdots$$

$$x_k = \sum_{j\in J} \langle x_k|e_j \; e_j + x_{k+1}$$

où $\|x_{k+1}\| \leqq r\|x_k\|$.

L'algorithme (4.4) converge et en additionnant membre à membre il vient

$$x = \sum_{j\in J} \langle y|e_j\rangle \; e_j \quad \text{où } y = \mathcal{T}^{-1}(x).$$

On a donc rapidement une décomposition exacte a l'aide des décompositions approchées (4.4) fournies par le frame.

Venons-en aux exemples importants. Il est possible de donner un critère simple portant sur la transformée de Fourier $\hat{\psi}$ d'une fonction ψ (appartenant à $L^2(\mathbb{R})$, à $L^1(\mathbb{R})$ et d'intégrale nulle) pour que les fonctions $2^{j/2}\ \psi(2^j x-k)$, $j = 0, \pm 1, \pm 2, \ldots, k = 0, \pm 1, \pm 2, \ldots$ forment un frame. I. Daubechies a écrit un premier énoncé (sous forme de condition suffisante) énoncé ensuite amélioré par P. Tchamitchian (toujours sous forme de condition suffisante). La condition de Daubechies utilise la fonction auxiliaire $\beta(s)$ définie par

$$\beta(s) = \sup_{1 \leq x \leq 2} \sum_{-\infty}^{\infty} |\hat{\psi}(2^j x)|\ |\hat{\psi}(2^j x + s)|.$$

Alors la condition suffisante sur ψ est

$$\inf_{1 \leq x \leq 2} \sum_{-\infty}^{\infty} |\hat{\psi}(2^j x)|^2 > 2 \sum_{1}^{\infty} (\beta(2k\pi)\beta(-2k\pi))^{1/2} \tag{4.5}$$

Armeé de ce critère et d'un petit ordinateur, I. Daubechies part de l'ondelette $\psi(x) = \frac{2}{\sqrt{3}}\ \frac{1}{\pi^{1/4}}\ (1-x^2)e^{-x^2/2}$ (la normalisation est telle que $\| \psi \|_2 = 1$) et démontre que la collection $2^{j/2}\ \psi(2^j x-k)$, $j = 0, \pm 1, \pm 2, \ldots, k = 0, \pm 1, \pm 2, \ldots$ est un frame. La méthode employée permet de calculer la rapidité de la convergence de l'algorithme (4.4). L'importance de cet exemple vient de son utilisation dans les calculs numériques réalisés par les ingénieurs de Elf-Aquitaine. Dans la pratique on "suréchantillonne largement" et avec le même choix de ψ, on forme le frame $\psi^{\#}_{j,k}(x) = \gamma\, 2^{j/4}\ \psi(2^{j/4}x-k/2)$, $j = 0, \pm 1, \pm 2, \ldots, k = 0, \pm 1, \pm 2, \ldots$ Alors $r < 10^{-2}$. On peut donc, dans une première approximation décomposer f à l'aide des $\psi^{\#}_{j,k}$ comme s'il s'agissait d'une base orthonormeé. Si l'on n'est pas satisfait de la précision, il suffit d'itérer une ou deux fois.

Il importe de souligner que ces calculs de vitesse de convergence ne concernent que la norme $L^2(\mathbb{R})$. Nous examinerons plus loin les problèmes que l'on rencontre si l'on s'intéresse à d'autres cadres fonctionnels.

Une autre application remarquable du critère d'I. Daubechies a été le perfectionnement des algorithmes de décomposition atomique des fonctions de l'espace de Hardy complexe $\mathbb{H}^2$ dont nous allons rappeler la définition.

On appelle Ω le demi-plan ouvert supérieur $y > 0$ du plan complexe et on désigne par $\mathbb{H}^2$ l'espace de Hilbert des fonctions $f(z)$ holomorphes dans Ω et vérifiant $\sup_{y>0} \int_{-\infty}^{\infty} |f(x + iy)|^2 \, dx < + \infty$. Si cette condition est réalisée, alors f(z) a une trace sur l'axe réel, notée f(x + i0) et définie par $\lim_{y \downarrow 0} f(x + iy) = f(x + i0)$. Cette limite existe au sens $L^2(\mathbb{R})$ et aussi presque-partout. Finalement l'espace de Hilbert $\mathbb{H}^2$ se plonge isométriquement dans le sous espace $H^2(\mathbb{R}) \subset L^2(\mathbb{R})$ des traces sur l'axe reel des fonctions f(z). Le théorème de Paley-Wiener caractérise $H^2(\mathbb{R})$ par la condition (nécessaire et suffisante): $f \in H^2(\mathbb{R}) \iff \hat{f}(\xi) = 0$ sur $\xi < 0$.

Nous ne savons pas encore s'il existe une fonction ψ appartenant à la classe $\mathcal{S}(\mathbb{R})$ de Schwartz et à $H^2(\mathbb{R})$ telle que les fonctions $2^{j/2} \psi(2^j x-k)$, $j = 0, \pm 1, \pm 2, \ldots, k = 0, \pm 1, \pm 2, \ldots$ forment une base hilbertienne de $H^2(\mathbb{R})$ ou meme une base inconditionnelle de $H^2(\mathbb{R})$. Rappelons qu'une base inconditionnelle est l'image d'une base hilbertienne par un isomorphisme (non nécessairement isométrique).

En revanche le critère d'I. Daubechies fournit aisément des frames de $H^2(\mathbb{R})$. L'existence théorique de tels frames résultait des travaux de G. Weiss et de ses collaborateurs. L'exemple cidessous est le premier qui soit explicite en ce qui concerne les constantes numériques.

On part de $\psi(x) = (x + i)^{-2}$ qui appartient évidemment à $H^2(\mathbb{R})$. Alors S. Jaffard, en appliquant le critere d'I. Daubechies montre que $2^{j/2} (2^j x-2k+i)^{-2}$, $j = 0, \pm 1, \pm 2, \ldots, k = 0, \pm 1, \pm 1, \ldots$ est un frame de $H^2(\mathbb{R})$. On montre par ailleurs l'existence de $q_0 > 1$ tel que si $q \geqq q_0$, les fonctions $2^{j/2}(2^j x-kq+i)^{-2}$, $j = 0, \pm 1, \pm 2, \ldots, k = 0, \pm 1, \pm 2, \ldots$ forment une base inconditionnelle d'un sous-espace fermé $\mathcal{F}$ inclus dans $H^2(\mathbb{R})$ mais différent de $H^2(\mathbb{R})$. Existe-t-il une valeur critique q_0 telle que si $q > q_0$, cette denière propriété ait lieu (base inconditionnelle pour un sous-espace propre) alors que si $1 \leqq q < q_0$ les fonctions $2^{j/2} (2^j x-kq + i)^{-2}$ forment un frame? Nous ne savons pas répondre à cette question.

Un autre problème que nous ne savons pas résoudre est de savoir si l'algorithme de décomposition atomique des fonctions $f \in H^2$ sous la forme

$$f(x) = \sum_{-\infty}^{\infty} \sum_{-\infty}^{\infty} \alpha(j,k) 2^{j/2} (2^j x - 2k + i)^{-2} \qquad (4.6)$$

fonctionne encore pour la norme $H^1(\mathbb{R})$.

Rappelons que $\mathbb{H}^1$ se compose des fonctions holomorphes f(z) dans Ω telles que $\sup_{y>0} \int_{-\infty}^{\infty} |f(x + iy)|dx$ soit finie. Si f appartient à $H^2(\mathbb{R})$, l'algorithme (4.4) s'écrit sous la forme (4.6) et l'on a $\left(\sum_{-\infty}^{\infty} \sum_{-\infty}^{\infty} |\alpha(j,k)|^2\right)^{1/2} \leq C \|f\|_{H^2}$.

Réciproquement si la série double $\sum_{-\infty}^{\infty} \sum_{-\infty}^{\infty} |\alpha((j,k)|^2$ converge, alors le second membre de (4.6) représente toujours cerie fonction de $H^2(\mathbb{R})$.

On pourrait espérer le'résultat suivant. Si f appartient à H^1, l'algorithme (4.4) convergerait pour la norme H^1 et s'écrirait sous la forme (4.6) avec en outre

$$\left(\sum\sum_{(j,k) \in M(x)} |\alpha(j,k)|^2 2^j\right)^{1/2} = g(x) \in L^1(\mathbb{R}) \tag{4.7}$$

où $M(x)$ est l'ensemble des $(j,k) \in \mathbb{Z}^2$ tels que $k2^{-j} \leq x < (k+1)2^{-j}$. Réciproquement si cette condition est satisfaite, alors la série (4.6) converge, en norme H^1, vers une fonction f appartenant à H^1.

Rappelons que les difficultés rencontreés vicnnent ici de l'analyse complexe. En situation d'analyse réelle, on remplace l'espace de Hardy holomorphe H^1 par sa version réelle $\mathcal{H}^1$ (introduite par Stein et Weiss) et les ondelettes D.L.M. réalisent le programme désiré.

Une façon équivalente d'écrire que les fonctions $2^{j/2}(2^j x - 2k + i)^{-2}$ forment un frame pour $H^2(\mathbb{R})$ est d'écrire que les normes $\|f\|_{\mathbb{H}^2}$ et $\left(\sum_{-\infty}^{\infty} \sum_{-\infty}^{\infty} 8^{-j}|f'(2^{-j}(2k + i))|^2\right)^{1/2}$ soient équivalentes; ici $f'(z)$ est la dérivée de la fonction holomorphe f dans Ω.

5. La théorie des opérateurs

Les difficultés aux quelles nous venous de faire allusion s'éclairent en considérant les opérateurs associés aux frames. En langage abstrait, il s'agit de l'opérateur $T : H \to H$ défini par $T(x) = \sum_{j \in J} \langle x|e_j\rangle e_j$ si l'on suit les notations de (4.2). L'algorithme (4.4) signifie que l'on écrit $T = 1 + R$ où la norme de l'opérateur $R : H \to H$ est inférieure à 1. On inverse alors T par une série de Neumann.

En retournant aux situations concrètes que nous avons étudiées, on peut calculer le noyau-distribution $K(x,y)$ de l'opérateur T. C'est $K(x,y) = \sum_{-\infty}^{\infty} \sum_{-\infty}^{\infty} 2^j\psi(2^j x-k)\bar{\psi}(2^j y-k)$. Si la fonction ψ appartient à la classe $\mathcal{S}(\mathbb{R})$ de Schwartz et si tous les moments de ψ sont nuls, alors T appartient à l'algèbre A_∞ de Lemarié. Cela signifie que l'on a (en dimension 1), les trois conditions suivantes

$$|(\frac{\partial}{\partial x})^p (\frac{\partial}{\partial y})^q K(x,y)| \leqq C(p,q)|x-y|^{-1-p-q} \tag{5.1}$$

pour tout $p \geqq 0$ et tout $q \geqq 0$

$$T \text{ est borne sur } L^2(\mathbb{R}) \tag{5.2}$$

En designant par $S_0(\mathbb{R} \subset S(\mathbb{R})$ le sous-espace des fonctions de la classe de Schwartz dont tous les moments sont nuls, alors T et son adjoint T^*-envoient continûment $S_0(\mathbb{R})$ dans $S_0(\mathbb{R})$. (5.3)

Les opérateurs de l'algèbre A_∞ sont des opérateurs pseudo-différentiels dont les symboles $\sigma(x,\xi)$ vérifient les estimations

$$|(\frac{\partial}{\partial x})^p (\frac{\partial}{\partial \xi})^q \sigma(x,\xi)| \leqq C(p,q)\ |\xi|^{p-q}. \tag{5.4}$$

Si, de façon plus modeste, $|\psi(x)| \leqq C(1+x^2)^{-1}$, $\int_{-\infty}^{\infty} \psi(x)dx = 0$ et si $|\psi'(x)| \leqq C(1+x^2)^{-3/2}$, alors l'opérateur associé T est un opérateur de Calderón-Zygmund dont le noyau vérifie (5.1) pour $p + q \leqq 1$. Cet opérateur T est borné sur $L^2(\mathbb{R})$ et (5.3) est remplacé par la condition que T soit borné sur l'éspace H^1 de Stein et Weiss et aussi sur son dual BMO (l'espace de John et Nirenberg).

Le probleme fondamental posé par l'utilisation des frames pour d'autres espaces fonctionnels que $L^2(\mathbb{R})$ est celui du calcul symbolique pour l'algèbre A_∞ de Lemarié. Est-il vrai que si T appartient à A_∞ et si $T : L^2(\mathbb{R}) \to L^2(\mathbb{R})$ est un isomorphisme, alors T^{-1} appartienne à A_∞? P. Tchamitchian puis P.G. Lemarié ont trouvré un contre-exemple ruinant toute ideé simple. Pour tout $p \neq 2$, il existe un opérateur $T \in A_\infty$ qui est un isomorphisme sur $L^2(\mathbb{R})$ mais dont l'inverse T^{-1} n'est pas borné sur $L^p(\mathbb{R})$. En particulier T^{-1} ne peut être un opérateur de Calderón-Zygmund ni, a fortiori, un opérateur de l'algèbre de Lemarié (ou de l'algèbre plus modeste des opérateurs de Calderón-Zygmund bornés sur H^1 et sur BMO).

Faute de pouvoir résoudre cet important problème du calcul symbolique pour l'algèbre A_∞, nous ne savons pas si l'usage des frames se limite à l'espace $L^2(\mathbb{R})$. Nous disposons cependant d'un exemple. Pour tout p appartenant à $]1/2[$, il existe une fonction ψ appartenant à $S_0(\mathbb{R})$ telle que les fonctions $2^{j/2}\ \psi(2^j x-k)$, $j = 0, \pm1, \pm2,\ldots,k = 0, \pm1, \pm2,\ldots$ forment un frame et même une base inconditionnelle de $L^2(\mathbb{R})$ mais telle que la décomposition qui fonctionne si bien en norme L^2 cesse d'avoir la moindre signification si f appartient à L^p, tout simplement parceque au moins l'un des coefficients devient infini.

Cette pathologie est-elle exceptionnelle? L'évite-t-on en utilisant des ondelettes analysantes raisonnables comme $\psi(x) = (1-x^2)e^{-x^2/2}$ pour $L^2(\mathbb{R})$ et $\psi(x) = (x + i)^{-2}$ pour H^2? Nous ne savons pas répondre à ces questions, pourtant essentielles.

Voici quelques références. La plupart des textes traitant des ondelettes sont en cours de publication. Je recommande tout particulièrement [4] qui est d'une lecture agréable et stimulante.

Références

1. I. Daubechies, A. Grossmann and Y. Meyer. "Painless non-orthogonal expansions" J. Math. Phys. 27 (1986) 1271-1283.
2. P.G. Lemarié et Y. Meyer. "Ondelettes et bases hilbertiennes" à paraître dans la Revista Matematica Ibero Americana (publiée par l'université Autonoma de Madrid).
3. Y. Meyer. Principe d'incertitude, bases hilbertiennes et algèbres d'opérateurs. Séminaire Bourbaki. 1985-1986 no. 662.
4. P. Tchamitchian. Calcul symbolique sur les opérateurs de Calderón-Zygmund et bases inconditionnelles de $L^2(\mathbb{R})$. C.R. Acad. Sci. Paris, t. 303, Série I, no. 6, 1986 p. 215-217.
5. I. Daubechies. Discrete sets of coherent states and their use in signal analysis. Courant Institute of Mathematical Sciences. A paraître.

Y. Meyer
CEREMADE
Université Paris IX - Dauphine
Place de Lattre de Tassigny
75775 Paris CEDEX 16
France.

X MORA

Finite–dimensional attracting invariant manifolds for damped semilinear wave equations

The purpose of this paper is to show that, when the damping is sufficiently large, the dynamical system generated by certain damped semilinear wave equations has the property that its global attractor (i.e. the attractor in the sense of Babin, Vishik [1983] and Hale [1985]) is contained in a finite-dimensional local invariant manifold of class C^1. The exact result is stated in Theorem 2 and its Corollary.

A previous result in this direction has been obtained by Solà-Morales, València [1986]. These authors study a similar problem with Neumann boundary conditions and they give sufficient conditions on the coefficients, in order that all the flow be attracted by the invariant subspace formed by the spatially homogeneous states.

Our proof of Theorem 2 will be crucially based upon a special choice of the norm on the state space. They way of choosing this norm is inspired by the work of Solà-Morales, València [1986] (see also Haraux [1985]).

1. Introduction and preliminaries

We consider evolution problems of the following form, where the unknown u is a function of $x \in [0,\pi]$ and $t \in \mathbb{R}$ with values in $\mathbb{R}$:

$$\sigma^2 u_{tt} + 2\alpha u_t = u_{xx} + f(u) \tag{1}$$

$$u|_{x=0} = u|_{x=\pi} = 0 \tag{2}$$

$$u|_{t=0} = u_0, \quad \sigma u_t|_{t=0} = \sigma v_0 . \tag{3}$$

Here, σ and α are non-negative real parameters, f is a function $\mathbb{R} \to \mathbb{R}$, and u_0 and v_0 are functions of $x \in [0,\pi]$. We assume that f satisfies the following conditions:

$$f \text{ is of class } C^{1+\eta} \text{ for some } \eta \in (0,1) \tag{4}$$

$$f(0) = 0 \text{ and } \limsup_{|u| \to \infty} \operatorname{sgn}(u) f(u) < 0 \tag{5}$$

In the singular limit $\sigma = 0$, we have a reaction-diffusion problem, for which the existence of finite-dimensional invariant manifolds which contain the global attractor has been treated by several authors: Henry [1974-81] (p. 166), Mañé [1977], Kamaev [1981], and Mora [1983].

We shall here deal with the case $\sigma > 0$. In this case, the time variable can always be rescaled so as to reduce the problem to the case $\sigma = 1$ (take $t_{new} = \sigma^{-1} t_{old}$ and $\alpha_{new} = \sigma^{-1}\alpha_{old}$). According to this, in the sequel we assume always $\sigma = 1$.

We shall consider (1)-(2) as a first-order system for the pair $(u,u_t) \equiv (u,v)$. The resulting problem will be regarded as an evolution problem on the function space $H^1_o \times L_2 \equiv E$. Here, L_2 and H^1_o have the usual meaning as Hilbert spaces of functions on $[0,\pi]$; the inner product and the norm of L_2 are taken as usual, and they will be denoted respectively as $\langle\cdot,\cdot\rangle_{L_2}$, or simply $\langle\cdot,\cdot\rangle$, and $\|\cdot\|_{L_2}$; the inner product on H^1_o will be taken as $\langle u,y\rangle_{H^1_o} \equiv \langle u_x,y_x\rangle_{L_2}$ which results in the norm $\|u\|_{H^1_o} = \|u_x\|_{L_2}$. In the sequel a generic element of E will be denoted by $U = (u,v)$. Let us now define A as the linear operator on E given by

$$AU \equiv (v,\ u_{xx} - 2\alpha v) \tag{6}$$

with domain $D = (H^2 \cap H^1_o) \times H^1_o$, and let F be the non-linear operator given by

$$F(U) \equiv (0,\ f\circ u) \tag{7}$$

with these definitions, problem (1)-(3) can be rewritten in the form

$$\dot{U} = AU + F(U) \tag{8}$$

$$U(0) = U_0 \tag{9}$$

The operator A defined above is the generator of a semigroup, in fact a group, on the space E. On the other hand, it is easily verified that the condition that $f : \mathbb{R} \to \mathbb{R}$ be of class $C^{1+\eta}$ implies that $F : E \to E$ is also of class $C^{1+\eta}$ and F, DF, and the Hölder seminorm of DF are uniformly bounded on bounded sets of E. With this, problem (8)-(9) fits in the standard theory of semilinear evolution equations, which establishes the existence and uniqueness of (mild) solutions, and its $C^{1+\eta}$ dependence on the initial state (see Segal [1963]).

It is a well-known fact that this dynamical system has a functional that is strictly decreasing along non-stationary trajectories, namely

$$\Phi(U) \equiv \int_0^\pi [\frac{v^2}{2} + \frac{u_x^2}{2} - \phi(u)] \text{ where } \phi(u) \equiv \int_0^u f \qquad (10)$$

Moreover, the hypotheses on f imply that the level sets of Φ, $\{U|\Phi(U) \leqq c\}(c \in \mathbb{R})$ are bounded in $H_0^1 \times L_2$. These two properties imply that the positive semi-orbits of the dynamical system remain bounded, and in consequence they are defined in the whole interval $0 \leqq t < +\infty$. In fact, it can be shown that the positive semi-orbits are not only bounded but even precompact (Webb [1979], Thm. 2.2), from which it follows that every positive semiorbit is attracted by the set of stationary states (ibidem, Thm. 3.4). Using these facts, one can establish the existence of a compact global attractor which has a finite Haussdorff dimension (see Babin, Vishik [1983], §6, Hale [1985], Thm. 6.1, and Ghidaglia, Temam [1986]).

2. Abstract result

The globally attracting invariant manifolds that we are looking for will be obtained by the method of Perron-Bogolyubov-Mitropol'skii as it appears in Henry [1974-81] (Chap. 6). This method can be applied to general evolution problems of the form (8)-(9), where A is the generator of a linear semigroup on a Banach space E, and F is a non-linear perturbation. In this method, the attracting invariant manifolds are obtained in connection with a partition of the spectrum Σ of A into two parts Σ_1, Σ_2 such that

$$\text{Sup Re } \Sigma_2 < \min (\inf \text{Re } \Sigma_1, 0) \qquad (11)$$

In addition to this, we will assume that Σ_1 is bounded; in this case Σ_1 gives rise to a projection P_1 by means of a Dunford's integral. By defining $P_2 \equiv I - P_1$, $E_i \equiv P_i E$, and $A_i \equiv A|_{E_i}$ (i = 1,2), one has the decompositions $E = E_1 \oplus E_2$, $A = \text{diag}(A_1, A_2)$, and $e^{At} = \text{diag}(e^{A_1 t}, e^{A_2 t})$. Moreover, the fact that Σ_1 is bounded implies that A_1 is a bounded linear operator, and therefore $e^{A_1 t}$ is a group. In order to obtain the attracting invariant manifolds, one needs not only (11) but actually

$$\xi_2 < \min(\xi_1, 0) \tag{12}$$

where ξ_1, ξ_2 are the exponents appearing in a pair of bounds of the form

$$\|e^{-A_1 t}\| \leq M_1 e^{-\xi_1 t} \quad (\forall t \geq 0) \tag{13}$$

$$\|e^{A_2 t}\| \leq M_2 e^{\xi_2 t} \quad (\forall t \geq 0) \tag{14}$$

In addition to this, $\xi_1 - \xi_2$ and $-\xi_2$ are required to be large enough compared with some positive quantities which grow with the "size" of the non-linearity F.

To take care of our specific problem (1)-(3), it will suffice the particular result stated next, where the condition relative to the size of the non-linearity is especially simple.

THEOREM 1: Let A be the generator of a linear semigroup on the Banach space E, and let $F : E \to E$ be uniformly bounded in $C^{1+\eta}(E,E)$ for some $\eta \in (0,1)$, with Lipschitz constant L. Assume also that A has a spectral decomposition as described above, with $\xi_2 < \xi_1 < 0$ and $M_1 = M_2 = \|P_1\| = \|P_2\| = 1$. Suppose finally that

$$\xi_1 - \xi_2 > 4L \tag{15}$$

and let ζ be any real number in the interval $0 < \zeta < \min(\eta, \xi_2^{-1}(\xi_1 - \xi_2 - 4L))$. Then the dynamical system determined by (8)-(9) has a globally and exponentially attracting invariant submanifold M $C^{1+\zeta}$-diffeomorphic to E_1 (in fact, M is given by the graph of a mapping $E_1 \to E_2$ which is uniformly bounded in $C^{1+\zeta}(E_1, E_2)$). If, moreover, dim $E_1 < \infty$, then M attracts with asymptotic phase. □

From the properties of M it follows that it must contain the global attractor of the dynamical system. The proof of Theorem 1 can be obtained by following the lines of Henry [1974-81] (Thms. 6.1.2, 6.1.4, 6.1.5, 6.1.7).

3. Application to problem (1)-(3)

In the following, we shall apply Theorem 1 to the problem described in §1. Henceforth E, A, F have again the particular meaning that they had in §1.

If the function f is uniformly bounded in $C^{1+\eta}(\mathbb{R},\mathbb{R})$, then F is also uniformly bounded in $C^{1+\eta}(E,E)$, so that Theorem 1 can be applied directly. In the contrary case, we can still obtain a related result by making use of the fact that the dynamical system on E determined by (8)-(9) has the following property:

(P) There is a bounded positively invariant set B such that every positive semi-orbit goes eventually in B.

Having property (P), we will modify F outside B in such a way that the modified F be uniformly bounded in $C^{1+\eta}(E,E)$. Obviously, the modified flow will coincide with the original one within the set B, which we know that contains the global attractor of the original flow. Therefore, If M is the manifold of Theorem 1 for the modified flow, then $M \cap B$ will be a local invariant manifold for the original flow with the property of containing its global attractor.

In the case of problem (1)-(3), property (P) can be obtained by making use of the functional (10). According with what we have said in §1, it will suffice to take $B = \{U \in E | \Phi(U) \leqq c_1\}$, with $c_1 < \infty$ large enough so that B contains all the stationary states. The existence of such a $c_1 < \infty$ follows from the hypotheses of f.

The modification of F outside B mentioned above can be easily accomplished by adequately modifying f outside some interval of the form $(-c_2, c_2)$, where c_2 is an upper bound on the sup norm of u when $U = (u,v)$ varies over B. Such a bound is a straightforward consequence of the fact that B is bounded in $H^1_o \times L_2$.

Let us remark here that, after this modification, $f : \mathbb{R} \to \mathbb{R}$ is globally Lipschitzian, which implies that the mapping $u \mapsto f \circ u$ takes L_2 to L_2 and satisfies also a global Lipschitz condition with the same constant as $f:\mathbb{R} \to \mathbb{R}$. This fact will be useful below in connection with estimating the Lipschitz constant of $F:E \to E$.

Let us now consider the spectrum of the linear operator (6). This is given by $\Sigma = \{\lambda^{\pm}_k | k = 1,2,3...\}$, where

$$\lambda_k^{\pm} \equiv -\alpha \pm \sqrt{\alpha^2 - k^2} \tag{16}$$

For $k \leq \alpha$ the numbers $\lambda_k^{\pm}$ are real, with λ_k^{+} decreasing and λ_k^{-} increasing as k increases: for $k > \alpha$ they are complex with exactly the same real part $-\alpha$. Therefore, in order to look for a finite-dimensional invariant manifold, we are bound to take $\Sigma_1 = \{\lambda_k^{+} | 1 \leq k \leq n\}$, where n should be less than α. For simplicity we shall assume that

$$n + 1 < \alpha \tag{17}$$

so that λ_{n+1}^{+} is still real and greater than $-\alpha$.

In order to put ourselves in the hypotheses of Theorem 1, we shall introduce a new norm on E, related to the more detailed spectral partition $\Sigma = \Sigma_1 \cup \Sigma_{22} \cup \Sigma_{-1}$, where Σ_2 has been divided into two parts, namely $\Sigma_{-1} \equiv \{\lambda_k^{-} | 1 \leq k \leq n\}$ and $\Sigma_{22} \equiv \{\lambda_k^{\pm} | k > n\}$. Finally, we define also $\Sigma_{11} \equiv \Sigma_1 \cup \Sigma_{-1}$. For each of the spectral sets $\Sigma_i (i = 1,2,-1,11,22)$ we shall denote the corresponding projection by P_i, the corresponding subspace of E by E_i, and the corresponding part of A by A_i. These objects are easily identified by using the sine series representation of u and v, whose coefficients will be denoted respectively by $\hat{u}_k$ and $\hat{v}_k$. To specify an element of L_2 or E through these coefficients, we shall write simply $u = (\hat{u}_k)_{k \geq 1}$ or $U = (\hat{u}_k, \hat{v}_k)_{k \geq 1}$. In terms of this representation, the spaces $E_1, E_{-1}, E_{11}, E_{22}$ can be described respectively as consisting of the elements of the form $(\hat{u}_k, \lambda_k^{+}\hat{u}_k)_{k \leq n}$, $(\hat{u}_k, \lambda_k^{-}\hat{u}_k)_{k \leq n}$, $(\hat{u}_k, \hat{v}_k)_{k \leq n}$, $(\hat{u}_k, \hat{v}_k)_{k > n}$, where the subscript indicates the values of k for which the coefficients do not vanish.

4. The inner product on E

The new norm on E will derive from an inner product obtained by direct sum of a pair of inner products on E_{11} and E_{22}. In the following, $U = (u,v)$ and $Y = (y,w)$ denote two generic elements of E or its subspaces E_{11} and E_{22}. The inner products on E_{11} and E_{22} are chosen respectively as follows:

$$\langle U,Y \rangle_{E_{11}} \equiv \alpha^2 \langle u,y \rangle - \langle u_x, y_x \rangle + \langle \alpha u + v, \alpha y + w \rangle \tag{18}$$

$$\langle U,Y \rangle_{E_{22}} \equiv \langle u_x, y_x \rangle + (\alpha^2 - 2(n+1)^2) \langle u,y \rangle + \langle \alpha u + v, \alpha y + w \rangle \tag{19}$$

The corresponding norms will be denoted respectively by $\|\cdot\|_{E_{11}}$ and $\|\cdot\|_{E_{22}}$. The resulting inner product on E is then the following

$$\langle U,Y\rangle_E \equiv \langle U_{11},Y_{11}\rangle_{E_{11}} + \langle U_{22},Y_{22}\rangle_{E_{22}} \tag{20}$$

where U_i, Y_i $(i = 11,22)$ denote the corresponding projections of U and Y. The norm corresponding to this inner product will be denoted as $\|\cdot\|_E$.

The fact that (18) and (19) are positive-definite follows from assumption (17) together with the implications

$$U \in E_{11} \Rightarrow u = (\hat{u}_k)_{k\leq n} \Rightarrow \|u_x\|_{L_2} \leq n\ \|u\|_{L_2} \tag{21}$$

$$U \in E_{22} \Rightarrow u = (\hat{u}_k)_{k>n} \Rightarrow \|u_x\|_{L_2} \geq (n+1)\ \|u\|_{L_2} \tag{22}$$

In fact, these relations and assumption (17) imply that the inner product (20) is equivalent to the one given by the direct sum of those of H_0^1 and L_2, i.e. $\langle U,Y\rangle_{H_0^1 \times L_2} = \langle u_x,y_x\rangle + \langle u,w\rangle$. Another consequence of (21)-(22) which will be used below is the inequality

$$\|U\|_E \geq \sqrt{\alpha^2-(n+1)^2}\ \ \|u\|_{L_2} \quad (\forall U \in E) \tag{23}$$

We shall also use the fact that

$$u = 0 \Rightarrow \|U\|_E = \|v\|_{L_2} \tag{24}$$

which follows immediately from the definitions and the fact that E_{11} and E_{22} are orthogonal to each other with respect to the inner product $\langle\cdot,\cdot\rangle_{H_0^1 \times L_2}$.

The inner product on E_{11} has the remarkable property of making the subspaces E_1 and E_{-1} orthogonal to each other. This orthogonality is easily verified by using the sine series representation, in terms of which the generic elements of E_1 and E_{-1} have respectively the form $(\hat{u}_k,\lambda_k^+\hat{u}_k)_{k\leq n}$ and $(\hat{u}_k,\lambda_k^-\hat{u}_k)_{k\leq n}$; by introducing this into (18) one obtains

$$\langle U,Y\rangle_{E_{11}} = \frac{\pi}{2} \sum_{k=1}^{n} [\alpha^2-k^2+(\alpha+\lambda_k^+)(\alpha+\lambda_k^-)]\ \hat{u}_k\hat{y}_k = 0$$

where we have used the equality $(\alpha+\lambda_k^+)(\alpha+\lambda_k^-) = -(\alpha^2-k^2)$.

Moreover, by using the sine series representation it is also easily verified that

$$\|e^{-A_1 t}\| = e^{-\lambda_n^+ t} \quad (\forall t \geq 0)$$

$$\|e^{A_{-1} t}\| = e^{\lambda_n^- t} \quad (\forall t \geq 0)$$

On the other hand, the inner product on E_{22} has been chosen in such a way that it gives exactly

$$\|e^{A_{22} t}\| = e^{\lambda_{n+1}^+ t} \quad (\forall t \geq 0)$$

Of course, the inequality $\|e^{A_{22}t}\| \geq e^{\lambda_{n+1}^+ t}$ is obtained by considering the effect of $e^{A_{22}t}$ on an element of the form $U = (\hat{u}_k, \lambda_k^+ \hat{u}_k)_{k=n+1}$. In order to prove the reverse inequality, it suffices to show that, for any $U_0 \in E_{22} \cap D$, $\psi(t) \equiv \|e^{A_{22}t} U_0\|_E^2$ satisfies the differential inequaltiy $\psi_t \leq 2\lambda_{n+1}^+ \psi$.

Let us call $e^{A_{22}t} U_0 \equiv U(t) = (u(t), v(t))$. Then

$$\psi = -\langle u_{xx}, u\rangle + 2(\alpha^2-(n+1)^2)\langle u,u\rangle + 2\alpha\langle u,v\rangle + \langle v,v\rangle,$$

and using the relations $u_t = v$ and $v_t = u_{xx} - 2\alpha v$ we obtain

$$\psi_t = 2\alpha \langle u_{xx}, u\rangle - 4(n+1)^2\langle u,v\rangle - 2\alpha\langle v,v\rangle$$

From these two equalities it follows that

$$\psi_t + 2\alpha\psi = 4(\alpha^2-(n+1)^2)\langle u, \alpha u + v\rangle$$
$$= 4\sqrt{\alpha^2-(n+1)^2}\ \langle\sqrt{\alpha^2-(n+1)^2}\ u, \alpha u + v\rangle$$

Now, inequality $|\langle a,b\rangle| \leq \frac{1}{2}(\|a\|^2 + \|b\|^2)$ applied to the last term of the previous relation gives

$$\psi_t + 2\alpha\psi \leq 2\sqrt{\alpha^2-(n+1)^2}\ [(\alpha^2-(n+1)^2)\langle u,u\rangle + \langle \alpha u+v, \alpha u+v\rangle],$$

and (22) allows us to conclude that

$$\psi_t + 2\alpha\psi \leqq 2\sqrt{\alpha^2-(n+1)^2}\,\psi$$

which is the desired result.

5. Final results

We can now proceed to the application of Theorem 1. We first notice that the orthogonality of E_1 and E_{-1} together with the obvious orthogonality of E_{11} and E_{22} imply that E_1 and E_2 are also orthogonal to each other, so that $\|P_1\| = \|P_2\| = 1$. On the other hand, from (26) and (27), follows that

$$\|e^{A_2 t}\| = e^{\lambda^+_{n+1} t} \qquad (\forall t \geqq 0) \tag{28}$$

Relations (25) and (28) give us (13) and (14) with $M_1 = M_2 = 1$ and

$$\xi_1 = \lambda^+_n, \quad \xi_2 = \lambda^+_{n+1} \tag{29}$$

Let us now estimate the Lipschitz constant of F. For this we shall take profit of the fact that, after the modification of f, the mapping $u \mapsto f\circ u$ takes L_2 to L_2 and satisfies a global Lipschitz condition with the same constant as $f : \mathbb{R} \to \mathbb{R}$. In the following this constant will be denoted by ℓ. By using this fact, together with relations (23) and (24), we obtain that

$$\|F(U) - F(Y)\|_E = \|f\circ u - f\circ y\|_{L^2}$$

$$\leqq \ell\,\|u-y\|_{L^2} \leqq \frac{\ell}{\sqrt{\alpha^2-(n+1)^2}}\,\|U-Y\|_E$$

which gives

$$L = \ell/\sqrt{\alpha^2-(n+1)^2} \tag{30}$$

By introducing the obtained values of ξ_1, ξ_2 and L into (15), this condition takes the form

$$\sqrt{\alpha^2-n^2} - \sqrt{\alpha^2-(n+1)^2} > \frac{4\ell}{\sqrt{\alpha^2-(n+1)^2}} \quad (31)$$

The pair of conditions (17) and (31) is equivalent to the following:

$$2n+1 > 8\ell \quad (32)$$

$$\left(\frac{\alpha}{\sigma}\right)^2 > (n+1)^2 + \frac{16\ell^2}{(2n+1)-8\ell} \quad (33)$$

where we have substituted α/σ for α, which takes care of problem (1)-(3) in the general case of σ not necessarily equal to 1.

By virtue of Theorem 1, we can therefore conclude the following

Theorem 2: Let us consider the dynamical system on $H_o^1 \times L_2$ determined by (1)-(3) with σ and α positive, and assume that f satisfies conditions (4)-(5). Then there exists a finite constant ℓ (equal to the Lipschitz constant of f on some subinterval of $\mathbb{R}$) such that, for every integer n satisfying (32) and (33) there is a local invariant submanifold of class C^1 and dimension n which contains the global attractor (in fact the obtained manifold is $C^{1+\zeta}$ for some $\zeta > 0$). □

COROLLARY: There exists a finite constant d* such that for $\frac{\alpha}{\sigma} > d^*$ the global attractor lies in a finite-dimensional local invariant submanifold of class C^1. □

Indeed, the range of values of $\frac{\alpha}{\sigma}$ and n determined by conditions (32) and (33) looks like the shaded area in Figure 1. When $\frac{\alpha}{\sigma}$ grows towards infinity, the range of values of n for which (32) and (33) hold spreads to the right and also to the left. Asymptotically, this range is given by $4\ell - \frac{1}{2} < n < \frac{\alpha}{\sigma} - 1$. This is in accordance with what happens in the parabolic case ($\sigma = 0$), where the range of admissible values of n is bounded from below but not from above.

Finally, let us announce that in a joint paper with J. Solà-Morales, it will be shown that, for small values of the damping ratio $\frac{\alpha}{\sigma}$ one can give examples where the global attractor is not contained in any finite-dimensional local invariant manifold of class C^1.

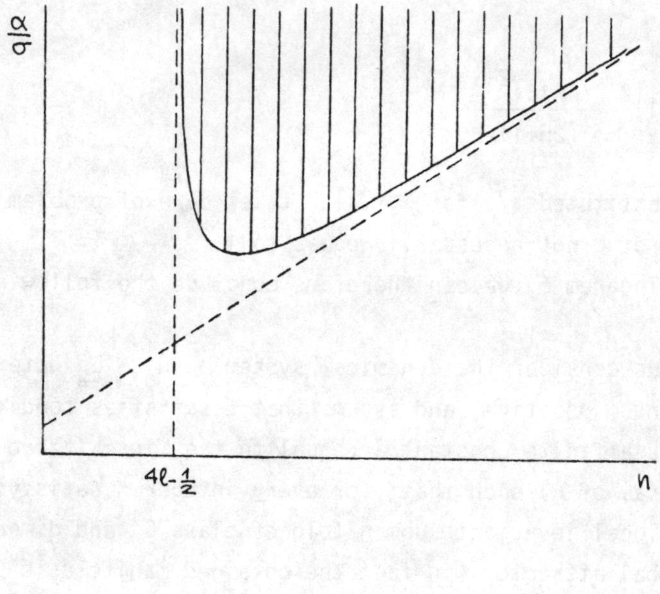

Figure 1

References

1. Babin, A.V., Vishik, M.I., 1983, Regular attractors of semigroups and evolution equations. J. Math. Pures Appl. 62: 441-491.
2. Ghidaglia, J.M., Temam, R., 1986. Attractors for damped nonlinear hyperbolic equations. J. Math. Pures Appl. (in press).
3. Hale, J.K., 1985. Asymptotic behavior and dynamics in infinite dimensions. Res. Notes in Math. 132: 1-42.
4. Haraux, A., 1985. Two remarks on hyperbolic dissipative problems. Publications du Laboratoire d'Analyse Numérique (Univ. Paris VI) No. 85029.
5. Henry, D., 1974-81. Geometric Theory of Semilinear Parabolic Equations. Univ. Kentucky Lecture Notes (1974). Lecture Notes in Math. 840 (1981).
6. Kamaev, D.A., 1981. On the Hopf conjecture for a class of equations of chemical kinetics. Zap. Nauchn. Sem. Leningrad. Otdel. Mat. Inst. Steklov. 110: 57-73 (Russian). Transl.: J. Soviet Math. 25 (1984).

7. Mañé, R., 1977. Reduction of semilinear parabolic equations to finite dimensional C^1 glows. Lecture Notes in Math. 597: 361-378.

8. Mora, X., 1983. Finite-dimensional attracting manifolds in reaction-diffusion equations. Contemporary Math. 17: 353-360.

9. Segal, I., 1963. Non-linear semi-groups. Ann. Math. 78: 339-364.

10. Solà-Morales, J., València, M., 1986. Trend to spatial homogeneity for solutions of semilinear damped wave equations. Proc. Roy. Soc. Edinburgh (in press).

11. Webb, G.F., 1979. A bifurcation problem for a nonlinear hyperbolic partial differential equation. SIAM J. Math. Anal. 10: 922-932.

X. Mora
Departament de Matemàtiques
Universitat Autònoma de Barcelona
Barcelona
Spain.

J–M MOREL & L OSWALD

Remarks on the equation $-\Delta u = \lambda f(u)$ with f nondecreasing

I. Introduction

Let Ω be a smooth bounded open subset of $\mathbb{R}^N$, and let f be a nondecreasing, strictly convex function of class C^2 with $f(0) > 0$ (e.g. $f(u) = e^u$, $f(u) = 1 + u^2$). We consider the following problem:

$$(P_\lambda) \begin{cases} -\Delta u = \lambda f(u) \\ u \in C_0^1(\Omega) \\ u \geq 0 \text{ in } \Omega. \end{cases}$$

Solutions for (P_λ) can be defined in at least one of the three following ways:

(1) the solutions $u(\lambda)$ obtained by continuation from the solution 0 of (P_0).

(2) the minimal solutions $\underline{u}_\lambda$ obtained by iteration from the subsolution 0 by means of the algorithm A , A being defined by $Au = \lambda(-\Delta^{-1})f(u)$.

(3) the solutions obtained by a local minimization of the functional $J_\lambda(u) = \frac{1}{2}\int |\nabla u|^2 - \lambda \int F(u)$ where $F(u) = \int_0^u f(s)\, ds$.

The first aim of this note is to give a synthetic presentation of the equivalence of the three definitions. To do this, we refer, among others, to the works of Crandall-Rabinowitz [3], Mignot-Puel [9], Amann [1], Laetsch [6], Keener-Keller [5], Cohen [2], Lions [7].

Our second aim is to examine which part of the preceding results remains true when the convexity hypothesis is removed. We focus on the minimal solution $\underline{u}(\lambda)$ obtained by iteration and analyse its stability properties. We point out that, as in the convex case, algorithm A provides a way for deciding whether (P_λ) has a solution. Moreover, the solution obtained by algorithm A is, in general, a strict local minimum of the energy. Part of our approach follows and simplifies results of Puel [10] and is strongly related to a stability analysis due to Lions [8].

II. The convex case

In this part we assume that f is strictly convex.

(1) Definition of the solutions obtained by continuation

(P_λ) can be reset as an integral equation:

$$\begin{cases} T(\lambda,u) = u + \lambda\, \Delta^{-1} f(u) = 0 \\ u \in C_0^1(\Omega) \\ u \geqq 0 \end{cases}$$

It is easy to check that the operator T is C^1, maps $\mathbb{R} \times C_0^1$ into C_0^1 and that $T(0,0) = 0$ and $T_u'(0,0) = Id$. So applying the implicit function theorem to T in a neighbourhood of (0,0) in $\mathbb{R}^+ \times C_0^1$, we define a curve $u(\lambda)$ in C_0^1 of solutions of (P_λ). For λ small, $u(\lambda)$ is small: since $f(0) > 0$ we have $f(u\,(\lambda)) > 0$ and therefore $u(\lambda) > 0$. Moreover, differentiating (P_λ) with respect to λ we find:

$$- \Delta u'(\lambda) - \lambda f'(u(\lambda))u'(\lambda) = f(u(\lambda)).$$

This shows that as far as $u > 0$ and $\lambda_1(-\Delta\cdot - \lambda f'(u)\cdot) > 0$, we have $u'(\lambda) > 0$. Consequently, $u(\lambda)$ increases and $\lambda_1(-\Delta\cdot - \lambda f'(u(\lambda))\cdot)$ decreases with λ as far as $\lambda_1(-\Delta\cdot - f'(u(\lambda))\cdot)$ remains positive. Thus we have defined $u(\lambda)$ on a maximal interval $[0,\lambda^*)$. Note that since f' is nondecreasing and $u(\lambda)$ is nondecreasing, $\lambda^* < \infty$. Therefore we have the following diagram:

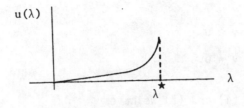

(2) Definition of the solutions obtained by iteration

The algorithm A_λ is defined from $C_o^1(\Omega)$ into $C_o^1(\Omega)$ by $A_\lambda u = -\lambda\Delta^{-1}f(u)$. For every λ, let us consider the sequence $u_n = A_\lambda^n 0$. Since f is nondecreasing, the sequence u_n increases with n. Set $\lambda^{**} = \sup\{\lambda: A_\lambda^n 0$ converges in $C_o^1(\Omega)\}$. Then $A_\lambda^n 0$ converges for every λ in $[0,\lambda^{**})$. Indeed, if $\lambda > \lambda'$ then $A_\lambda^n 0 > A_{\lambda'}^n 0$. Moreover, when the sequence $A_\lambda^n 0$ converges, it converges to the smallest solution of (P_λ). Therefore we have defined the minimal solution $\underline{u}_\lambda$ of (P_λ) for λ in $[0,\lambda^{**})$.

(3) Equivalence of the three kinds of solutions

Let us detail these equivalences in a theorem:

THEOREM 1: (a) $\lambda^* = \lambda^{**}$ and (P_λ) has no solution for $\lambda > \lambda^*$.

(b) Moreover the following propositions are equivalent:

u is obtained by continuation (i.e. $u = u(\lambda), \lambda < \lambda^*$). (1)

u is obtained by the algorithm A_λ (i.e. $u = \underline{u}_\lambda, \lambda < \lambda^*$). (2)

u is a strict local minimum for the energy J_λ. (3)

The proof of Theorem 1 is based upon the following lemma (c.f. Crandall-Rabinowitz):

LEMMA 1: Let u,v be in $C_o^1(\Omega)$ and satisfy

$$-\Delta u - f(u) = 0 \text{ with } \lambda_1(-\Delta\cdot - f'(u)\cdot) \geq 0 \quad (4)$$

$$-\Delta v - f(v) \geq 0 \quad (5)$$

then

(i) $\lambda_1 > 0$ implies $v \geq u$

(ii) $\lambda_1 = 0$ implies $v = u$.

PROOF OF LEMMA 1: From (4) and (5) we have

$$-\Delta(v-u) \geq f(v) - f(u) \geq f'(u)(v-u)$$

hence

$$- \Delta(v-u) - f'(u)(v-u) \geqq 0. \qquad (6)$$

(i) If $\lambda_1(-\Delta\cdot - f'(u)\cdot) > 0$, the conclusion follows from (6) by the maximum principle.

(ii) If $\lambda_1(-\Delta\cdot - f'(u)\cdot) = 0$, let ϕ_1 be the corresponding eigenfunction; multiplying (6) by ϕ_1 we have:

$$\int (-\Delta(v-u) - f'(u)(v-u))\phi_1 = \int (-\Delta\phi_1 - f'(u)\phi_1)(v-u) = 0.$$

So using (6) and since $\phi_1 > 0$ we find that:

$$- \Delta(v-u) - f'(u)(v-u) = 0. \qquad (7)$$

By Taylor formula, using (4) and (5) we obtain:

$$- \Delta(v-u) - (f(v)-f(u)) = - \Delta(v-u) - f'(u)(v-u) - \frac{1}{2} f''(u+\theta(v-u))(v-u)^2$$

$$= - \frac{1}{2} f''(u+\theta(v-u))(v-u)^2 \geqq 0.$$

Since f is strictly convex, we conclude that u = v.

PROOF OF THEOREM 1:

STEP 1: Proof of (a).

Let $\lambda \geqq \lambda^*$ and u be a solution of (P_λ). We have

$$- \Delta u - \lambda f(u) = 0$$

and so

$$- \Delta u - \mu f(u) \geqq 0 \text{ for every } \mu < \lambda^*.$$

Therefore we can apply Lemma 1(i) to $u(\mu)$ and u and we obtain $u \geqq u(\mu)$ for every $\mu < \lambda^*$. We derive from this estimate that $u(\lambda^*) = \lim_{\mu \uparrow \lambda^*} u(\mu)$ exists, is a solution of P_{λ^*} and that $\lambda_1(-\Delta\cdot - \lambda^* f'(u(\lambda^*))\cdot) = 0$. Then, applying Lemma 1(ii) to $u(\lambda^*)$ and u we find that $u = u(\lambda^*)$. This proves the second part of assertion (a) and in particular that $\lambda^{**} \leqq \lambda^*$. Since the converse inequality obviously holds, we conclude that $\lambda^* = \lambda^{**}$.

STEP 2: Proof of the equivalence of (1) and (3).

The fact that (1) implies (3) is obvious. Indeed, if $u = u(\lambda)$ is obtained by continuation, then $\lambda_1(-\Delta\cdot\lambda f'(u(\lambda))\cdot) > 0$ for $\lambda < \lambda^*$. Thus it remains to prove that (3) implies (1). Let u be a local minimum of J_λ, then $\lambda_1(-\Delta\cdot-\lambda f'(u)\cdot) \geqq 0$; so applying Lemma 2 to u and $u(\lambda)$ and then to $u(\lambda)$ and u we find $u = u(\lambda)$. Therefore if u is a strict local minimum we have $u = u(\lambda)$.

STEP 3: Proof of the equivalence of (1) and (2).

The implicit function theorem says that in a neighbourhood of (0,0) in $R \times C^1_o(\Omega)$, $(\lambda,u(\lambda))$ is the only solution of P_λ. Since $0 \leqq \underline{u}_\lambda \leqq u(\lambda)$ and since the set of the solutions of (P_λ) is a manifold in a neighbourhood of (0,0) we have $\underline{u}_\lambda = u(\lambda)$ for λ small enough. Set $\lambda_{max} = \inf\{\lambda \geqq 0;\ u(\lambda) \neq \underline{u}_\lambda\}$. Let us suppose by contradiction that $\lambda_{max} < \lambda^*$. According to the inequalities

$$u(\lambda) = \underline{u}_\lambda \leqq \underline{u}_{\lambda_{max}} \leqq u(\lambda_{max}) \text{ for } \lambda < \lambda_{max}$$

and to the continuity of the curve $\lambda \to u(\lambda)$ we have $\underline{u}_{\lambda_{max}} = u(\lambda_{max})$. Let λ_n be a sequence such that $\lambda_n \downarrow \lambda$ and $u(\lambda_n) \neq \underline{u}_{\lambda_n}$. The following inequalities hold:

$$u(\lambda_{max}) = \underline{u}(\lambda_{max}) \leqq \underline{u}_{\lambda_n} \leqq u(\lambda_n), \text{ therefore } \underline{u}_{\lambda_n} \to u(\lambda_{max}) \text{ as } n \to \infty .$$

Since the set of the solutions of (P_λ) is a manifold in a neighbourhood of $(\lambda,u(\lambda))$ and since $\underline{u}(\lambda_n)$ does not belong to that manifold we obtain a contradiction.

II. The case f nonconvex and nondecreasing

(1) Properties of the algorithm A

LEMMA 2: Assume f is C^1 nondecreasing. Then algorithm A defined in §1 is a gradient algorithm. In particular $J(Av) \leqq J(v)$.

If u is such that $\underline{u}_0 < u < \bar{u}_0$, with $\underline{u}_0, \bar{u}_0 \in C^1_o(\Omega)$, $A^n\underline{u}_0 \uparrow u$ and $A^n\bar{u}_0 \downarrow u$ then u is a strict local minimum for J in $C^1_o(\Omega)$.

If one only has $A^n\underline{u}_0 \uparrow u$, $u \in C^1_o(\Omega)$ (or $A^n\bar{u}_0 \downarrow u$) then $\lambda_1(-\Delta\cdot-\lambda f'(u)\cdot) \geqq 0$.

PROOF: Let us first notice that $Av = \nabla v - \nabla J(v)$, where $J(v) = v - Kf(v)$ is the gradient of J in $H_0^1(\Omega)$. Assume $v \in C_0^1(\Omega)$, then $Av \in C_0^1(\Omega)$ and applying Taylor formula we get:

$$J(Av) = J(v-\nabla J(v)) =$$

$$= J(v) - J'(v)(\nabla J(v)) + \frac{1}{2} J''(v-\theta\nabla J(v))(\nabla J(v),\nabla J(v))$$

Thus

$$J(Av) = J(v) - \frac{1}{2} \|\nabla J(v)\|^2_{H^1} - \frac{1}{2} \int_\Omega f'(v-\theta\nabla J(v))|\nabla J(v)|^2$$

$$J(Av) \leq J(v) - \frac{1}{2} \|\nabla J(v)\|^2_{H_0^1} \tag{8}$$

Assume now that $A^n\underline{u}_0 \uparrow u$ and $A^n\bar{u}_0 \downarrow u$. Then for $\underline{u}_0 \leq v \leq \bar{u}_0$, $v \neq u$, the comparison principle gives $A^n\underline{u}_0 \leq A^n u \leq A^n\bar{u}_0$. Thus $A^n v \to u$ in $C_0^1(\Omega)$. From (8) we obtain $J(A^n v) < J(v)$, on the other hand $J(A^n v) \to J(u)$, thus $J(u) < J(v)$. Since $[\underline{u}_0, \bar{u}_0]$ is a neighbourhood of u in $C_0^1(\Omega)$ this proves that in this case u is a strict local minimum for the energy. Now, if $\underline{u}_0$ is a strict sub-solution, the same argument implies that $\underline{u} = \lim_{n\to\infty} A^n u_0$ as it exists satisfies: $J(\underline{u}) = \min_{\underline{u}_0 \leq v \leq u} J(v)$. Hence for every $\phi \geq 0$, $\phi \in C_0^1(\Omega)$, we have $J(u-t\phi) \leq J(u)$ for $t > 0$ small enough. Then it is easy to deduce that

$$\lambda_1(-\Delta\cdot - f'(u)\cdot) \geq 0.$$

(2) <u>An existence result for (P_λ) with f nondecreasing</u>

<u>THEOREM 2</u>: Assume f is nondecreasing and C^1. Then there exists $\lambda^* > 0$ such that for $\lambda < \lambda^*$, (P_λ) has a minimal solution $\underline{u}(\lambda)$ and for $\lambda > \lambda^*$, (P_λ) has no solution. Moreover, $\lambda_1(-\Delta\cdot - f'(u)\cdot) \geq 0$ and the function $\lambda \to \underline{u}(\lambda)$ is non-decreasing.

The proof of Theorem 2 is a straightforward consequence of Lemma 2 and the use of algorithm A.

<u>REMARK 1</u>: Assume $\underline{u}(\lambda)$ is a continuity point for $\lambda \to \underline{u}(\lambda)$. Then using the monotonicity of this function and Lemma 2 it is easy to see that

either $u(\lambda)$ is a local strict minimum for J_λ

or there exists a sequence $u_n(\lambda) \downarrow u(\lambda)$ of solutions of (P_λ)

(Set $u_n(\lambda) = \lim_{k\to\infty} A^k u(\lambda + \frac{1}{n})$).

This alternative is shown in a slightly different framework in Lions [8].

<u>REMARK 2</u>: Jump points for $\lambda \to \underline{u}(\lambda)$ may happen, as it is shown by the following example.

Let f be a C^2 convex positive and nondecreasing function such that (P_λ) has the following bifurcative diagram

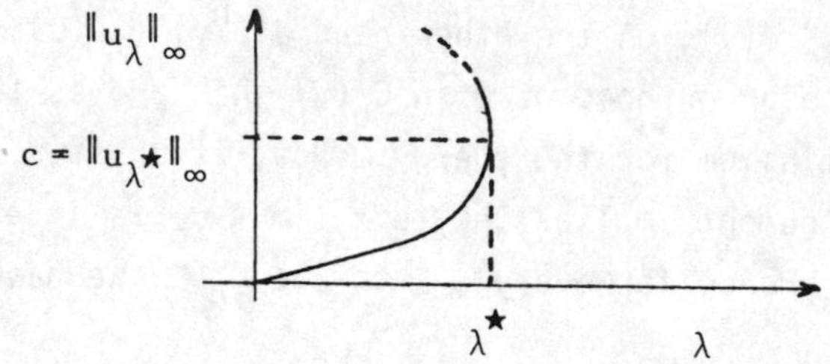

For $\lambda > \lambda^*$, (P_λ) has no solution. We denote by u_λ the minimal solution of (P_λ). Such functions exist (cf. [9] Theorem 3 p. 814). Then let g be a function satisfying:

$$\begin{cases} g \text{ is nondecreasing and } C^2 \\ g(u) = f(u) \text{ for } u \leq c+1 \\ g(u) \leq f(c+1) + 1 \end{cases}$$

We denote by (Q_λ) the problem corresponding to g. Let v_λ be the smallest solution of (Q_λ). v_λ exists for every λ in $\mathbb{R}$ because g is bounded.

(1) It is easy to check that for every $\lambda \leqq \lambda^*$, $v_\lambda = u_\lambda$.

(2) Let $\lambda > \lambda^*$. If $\|v_\lambda\|_\infty < c+1$ then v_λ would be a solution of (P_λ) with $\lambda > \lambda^*$ which is impossible. So

$$\|v_\lambda\|_\infty \leqq c \quad \text{if } \lambda \leqq \lambda^*$$

$$\|v_\lambda\|_\infty \geqq c+1 \text{ if } \lambda > \lambda^*$$

Therefore λ is a jump point for the function $\lambda \to v_\lambda$.

REMARK 3: Let us consider the equation:

$$(P'_\lambda) \begin{cases} -\Delta u = \lambda(f(u) + h) \\ u \in C_0^1(\Omega),\ u \geqq 0 \text{ in } \Omega \end{cases}$$

with $f \in C_0^2(\Omega)$, $h \in L^\infty(\Omega)$. The previous study applies to this equation if f is nondecreasing. If h is small (P'_λ) can be considered as a perturbation of (P_λ). Using Smales' density theorem [11] and following ([4], Section 1) it is possible to prove that for "almost every" h in $L^\infty(\Omega)$ (i.e. for h in a dense G_δ), the solution (u,λ) of (P_λ) form a C^1 manifold. This excludes the second case in the alternative of Remark 1. Thus generically, i.e. for almost every h, one can say

either $\underline{u}(\lambda)$ is a strict local minimum

or $\underline{u}(\lambda)$ is a jump point of $\lambda \to \underline{u}(\lambda)$.

References

1. Amann, H., On the existence of positive solutions of nonlinear elliptic boundary value problems. Indiana Univ. Math. J. 21, 125-146 (1971).
2. Cohen, D.S., Positive solutions of a class of nonlinear eigenvalue problems. J. Math. Mech. 17, 209-215 (1967).
3. Crandall, M.G., Rabinowitz, P.H., Some Continuation and Variational Methods for Positive Solutions of Nonlinear Elliptic Eigenvalue Problems. Arch. Rat. Mech. Anal. 58 - 3 - p. 207-218 (1975).

4. Devys, C., Morel, J.M., Witomski, P., A homotopy method for solving an equation of the type $-\Delta u = F(u)$. Annales de l'I.H.P., Analyse Nonlinéaire, 1 - 4, 205 (1984).
5. Keener, J.P., Keller, H.B., Positive solutions of nonlinear eigenvalue problems, J. Diff. Eq. 16, 103 (1974).
6. Laetsch, T., On the number of solutions of boundary value problems with convex nonlinearities. J. Math. Anal. Appl. 35, 389, (1971).
7. Lions, P.L., On the existence of positive solutions of semilinear elliptic equations, SIAM Review 24, p. 441-467 (1982).
8. Lions, P.L., Structure of the Set of Steady-State Solutions and Asymptotic Behaviour of Semilinear Heat Equations, J. of Diff. Equ. 53, p. 362 (1984).
9. Mignot, F., Puel, J.P., Sur une classe de problèmes non linéaires avec une nonlinéarité positive, croissante, convexe. Comm. P.D.E. 5, p. 791-836 (1980).
10. Puel, J.P., Contrôle optimal sur les solutions minimum ou maximum de certains problèmes semilinéaires. Rapport 75018, Analyse Numérique, Paris VI (1975).
11. Smale, S., An infinite dimensional version of Sard's theorem. American Journal of Mathematics 87 p. 861-866 (1965).

J.-M. Morel
CEREMADE
Université Paris IX - Dauphine
Place de Lattre de Tassigny
75775 Paris CEDEX 16
France.

L. Oswald
Laboratoire d'Analyse Numérique
Université Pierre et Marie Curie
75005 Paris CEDEX 05
France.

I PERAL ALONSO

Some remarks on semilinear wave equations in $\mathbf{R}^n$

Abstract

We consider wave equations

$$\Box u = F(u)$$

where

$$\Box u = u_{tt} - \sum_{i=1}^{n} u_{x_i x_i}.$$

Preferential attention will be given to n = 3. The Cauchy Problem and the energy method are the subject of Section 1. A remark on monotony method and global existence of solution is the tool in Section 2. Finally, Section 3 contains the calculus of self-similar solutions.

1. The energy method

It is well known that to prove existence and uniqueness of global solutions for

$$\Box\, u = F(u)$$

it is necessary to consider the energy,

$$\varepsilon(t) = \frac{1}{2}\int_{\mathbb{R}^n} \{|u_t|^2 + |\nabla_x u|^2\}dx + \int_{\mathbb{R}^n} G(u(x))dx$$

where

$$G(u) = \int_0^u F(s)ds.$$

The proof of existence of a global weak solution depends on $\varepsilon(t)$ being "always positive". A condition for this is the following hypothesis:

$$uF(u),\ G(u) \text{ is lower bounded by } -c(1 + u^2).$$

If $F(u) \geqq 0$, $F'(u) \geqq 0$, $F''(u) \geqq 0$ and

$$\int^{\infty} |G(s)|^{-\frac{1}{2}}\, ds < +\infty$$

(superlinear case) then there is a blow-up.

Proofs of uniqueness need some hypothesis on the asymptotic behaviour of $F(u)$ for $|u| \to +\infty$. This is a natural condition because it is necessary to estimate the term

$$\int_{\mathbb{R}^n} G(u)dx$$

by the energy. This fact is essentially the Sobolev inequality.

Let us assume that

$$F(u) = |u|^{k-1}u$$

and, in this way the analysis of the energy behaviour is reduced to elementary observations on homogeneity.

We consider the Cauchy problem

$$(P_k) \quad k > 1 \quad \begin{cases} \Box u + |u|^{k-1}u = 0 & (x,t) \in \mathbb{R}^n \times \mathbb{R} \\ u(x,0) = f(x) \\ u_t(x,0) = g(x) \end{cases}$$

where we assume $f, g \in C_0^{\infty}(\mathbb{R}^n)$.

Let $u(x,t)$ be a global solution of (P_k) with finite energy. Consider

$$u_s(x,t) = s^{2/(k-1)}u(sx,st)$$

the family of dilation of u by a rescaling. Then

$$\begin{cases} \Box u_s + |u_s|^{k-1}u_s = 0 \\ u_s(x,0) = f_s(x) \\ u_{s,t}(x,0) = g_s(x) \end{cases}$$

where

$$f_s(x) = s^{2/(k-1)}f(sx)$$

and

$$g_s(x) = s^{(k+1)/(k-1)}g(sx).$$

Now, we have:

$$\varepsilon_s(0) = \varepsilon_s(t) + \varepsilon_1(0)s^{(4/(k-1))+2-n},$$

where ε_s is the total energy of u_s and, obviously, $\varepsilon_1(0)$ is the initial energy of u.

On the other hand, we have

$$\|u_s\|_\infty = s^{2/(k-1)} \|u\|_\infty$$

and then

$$\varepsilon_s(0) = c \|u_s\|^{2/(k-1)+1-(n/2)}.$$

It is observed that the exponent

$$\alpha(k,n) = \frac{4}{k-1} + 2 - n$$

for a fixed value of n, is positive if

$$1 \leqq k < 1 + \frac{4}{n-2} = \frac{2n}{n-2} - 1 = \alpha_0(n)$$

But, $\alpha_0(n) + 1$ is the critical value for the Sobolev inclusion of $H^1(\mathbb{R}^n)$ in L^q. This situation can be represented by the following graph:

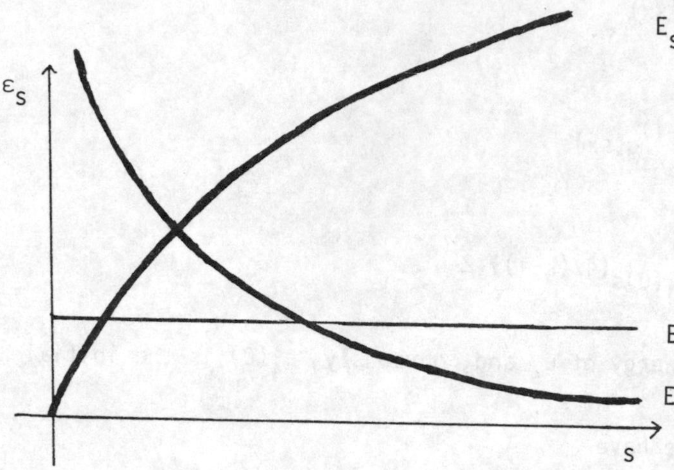

The uniqueness result for $1 \leqq n \leqq 3$ is due to Jörgens [1961], and is obtained, for $n \geqq 4$. By Ginibre-Velo [1983]. These results can be formulated in the following way:

"Suppose $1 \leqq k < \alpha_0(n)$. Then the problem (P_k) has a unique solution".

W. Strauss [1968] proved that for sufficiently small initial data, the problem (P_k) also has a unique solution if $k \geqq 5$. Using the same type of techniques of W. Strauss, we will go on to prove some qualitative results for the problem (P_k), $k \geqq 5$ and $n = 3$. The idea is to use the fact observed in the graph for $k \geqq 5$ in $n = 3$, i.e. if $s > 0$ the energy $\varepsilon_s(0) \to +\infty$ and for $s \to +\infty$ $\varepsilon_s(0) \to 0$. The result is the following:

<u>THEOREM 1</u>: Let $k \geqq 5$, $f,g \in C_0^\infty(\mathbb{R}^3)$ in the problem (P_k). Then $\exists s_0 > 1$ such that if $s > s_0$ and

$$\|u_s(\cdot,t)\|_\infty \leqq \rho\, s^{2/(k-1)} \text{ in } 0 \leqq t \leqq \frac{\log s}{s} = T_0$$

implies that

$$\|u_s(\cdot,t)\|_\infty \leqq \frac{c(s)}{((1/s)+t)^\alpha} \text{ for some } \alpha \in (0,1)$$

and for all $t \in [T_0,\infty)$.

For the proof, let us remember certain results.

LEMMA 1: Let $s > 1$ and $k \geq 5$. Let u_s be the solution of the Cauchy Problem

$$(P_s)\begin{cases} \square v = 0 \\ v(x,0) = s^{2/(k-1)} f(sx) \\ v_t(x,0) = s^{(k+1)/(k-1)} g(sx) \quad f,g \in C_0^\infty(\mathbb{R}^3) \end{cases}$$

Then $\exists\rho = \rho(f,g) > 0$ such that

$$\|u_s(\cdot,t)\|_\infty \leq \rho \frac{s^{(3-k)/(k-1)}}{((1/s)+t)} \tag{1.1}$$

LEMMA 2: If u is the solution of the problem

$$\begin{cases} \square u = F(x,t) \qquad x \in \mathbb{R}^3 \\ u(x,0) = 0 \\ u_t(x,0) = 0 \end{cases}$$

then

$$\|u(\cdot,t)\|_\infty \leq C \int_0^t (t-\tau)^{2-(3/q)} \|\nabla F(\cdot,\tau)\|_q \, d\tau \text{ for } 1 < q < 3 \tag{1.2}$$

LEMMA 3: Let $1 \geq a > 0$, $0 < \frac{1}{\gamma} < \sigma < 1$.

Then

$$I(a,t) = \int_0^t (t-\tau)^{-\sigma}(a+\tau)^{-\gamma\sigma} \, d\tau \leq C_{\gamma\tau}\, a^{1-\gamma\sigma}(a+t)^{-\sigma} \tag{1.3}$$

PROOF OF THE THEOREM: We call $F(u) = -|u|^{k-1}u$. Suppose that t_0 is such that

$$\|u_s(\cdot,t)\| \leq \mu\, s^{2/(k-1)} \quad \text{in } t \in [0,t_0],$$

then

$$\|u_s(\cdot,t)\|_\infty \leq \rho \frac{s^{(3-k)/(k-1)}}{t + (1/s)} + \bar{c} \int_0^t (t-\tau)^{2-(3/q)} \|\nabla F(u_s(\cdot,\tau))\|_q dz \leq$$

$$\leq \rho \frac{s^{(3-k)/(k-1)}}{(t + (1/s))} + \bar{c} \int_0^t (t-\tau)^{2-(3/q)} \|\nabla u_s\|_2 F'(u_s)\|_r \, d\tau \tag{1.4}$$

where $1 < q < 2$ and $r = \frac{2q}{2-q}$.

Let $\alpha(h) = \sup_{|u| \leqq h} |\frac{F'(u)}{|u|^4}| = \bar{c}\, h^{k-5}$.

Then

$$\|F'(u_s)\|_r \leqq \alpha(s^{2/(k-1)}\mu) \| u_s^4(\cdot,\tau)\|_r \leqq$$

$$\leqq \alpha(s^{2/(k-1)}\mu) \|u_s\|_r^{6/r} \|u_s(\cdot,\tau)\|_\infty^{(4r-6)/r} \leqq$$

$$\leqq \alpha(s^{2/(k-1)}\mu) \|\nabla u_s\|_2^{6/r} \| u_s(\cdot,\tau)\|_\infty^{4-(6/r)} \leqq$$

$$\leqq \alpha(s^{2/(k-1)}\mu) |\varepsilon_s(0)|^{3/r} \|u_s\|_\infty^{4-(6/r)}.$$

If we take q close to 1 then r is close to 2 but $r > 2$, so that

$$\frac{4r - 6}{r} > 1.$$

To estimate (1.4) we use a convex inequality. In (1.4) we have

$$\|u_s(\cdot,t)\|_\infty \leqq \rho \frac{s^{(3-k)/(k-1)}}{(t + (1/s))} + c_1 (s^{2/(k-1)}\mu)^{k-5} |\varepsilon_s(0)|^{(3/r)+(1/2)} \cdot$$

$$\cdot \int_0^t (t-\tau)^{2-(3/q)} \| u_s(\cdot,\tau)\|_\infty^{4-(6/r)} d\tau \tag{1.5}$$

and for $q = \frac{1}{(1 - (\varepsilon/4))}$, $\frac{1}{r} = \frac{1}{2} - \frac{\varepsilon}{4}$, (1.5) is now

$$\|u_s(\cdot,t)\|_\infty \leqq \rho \frac{s^{(3-k)/(k-1)}}{(t + (1/s))} + c_1\mu^{k-5} s^{3\varepsilon/4)(k-5/k-1)} \cdot \int_0^t (t-\tau)^{-1+(3\varepsilon/4)} \| u_s(\cdot,\tau)\|_\infty^{1+(3\varepsilon/2)} d\tau. \tag{1.6}$$

We put $\sigma = 1 - \frac{3\varepsilon}{4}$, $\gamma = 1 + \frac{3\varepsilon}{2}$, then

$$0 < \frac{1}{\gamma} < \sigma < 1$$

i.e. the hypothesis of Lemma 3. Then, if

$$M(t) = \sup_{0 \leqq \tau \leqq t} |(\frac{1}{s} + \tau)^\sigma \|u_s(\cdot,\tau)\|_\infty |$$

we obtain from Lemma 3

$$\|u_s(\cdot,t)\|_\infty \leqq \rho \frac{s^{(3-k)/(k-1)}}{((1/s) + t)} + c_1\mu^{k-5} s^{(3\varepsilon/4)(k-5/k-1)} \cdot c_{\gamma\sigma} (\frac{1}{s})^{1-\sigma\gamma} (\frac{1}{s} + t)^{-\sigma} [M(t)]^\gamma \tag{1.7}$$

and then

$$M(t) \leqq \rho \frac{s^{(3-k)/(k-1)}}{(t + \frac{1}{s})^{1-\sigma}} + c_2\mu^{k-5} s^{(3\varepsilon/4)(k-5/k-1)} (\frac{1}{s})^{1-\sigma\gamma} [M(t)]^\gamma \tag{1.8}$$

Now, if $t_0 > \frac{\log s}{s} = T_0$

$$M(t) \leqq \frac{s^{(3-k)/(k-1)} s^{1-\sigma}}{(1 + \log s)^{1-\sigma}} + c_2\mu^{k-5} s^{(3\varepsilon/4)(k-5/k-1)} (\frac{1}{s})^{1-\sigma\gamma} [M(t)]^\gamma$$

On the other hand, we have the following convex inequality:

<u>LEMMA 4</u>: Let $M(t)$ be a real function on $[0,T]$, such that

(i) $0 \leqq M(t) \leqq c_1 + c_2[M(t)]^\gamma$, $\gamma > 1$, $c_1, c_2 > 0$

(ii) $M(0) \leqq c_1$.

Then, if $c_1 c_2^{1/(\gamma-1)} \leq (1 - \frac{1}{\gamma}) \frac{1}{\gamma^{1/(\gamma-1)}}$, (*), we have

$$M(t) \leq \frac{c_1}{(1 - \frac{1}{\gamma})}$$

REMARK: The elementary idea in Lemma 4 is the convexity of the function

$$f(x) = c_2 x^\gamma + c_1 - x$$

With (*) f attains a negative minimum in x_0 (see Figure below). Then the point x_2 is an upper bound for x_1 which is the bound for M(t)

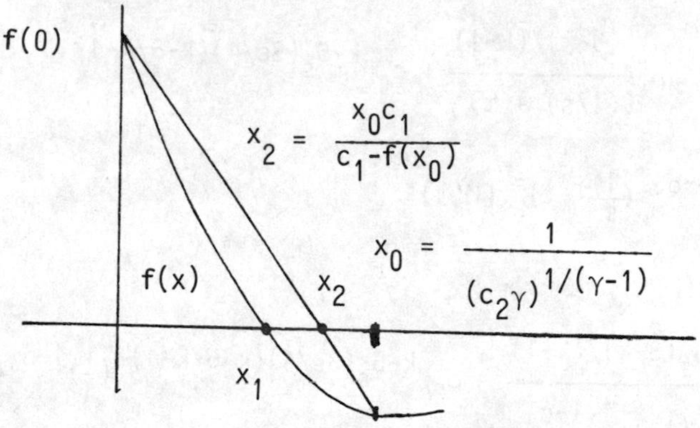

The direct application of Lemma 4 is possible in this case, under the hypothesis

$$\frac{\bar{c}(\mu)}{(\log s + 1)^{1-\sigma}} < (1 - \frac{1}{\gamma})(\frac{1}{\gamma})^{1/(\gamma-1)} = c(\varepsilon)$$

which holds for s sufficiently large. Then we have in this case

$$M(t) \leqq \rho \frac{s^{(3-k)/(k-1)} s^{1-\sigma}}{(\log s + 1)^{1-\sigma}} \cdot \frac{1}{1 - \frac{1}{\gamma}}$$

in other words

$$\|u(\cdot,t)\|_\infty \leqq \frac{c(s,\varepsilon)}{(t + \frac{1}{s})^\sigma}$$

The argument for ending is now standard.

The result in Theorem 1 may be reconsidered in the following way:

"For f, g $\in C_0^\infty(\mathbb{R}^3)$ in the (P_k)-problem there exists a $T_0(f,g) > 0$ such that, if a solution is defined and bounded in $\mathbb{R}^3 \times [0,T_0]$, it is then defined and bounded in $\mathbb{R}^3 \times [0,\infty)$".

For a dilation sufficiently large, s >> 1, the local result shows that the solution is defined and bounded in $\mathbb{R}^3 \times [0, \frac{1}{s \log \log s}]$ (for example). On the other hand, Theorem 1 is a regularity result in the hypothesis that the solution is bounded in $\mathbb{R}^3 \times [0, \frac{\log s}{s}]$.

2. Monotony method

The following argument is available in 1, 2, and 3 spatial dimensions and is based on the explicit form of fundamental solutions of wave equations in such dimensions.

Our exposition will be in dimension n = 3.

For the linear Cauchy problem

$$(P.L) = \begin{cases} u_{tt} - \Delta u = F(x,t) \\ u(x,0) = f(x) & x \in \mathbb{R}^3 \\ u_t(x,0) = g(x) \end{cases}$$

the explicit solution is

$$u(x,t) = \frac{1}{4\pi} \int_{|y|=1} g(x + ty)d\ (y) + \frac{\partial}{\partial t} (\frac{t}{4\pi} \int_{|y|=1} f(x+ty)d\sigma(y) +$$

$$+ \frac{1}{4\pi} \int_0^t \tau \left\{ \int_{|y|=1} F(x \pm \tau y, t-\tau)d\sigma(y)\right\} d\tau \qquad (2.1)$$

where $d\sigma(y)$ is the measure in S^2 and we assume sufficient regularity in the initial data.

The basic observation is that the fundamental solution is a positive measure in the light cone.

We will denote $v_0(x,t)$ the solution of (P,L) when F = 0, i.e. the solution of the homogeneous problem.

The last term will be denoted by

$$L_{(x,t)}(F)$$

hence, (2.1) can be written as

$$u(x,t) = v_0(x,t) + L_{(x,t)}(F). \qquad (2.2)$$

We have, obviously, two inequalities:

$$\text{(i)}\quad \|v_0(\cdot,t)\|_{L^\infty(\mathbf{R}^3)} \leq c(\|f\|_{W^{\infty,1}(\mathbf{R}^3)} + \|g\|_{L^\infty(\mathbf{R}^3)}\, t)$$

$$\text{(ii)}\quad \|L_{(,xt)}(F)\|_{L^\infty(\mathbf{R}^3)} \leq c \int_0^t (t-\tau)\, \|F(\cdot,\tau)\|_{L^\infty(\mathbf{R}^3)}\, d\tau$$

After this introduction, we shall consider the semilinear Cauchy Problem (P):

$$(P)\quad \begin{cases} u_{tt} - \Delta u = F(u) \\ u(x,0) = f(x) \\ u_t(x,0) = g(x) \qquad x \in \mathbf{R}^3 \end{cases}$$

If $F' \geqq 0$ and $F(u) \geqq cu^\alpha$, $\alpha > 1$ if $u > 0$, the results of F. John and T. Kato prove that there is a blow-up in finite time. Recently, L. Caffarelli and A. Friedman have proved that the blow-up takes place in a C^1 space-like surface, under natural regularity conditions.

Here, we will assume:

(1) $F \in C^1(\mathbf{R})$ and $F'(u) \leqq 0$

(2) $F(u) \leqq A \quad \forall u \in \mathbf{R}$

(3) $f \in W^{1,\infty}(\mathbf{R}^3)$, $g \in L^\infty(\mathbf{R}^3)$

<u>DEFINITION</u>: u is a weak solution of (P), if and only if, $u \in L^1_{loc}(\mathbf{R}^4)$, $F(u) \in L^1_{loc}(\mathbf{R}^4)$ and

$$u(x,t) = v_0(x,t) + L_{(x,t)}F(u)$$

(in the sense of (2.2)).

<u>THEOREM 2</u>: In hypotheses (1), (2) and (3) the Cauchy Problem (P) has a unique weak solution such that for each $T > 0$

$$u \in C([0,T], \; L^\infty(\mathbb{R}^3))$$

<u>REMARKS</u>: (1) Hypothesis (1) may be reduced to F non-increasing and locally Lipschitzian

(2) Hypothesis (2) may alternatively be substituted by

(2') $B \leq F(u) \quad \forall u \in \mathbb{R}$.

<u>PROOF</u>: Consider $u_0(x,t)$ to be the solution of the linear Cauchy Problem

$$(P_0) \equiv \begin{cases} v_{tt} - \Delta v = A \\ v(x,0) = f(x) \\ v_t(x,0) = g(x) \end{cases}$$

We have

$$u_0(x,t) = v_0(x,t) + L_{(x,t)}(A),$$

where v_0 is the solution of the homogeneous linear problem.

From (2), $F(u_0(x,t)) \leq A$. We obtain the problem

$$(P_1) = \begin{cases} v_{tt} - \Delta v = F(u_0(x,t)) \\ v(x,0) = f(x) \\ v_t(x,0) = g(x) \end{cases}$$

of which solution u_1, can be written as

$$u_1(x,t) = v_0(x,t) + L_{(x,t)}(F(u_0)).$$

But then

$$u_1(x,t) \leq u_0(x,t),$$

because $L_{(x,t)}$ is an operator with positive kernel and

$$F(u_0(x,t)) \leq A \text{ in all } (x,t) \in \mathbb{R}^3 \times \mathbb{R}.$$

Likewise, the solution u_2 of the Cauchy Problem

$$(P_2) = \begin{cases} v_{tt} - \Delta v = F(u_1(x,t)) \\ v(x,0) = f(x) \\ v_t(x,0) = g(x) \end{cases}$$

is $u_2(x,t) = v_0(x,t) + L_{(x,t)}(F(u_1))$

Hence

$$u_0(x,t) = v_0(x,t) + L_{(x,t)}(A) \geq u_2(x,t) \geq$$

$$\geq v_0(x,t) + L_{(x,t)}F(u_0)) = u_1(x,t)$$

and, therefore

$$F(u_0) \leq F(u_2) \leq F(u_1) \leq A.$$

In the same way, it can be estimated that $u_3(x,t)$ the solution of

$$(P_3) \equiv \begin{cases} v_{tt} - \Delta v = F(u_2) \\ v(x,0) = f(x) \\ v_t(x,0) = g(x) \end{cases}$$

verifies

$$u_1 = v_0 + L_{(x,t)}F(u_0) \leq u_3 \leq v_0 + LF(u_1) = u_2.$$

Therefore, we have

$$u_1 \leqq u_3 \leqq u_2 \leqq u_0.$$

By induction process, we obtain

$$u_1 \leqq u_3 \leqq \dots \leqq u_{2k+1} \leqq u_{2k} \leqq \dots \leqq u_2 \leqq u_0$$

where u_m is the solution of

$$(P_m) = \begin{cases} v_{tt} - \Delta v = F(u_{m-1}) \\ v(x,0) = f(x) \\ v_t(x,0) = g(x). \end{cases}$$

After this iteration method, we fix $T > 0$ and let

$$X_T = \{\phi \in C([0,T],L^\infty(\mathbb{R}^3)) : u_1(x,t) \leqq \phi(x,t) \leqq u_0(x,t),$$
$$(x,t) \in \mathbb{R}^3 \quad [0,T]\}$$

with the norm

$$|||\phi|||_{X_T,K} = \sup_{t\in[0,T]} e^{-Kt} \|\phi(t)\|_{L^\infty(\mathbb{R}^3)}$$

where $k > 0$ will be chosen in a convenient way.

Let us consider the operator on X_T

$$\mathcal{F}(\phi) = v_0(x,t) + L_{(x,t)}(F(\phi))$$

then we have the following properties:

(i) $u_1 \leqq \mathcal{F}(\phi) \leqq u_0$ if $\phi \in X_T$

(ii) $\|\mathcal{F}(\phi)(t) - \mathcal{F}(\psi)(t)\|_{L^\infty(\mathbb{R}^3)} \leqq$

$$\leqq HT \int_0^t \|\phi(\tau) - \psi(\tau)\|_{L^\infty(\mathbb{R}^3)} d\tau$$

where

$$H = \max\{|F'(s)| : s \in [-(\|u_1\| + \|u_0\|), (\|u_1\| + \|u_0\|)]\}$$

If we take $K > HT$ and the corresponding norm $|||\cdot|||_{X_T,K}$, the result is a consequence of Picard's fixed point theorem.

EXAMPLES:

(1) For $N \leq 0$ let

$$F_N(u) = \begin{cases} N^k, & u \leq N \\ -|u|u^{k-1}, & u \geq N \end{cases}$$

(2) For $F(u) = -(e^u-1)$, we also have the results. Here, the energy method is not available for $n = 3$.

Perhaps the principal interest of the theorem lies in the following consequences:

Let us assume a function $\tilde{F}(u)$ such that:

(1) $\tilde{F} \in C^1(\mathbf{R})$ and $\tilde{F}'(u) \leq 0 \quad \forall u \in \mathbf{R}$.

Consider the Cauchy Problem

$$(P) \equiv \begin{cases} \Box u = \tilde{F}(u) \\ u(x,0) = f(x) \in W^{1,\infty}(\mathbf{R}^3) \\ u_t(x,0) = g(x) \in L^\infty(\mathbf{R}^3) \end{cases}$$

COROLLARY: If (P) has a weak solution $u(x,t)$ verifying

$$C \leq u(x,t) \quad \forall (x,t) \in \mathbf{R}^3 \times [0,T]$$

then $u \in C([0,T], L^\infty(\mathbf{R}^3))$ and u is the unique weak solution of (P).

(In the same way we have the results for the condition $u(x,t) \leq C$).

The typical example is that of Section 1, i.e.

$$\tilde{F}(u) = -|u|u^{k-1}.$$

According to the corollary, with a unilateral condition there is uniqueness for the Cauchy Problem.

For the same type of reason, the theorem can read as follows:

"If u is a weak solution of (P) such that

$$u(x,t) \underset{t \to T}{\to} +\infty$$

then there exists $y \in \mathbb{R}^3$ such that

$$u(y,t) \underset{t \to T}{\to} -\infty$$

Finally, if the initial data are sufficiently smooth, then the asymptotic decay in L^∞ can be obtained for $t \to \infty$ because we have

$$u_1 \leqq u \leqq u_0$$

3. <u>On self-similar solutions in the case $F(u) = -|u|u^{k-1}$</u>

Let us consider the equation

$$\Box u = -|u|u^{k-1} \text{ in } \mathbb{R}^n \times \mathbb{R}.$$

We look for solutions $u(t,r) \equiv u(t,|x|)$. For this, the equation is

$$u_{tt} - u_{rr} - \frac{(n-1)}{r} u_r = -|u|u^{k-1} \quad r > 0 \ t \in \mathbb{R}.$$

We take $u = f(t^2 - r^2)$ and then, for $\xi = t^2 - r^2$ we have

$$4\xi f'' + (4 + 2(n-1))f' = -|f|^{k-1}f.$$

We can immediately find a solution of the form $f(\xi) = \dfrac{A}{\xi^\alpha}$ $\xi > 0$. The result for $n > 1$ and $k > \dfrac{n+1}{n-1}$ is:

$$f(\xi) = \frac{A_{nk}}{\xi^{1/k-1}} \quad \xi > 0.$$

Then, for the P.D.E. we have

$$E_{n,k}(x,t) = \begin{cases} \pm \dfrac{A_{n,k}}{(t^2-|x|^2)^{1/k-1}} & |x| < |t| \\ 0 & |t| > |x| \end{cases}$$

When fixing $x \in \mathbb{R}^n$ we have

$$|E_{n,t}(,xt)| \simeq \frac{A(x)}{t^{2/k-1}}$$

and if $1 \leqq p < k-1$

$$\|E_{n,t}(\cdot,t)\|_p \leqq C^p_{k,n} \, |t|^{(n/p)-(2/k-1)}$$

The type of behaviour on t for large t is critical in the following sense. We can formulate the results in $\mathbb{R}^3 \times \mathbb{R}$.

THEOREM: Let $u \in L^\infty(\mathbb{R}, H^1 \cap L^{k+1}(\mathbb{R}^3))$ be the solution of

$$\begin{cases} \Box u = -|u|u^{k-1} \\ u(x,0) = f(x) \\ u_t(x,0) = g(x) \qquad f,g \in C_0^\infty(\mathbb{R}^3) \text{ and } k > 5. \end{cases}$$

If $\|u(\cdot,t)\|_\infty \leqq \dfrac{c}{(1+|t|)^{2/(k-1)}}$ then $\forall\, \alpha \in (\frac{2}{k-1}, 1)$ there exists a constant $C_\alpha > 0$ such that

$$\|u(\cdot,t)\|_\infty \leqq \frac{C_\alpha}{(1+|t|)^\alpha}$$

PROOF: We use the same method as in Section 1, i.e.

$$\|u(\cdot,t)\|_\infty \leqq \frac{c}{1+|t|} + c \, \|\nabla u\|_2 \int_0^t (t-\tau)^{2-(3/q)} \|F'(u)\|_{2q/(2-q)} \, d\tau \tag{3.1}$$

$$1 < q < 2$$

We have sop $\{u\} \subset \{|x| \leqq R + |t|\}$ if (sop f) U (sop g) $\subset B(0,R)$ and, then

$$\|F'(u)\|_{2q/(2-q)} \leq \frac{H}{(1+\tau)^{2-3((2-q)/2q)}} .$$

In this way

$$\|u(\cdot,t)\|_\infty \leq \frac{c_1}{1+|t|} + c \quad \|\nabla u\|_2 \int_0^t (t-\tau)^{2-3/1} \frac{d\tau}{(1+\tau)^{2-3((2-g)/2g}}$$

$$1 < q < 2$$

If we take $\sigma = \frac{3}{q} - 2$, $\gamma\sigma = 2-3(\frac{2-q}{2q}) = \frac{3}{2} - \sigma$. If $\sigma > \frac{1}{2}$ we have $0 < \frac{1}{\gamma} < \sigma < 1$ and by Lemma 3 in Section 1 that implies

$$\|u(\cdot,t)\|_\infty \leq \frac{c_\alpha}{(1+|t|)^\alpha} \quad \text{for } \frac{2}{k-1} < \alpha < \frac{1}{2} .$$

For $k \geqq 6$ a new iteration in (3.1) with the obtained behaviour gives the results. For $k > 5$ a finite number of iterations is necessary.

References

[1968] Strauss, W. Jou. Func. Analysis Vol. 2, no. 4, p. 409-457.

[1977] Strauss, W. "Invariant Wave Equations". Springer Verlag, 1978, Lecture Notes in Physics No. 73.

[1961] Jörgens, G. Math. Z. 77 (1961) p. 295-307.

[1983] Ginibre-Velo. "Seminaire Gouloauic-Meyer-Schwartz 1983-84". Expose No. 2. Ecole Polytechnique à Paris.

[1979] John F. Manuscripta Math. 28 (1979), p. 235-268.

[1980] Kato, T. Comm. Pure Appl. Math. 33, 1980, p. 501-505.

[1984] Caffarelli-Friedman. Preprint.

This work was supported by: Comité Conjunto Hispano-Norteamericano, Proyecto No. CCB 8402/058. Comisión Asesora de Investigación Cientifica y Técnica, Proyecto No. 2805/83.

I. Peral Alonso
División de Matématicas
Facultad de Ciencias
Universidad Autónoma
28049 Madrid
Spain.

B PERTHAME

Recent results on the quasi–variational inequality of the impulsive control

Abstract

We study the quasi-variational inequality (QVI) associated with the Hamilton-Jacobi-Bellman equation

$$\text{Max}\,(\max_{1 \leq i \leq m} (A^i u - f^i),\; u - Mu) = 0 \text{ in } \Omega, \tag{1}$$

$$u|_{\partial\Omega} = \phi, \tag{2}$$

$$Mu(x) = k + \inf_{\xi \geq 0,\; x+\xi \in \bar{\Omega}} \{c_o(\xi) + u(x+\xi)\}. \tag{3}$$

In general, this Dirichlet problem has no solution and the boundary condition must be relaxed in

$$u|_{\partial\Omega} = \inf\,(\phi, Mu) \tag{2'}$$

Then the system (1), (2') has a unique solution in $C(\bar{\Omega})$. When (2) holds, we can prove that, under technical assumptions, $u \in W^{2,\infty}(\Omega)$. Finally we solve the ergodic problem for the Q.V.I. (m = 1 in (1)) with Neumann boundary condition.

Introduction

We present some recent results on the fully nonlinear equation

$$\text{Max}\,(\max_{1 \leq i \leq m} (A^i u - f^i),\; u - Mu) = 0 \text{ in } \Omega, \tag{1}$$

where A^i are uniformly elliptic second order operators and

$$Mu(x) = k + \inf_{\substack{\xi \geq 0 \\ x+\xi \in \bar{\Omega}}} \{c_o(\xi) + u(x+\xi)\}. \tag{2}$$

Here Ω is a smooth bounded domain of $\mathbb{R}^N$ (non necessarily convex), k is a positive constant and $\xi \geqq 0$ means that all the coordinates of the vector are non-negative.

We are interested in two kinds of questions. For the first one we consider (1) with a Dirichlet boundary condition. In general this problem has no solution and the boundary condition must be relaxed in

$$u|_{\partial\Omega} = \inf(\phi, Mu), \tag{3}$$

where ϕ is the Dirichlet data. In this case we introduce an assumption which ensures that the equation (1), (3) admits a unique weak solution (in a sense detailed in Section I). The difficulty is that M does not map $C(\bar{\Omega})$ into itself. Then we give some strong regularity results in $W^{1,\infty}(\Omega)$ and $W^{2,\infty}(\Omega)$. The second question we treat, is the general ergodic problem for the Quasi-Variational Inequality (Q.V.I. in short) with Neumann condition:

$$\left.\begin{aligned} &\text{Max}\,(-\Delta u_\alpha + \alpha u_\alpha - f,\; u_\alpha - Mu_\alpha) = 0 \text{ in } \Omega, \\ &\frac{\partial u_\alpha}{\partial n}\Big|_{\partial\Omega} = 0. \end{aligned}\right\} \tag{4}$$

Our main result shows that, denoting by $\langle\phi\rangle = \text{meas}\,(\Omega)^{-1}\int_\Omega \phi\,dx$, αu_α converges uniformly to a constant λ, $u_\alpha - \langle u_\alpha\rangle$ converges uniformly and in $H^1(\Omega)$ to a function u_0, solution of

$$\left.\begin{aligned} &\text{Max}\,(-\Delta u_0 + \lambda - f,\; u_0 - Mu_0) = 0 \text{ in } \Omega, \\ &\frac{\partial u_0}{\partial n}\Big|_{\partial\Omega} = 0. \end{aligned}\right\}$$

The solution of equation (1) or (4) represents the optimal cost function of a diffusion process controlled both impulsively and continuously ([19, 22]). We will not study the relation between (1) and control theory. Let us only point out that the impulsive control leads, via the dynamic programming principle, to the Q.V.I. (1) with m = 1). It has been introduced and studied by A. Bensoussan and J.L. Lions (see [3,4]) and regularity results are due to [5, 9, 10, 12]. Our results complete and generalize them. On the other hand, the Hamilton-Jacobi-Bellman (H.J.B. in short) equation has been studied by

[8, 15, 16, 17] where it is proved that they have a unique solution in $W^{2,\infty}(\Omega)$. Finally the theory of ergodic stochastic control is also a standard matter (see [2, 6, 11, 14, 24, 25]) but the specificity of the implicit obstacle Mu requires a particular treatment.

This paper is organized as follows. In the first section we present the assumptions which ensure the existence of weak solutions of (1) with boundary condition (3). In the second section we go back to the Dirichlet boundary condition and we give some stronger regularity results ($W^{2,\infty}(\Omega)$ up to the boundary), these results are, in some sense, optimal. Finally in Section III we treat the theory of ergodic impulsive control.

I. The Dirichlet problem

1. Setting the problem

In this section we consider the equation (1), (3) where:

$$A^i = - a^i_{jk} \frac{\partial^2}{\partial x_j \, \partial x_k} + b^i_j \frac{\partial}{\partial x_j} + c^i ,$$

with

$$a^i_{jk}(x),\ b^i_j(x),\ c^i(x),\ f^i(x) \in C^{2,\alpha}(\bar{\Omega}) \text{ for some } \alpha,\ \ 0 < \alpha < 1, \tag{5}$$

$$\exists \nu,\ \nu > 0,\ c^i(x) \geq \nu,\ a^i_{jk}\xi_j\xi_k \geq \nu|\xi|^2,\ \ \forall \xi=(\xi_1,\dots,\xi_N) \in \mathbb{R}^N, \tag{6}$$

$$\phi \in C(\bar{\Omega}), \tag{7}$$

and the implicit obstacle Mu is defined by (2) with

$$c_o \in C(\mathbb{R}^N), c_o \geq 0,\ c_o(0) = 0,\ c_o(\xi_1+\xi_2) \leq c_o(\xi_1) + c_o(\xi_2) \qquad \forall \xi_1,\xi_2 \geq 0. \tag{8}$$

We will look for a solution of (1), (3) which belongs to the space $C(\bar{\Omega})$ for three reasons. The first one is that there is no variational method associated with the H.J.B. part of (1) and the boundary data (3) is not in $H^{1/2}(\partial\Omega)$. Therefore we cannot look after a solution in $H^1(\Omega)$ as in [3, 4].

The second reason is that, as we will see later, the solution of (1), (3) is not regular in general and we need strong assumptions to obtain the $W^{2,\infty}$ regularity. The last reason is that the space $C(\bar{\Omega})$ is well adapted to make a link between the optimal cost function of the controlled diffusion and the solution of (1), (3). We refer the interested reader to [3, 4, 17, 19] for this subject.

Thus we begin by explaining what we mean by a continuous solution of equation (1). It is enough to deal with the obstacle problem

$$\left.\begin{aligned} &\operatorname{Max}\{ \operatorname*{Max}_{1 \leqq i \leqq m} (A^i v - f^i),\ v - \psi\} = 0 \text{ in } \Omega, \\ &v|_{\partial\Omega} = \phi, \end{aligned}\right\} \tag{9}$$

for the functions ϕ,ψ which satisfy,

$$\phi \in C(\bar{\Omega}), \tag{10}$$

$$\psi \in C(\bar{\Omega}), \quad \psi \geqq \phi \quad \text{on } \partial\Omega. \tag{11}$$

Since we have made the strong assumptions (5), (6) we can use the following definition of continuous solution of (9).

Definition. A function $v \in C(\bar{\Omega})$ is a solution of (9) if $v|_{\partial\Omega} = \phi$, $v \leqq \psi$ in $\bar{\Omega}$ and

$$A^i v \leqq f^i \text{ in } \mathcal{D}'(\Omega), \ \forall i,\ 1 \leqq i \leqq m,$$

and, on the open set $\mathcal{O} = \{v < \psi\}$, $v \in W^{2,\infty}_{loc}(\Omega)$ and

$$\operatorname*{Max}_{1 \leqq i \leqq m} (A^i v - f^i) = 0 \text{ a.e. in } \mathcal{O}.$$

As a matter of fact, the solution of (9) with this definition coincides with the so-called "viscosity solution" for degenerated H.J.B. equations. Let us recall that this notion was introduced, for first order equations, in [7] and was extended to the second order in [17]. The main properties which motivate the introduction of the viscosity solutions are the stability and the uniqueness of these solutions and the direct link with the control problem. Thus we will refer to the solutions of (9) in the sense of the above definition as viscosity solutions.

2. Existence results

Let us begin with a result on the obstacle problem (9). We have

THEOREM 1: Under assumptions (5), (6), (10), (11), the equation (9) has a unique viscosity solution.

For a proof of this theorem we refer to [15, 17, 20].

Now to treat the equation (1), (3) we need to be sure that the implicit obstacle is continuous, but we know that even if $u \in C^{\infty}(\bar{\Omega})$, Mu need not be continuous. Thus we introduce as in [20] an assumption which ensures the continuity of Mu. Namely, let $u_0 \in W^{2,\infty}(\Omega)$ be the solution (cf. [8, 16] for the existence) of

$$\left.\begin{aligned} &\operatorname*{Max}_{1 \leq i \leq m} (A^i u_0 - f^i) = 0 \text{ in } \Omega, \\ &u_0|_{\partial\Omega} = \phi. \end{aligned}\right\} \qquad (12)$$

We will assume that

$$Mu_0 \in C(\bar{\Omega}). \qquad (13)$$

Then we have

LEMMA 1 [20]: Under assumption (13), let $u \in C(\bar{\Omega})$ satisfy $u|_{\partial\Omega} = \phi$ and $u \leq u_0$, then $Mu \in C(\bar{\Omega})$.

This continuity result allows us to prove

THEOREM 2: Under assumptions (5) - (8), (13) the equation (1), (3) has a unique viscosity solution.

The proof of this theorem is detailed in [20, 22]. It uses a generalization of the decreasing process introduced by Bensoussan-Lions (see [3, 4]) and which is written

$$\left.\begin{aligned} &\mathrm{Max}\,(\max_{1\leq i\leq m}(A^i u_n - f^i),\ u_n - Mu_{n-1}) = 0,\\ &u_n|_{\partial\Omega} = \inf(\phi, Mu_{n-1}). \end{aligned}\right\} \tag{14}$$

The existence of solutions of (14) is proved in two steps. First we can prove, as in Lemma 1, that Mu_{n-1} is continuous, then we may apply Theorem 1. Finally, an argument due to Hanouzet-Joly [12] proves that the sequence u_n converges uniformly and one can pass to the limit using the stability of the viscosity solutions as in Crandall-Lions [7].

REMARK: The assumption (13) may be automatically satisfied if Ω has some geometric property. For example if Ω is convex Mu_0 always belongs to $C(\bar{\Omega})$. This also holds under the weakest assumption on Ω:

$$\forall x \in \bar{\Omega},\quad \{\xi/\xi \geqq 0,\ \xi \neq 0,\ x+\xi \in \partial\Omega,\ \exists\ \varepsilon > 0,\ \forall\ y \geqq 0,\quad x+y \notin \Omega \ \text{ if } |y-\xi| \leqq \varepsilon\} = \phi \tag{15}$$

II. Regularity results

First let us come back to boundary condition (3). A natural question is to ask if the true Dirichlet boundary condition holds, i.e. if

$$u|_{\partial\Omega} = \phi. \tag{3'}$$

This is wrong in general and a sufficient condition so that (3') holds, is

$$\begin{aligned} &\exists \underline{u} \in W^{2,\infty}_{loc}(\Omega) \cap C(\bar{\Omega})\\ &A^i\underline{u} \leqq f^i \ \text{ a.e. in } \Omega,\ \forall i \text{ and } \underline{u} \leqq M\underline{u} \text{ in } \Omega,\\ &\underline{u} \leqq \phi \leqq M\underline{u} \text{ on } \partial\Omega. \end{aligned} \tag{16}$$

Under this condition we can prove some regularity results.

THEOREM 3 [20, 23]: Under assumptions (5) - (8), (13), (16), if $c_0 \in W^{2,\infty}(\mathbb{R}^N)$, $\phi \in W^{3,\infty}(\Omega)$, and if $Mu_0 \in W^{1,\infty}(\Omega)$ (resp. $W^{1,\infty}_{loc}(\Omega)$), then the solution u of (1),

(3') belongs to $W^{1,\infty}(\Omega)$ (resp. $W^{1,\infty}_{loc}(\Omega)$). If $Mu_0 \in D^{2,+}(\Omega)$ (resp. $D^{2,+}_{loc}(\Omega)$) then $u \in W^{2,\infty}(\Omega)$ (resp. $W^{2,\infty}_{loc}(\Omega)$).

Here $D^{2,+}$ denotes the set of semi-concave functions:

$$D^{2,+}(\Omega) = \{u \in C(\bar{\Omega}), \frac{\partial^2 u}{\partial \chi^2} \leq C \text{ in } \mathcal{D}'(\Omega), \forall \chi, |\chi| = 1\}.$$

This theorem relies on a method initiated in [5] and on an extension of Lemma 1 which shows that $Mu \in W^{1,\infty}$ (resp. $W^{1,\infty}_{loc}$, $D^{2,+}$, $D^{2,+}_{loc}$) if $Mu_0 \in W^{1,\infty}$ (resp. $W^{1,\infty}_{loc}$, $D^{2,+}$, $D^{2,+}_{loc}$). The $W^{2,\infty}$ esimates inside Ω are deduced from the method of [16], cf. [23]. The $W^{2,\infty}$ estimates on the boundary use a technique introduced by Jensen [13] and the argument of [8]. Let us also point out that the assumptions of Theorem 3 are almost necessary. Some counter examples prove that the solution could even not belong to $W^{2,1}_{loc}$.

REMARKS: (1) Of course the difficulties in this problem come from the boundary. It is proved that in the whole space, (1) always has a unique solution in $W^{2,\infty}(\mathbb{R}^N)$ (cf. [21]).

(2) Another point of view on the boundary condition (3) or (3') can be found in [1].

III. Ergodic control

We consider now a simpler version of (1) with a Neumann boundary condition:

$$\left.\begin{aligned} &\text{Max } (-\Delta u_\alpha + \alpha u_\alpha - f, u_\alpha - Mu_\alpha) = 0 \text{ in } \Omega, \\ &\frac{\partial u_\alpha}{\partial n}\Big|_{\partial\Omega} = 0. \end{aligned}\right\} \quad (17)$$

To deal with this problem we use the variational formulation of (17)

$$\left.\begin{aligned} &\int_\Omega \nabla u_\alpha \cdot \nabla(v-u_\alpha)\, dx + \alpha \int_\Omega u_\alpha (v-u_\alpha)\, dx \geq \int_\Omega f(v-u_\alpha)\, dx, \\ &\forall v \in H^1(\Omega),\ v \leq Mu_\alpha \text{ a.e. } ;\ u_\alpha \in H^1(\Omega),\ u_\alpha \leq Mu_\alpha. \end{aligned}\right\} \quad (18)$$

We can deduce from [4, 12] that, under assumption (15), (18) has a unique solutions $u_\alpha \in H^1(\Omega) \cap C(\bar{\Omega})$ when f satisfies

$$f \in L^p(\Omega) \text{ for some } p > \frac{N}{2} \text{ if } N \geqq 1,\ p \geqq 1 \text{ if } N = 1. \tag{19}$$

In the following we are interested in the limit of u_α as α tends to zero (the interpretation of this limit is given in [18]). Let us denote by $\langle\phi\rangle = \text{meas}\,(\Omega)^{-1} \int_\Omega \phi(x)dx$. Then we have

THEOREM 4 [18]: We assume (15), (19). The αu_α converges uniformly on $\bar{\Omega}$ to some constant λ, $u_\alpha - \langle u_\alpha\rangle$ converges uniformly and in $H^1(\Omega)$ to v_0. And (λ, v_0) is the unique solution of

$$\left.\begin{aligned} &\int_\Omega \nabla v_0 \cdot \nabla(v - v_0)\,dx \geqq \int_\Omega (f-\lambda)(v-v_0)\,dx, \\ &\forall v \in H^1(\Omega),\ v \leqq Mv_0 \text{ a.e.},\ \lambda \in \mathbb{R},\ v_0 \in H^1(\Omega),\ v_0 \leqq Mv_0 \text{ a.e.} \end{aligned}\right\} \tag{20}$$

The proof of this Theorem is given in [18] and relies on various estimates. First we must prove that αu_α is bounded in $L^\infty(\Omega)$ (this is not a consequence of the maximum principle since we do not assume that $f \in L^\infty(\Omega)$). Then we prove that $u_\alpha - \langle u_\alpha\rangle$ is bounded in $H^1(\Omega) \cap L^\infty(\Omega)$ (this is the key estimate of the proof) and finally we can deduce from these estimates and Hanouzet-Joly's method that u_α is compact in $C(\bar{\Omega})$. In [18] we give also some extensions to more general second order operators than $-\Delta$.

References

1. L. Barthelemy and Ph. Benilan, Phénomène de couche limite pour la pénalisation d'Inéquations Quasi-Variationnelles elliptiques avec conditions de Dirichlet au bord. To appear.
2. A. Bensoussan and J.L. Lions, On the asumptotic behaviour of the solution of varitional inequalities. In Theory of nonlinear operators, Akademic Verlag, Berlin, 1978.
3. A. Bensoussan and J.L. Lions, Applications des Inéquations Variationnelles en Contrôle stochastique. Dunod, Paris, 1978.
4. A. Bensoussan and J.L. Lions, Contrôle impulsionnel et Inéquations Quasi-Variationnelles. Dunod, Paris, 1982.
5. L.A. Caffarelli and A. Friedman, Regularity of the solutions of the quasi-variational inequality for the impulse control. Comm. in P.D.E., 3 (8), (1978) pp. 745-753. II. Comm. in P.D.E., 4 (3), (1979), pp.279-291.

6. I. Capuzzo-Dolcetta and M.G. Garoni, Comportement asymptotique de la solution des problèmes non sous forme divergence avec condition de derivées obliques au bord. Comptes-Rendus de l'Acad. Sci., Paris, 1984.
7. M.G. Crandall and P.L. Lions, Viscosity solutions of Hamilton-Jacobi equations. Trans. Amer. Math. Soc., 277 (1983), pp. 1-42.
8. L.C. Evans and P.L. Lions, Résolution des équations de H.J.B. pour les équations uniformément elliptiques. Comptes-Rendus de l'Acad. Sci., Paris, 290 (1980), pp. 1049-1052.
9. J. Fhrese and U. Mosco, Irregular obstacles and Q.V.I. of stochastic impulse control. Annali di Pisa, t. 4, 91 (1982), pp. 105-157.
10. M.G. Garoni and G.M. Troianiello, Some regularity results and a priori estimates for solution of variational and quasi-variational inequalities, Proc. Conf. Recent method in nonlinear analysis and applications. Roma 1978. Ed. E.D. Giorgi, D. Mangenes, U. Mosco, Pitagora Ed. Bologna, 1979.
11. F. Gimbert, Problèmes de Neumann quasi-linéaires ergodiques. In Thèse de 3ème Cycle, Univ. Paris IX-Dauphine, 1984. J. of Funct. Anal.
12. B. Hanouzet and J.L. Joly, Convergence uniforme des itérés définissant la solution d'une I.Q.V.. Comptes-Rendus Acad. Sci., Paris, 286 (1978), pp. 735-738.
13. R. Jensen, Boundary Regularity for variational Inequalities. Indiana Univ. Math. J, 29 (1980) pp. 495-504.
14. J.M. Lasry, Contrôle stochastique ergodique. Thèse d'Etat. Univ. Paris IX-Dauphine, 1974.
15. S. Lehnart, Bellman equations for optimal stopping time problems, Indiana Univ. Math. J. 32 (3) (1983), pp. 363-375.
16. P.L. Lions, Résolution analytique des problèmes de Bellman-Dirichlet. Acta Mathematica, 146 (1981), pp. 151-166.
17. P.L. Lions, Optimal control and H.J.B. equations. Part 1, 2, Comm. in P.D.E. 8 (1983), pp. 1101-1174, pp. 1229-1276. Part 3, in "Nonlinear Partial Differential Equations and Applications", College de France Seminar, Vol. V, Pitman, London, 1983.
18. P.L. Lions et B. Perthame, Quasi-Variational Inequalities and ergodic impulse Control. SIAM J. on Control and Optimization, Vol. 24, No. 4, (1986).
19. B. Perthame, Continuous and impulsive control of diffusion process in $\mathbb{R}^N$. Nonlinear Analysis T.M.A. 8 (10), 1984, pp. 1277-1239.

20. B. Perthame, Q.V.I. and H.J.B. equation in a bounded region, Comm. in P.D.E. 9 (6) (1984), pp. 561-595.

21. B. Perthame, I.Q.V. et équation de H.J.B. dans $\mathbb{R}^N$, Ann. Toulouse 5 (1983), pp. 237-257.

22. B. Perthame, Some remarks on Q.V.I. and the associated impulsive control problem. Ann. Inst. H. Poincaré, Vol. 2, no. 3, (1984), pp. 237-260.

23. B. Perthame, On the regularity of the solutions of Q.V.I.. J. of Funct. Anal. 64 (2) (1985), pp. 190-208.

24. M. Robin, On some impulse control problems with long term average cost. SIAM J. Control Optim. 19 (1981) pp. 333-358.

25. M. Robin, Long term average cost control problems for continuous time Markov processes : a survey, Acta Appl. Math., to appear.

B. Perthame
CEREMADE
Université Paris-Dauphine
Place de Lattre de Tassigny
75775 Paris CEDEX 16
France.

Ecole Normale Superieure
Centre de Mathématiques Appliquées
45, rue D'ulm
75230 Paris - CEDEX 05
France.

M PIERRE

An L^1– method to prove global existence in some reaction–diffusion systems

I. Introduction

We consider here a class of reaction-diffusion systems for which global existence cannot be established by classical methods. A typical example is the following which has been suggested by R.H. Martin.

$$\left.\begin{aligned} &\frac{\partial u}{\partial t} - a\Delta u = - u\, f(v) \text{ on } (0,\infty) \times \Omega \\ &\frac{\partial v}{\partial t} - b\Delta v = u\, f(v) \text{ on } (0,\infty) \times \Omega \\ &\frac{\partial u}{\partial n} = \frac{\partial v}{\partial n} = 0 \qquad \text{on } (0,\infty) \times \partial\Omega \\ &u(0,\cdot) = u_0 \geqq 0, \quad v(0,\cdot) = v_0 \geqq 0 \end{aligned}\right\} \qquad (1)$$

where Ω is a bounded open set in $\mathbb{R}^N$ with a regular boundary, $a,b > 0$ and $f : [0,\infty[\to [0,\infty[$ a regular function with $f(0) = 0$.

When $u_0, v_0 \in L^\infty(\Omega)$, local existence of a bounded solution (u,v) for (1) is classical. Except in the case $a = b$, global existence is not so obvious. It has been established by Masuda in [5] under the assumption that $f(v)$ does not grow faster than a power of v. A new device based on L^p-regularity theory for the heat operator and a duality principle has been introduced by Hollis-Martin-Pierre [3]; it allows us to generalize Masuda's result to a quite general class of systems.

However both techniques rely strongly on the assumption of polynomial growth of $f(v)$.

Here we prove, in particular, global existence of a solution to (1) without restriction on f; moreover, we can treat the case when u_0, v_0 are only in $L^1(\Omega)$ and L^1-forcing terms can also be added. The technique is completely different and based on L^1-estimates coupled with estimates in some Orlicz-spaces. This method carries over to a general class of systems of two equations described below. Among them is the following version of the so-called "Bruxellator":

$$\left.\begin{aligned} &\frac{\partial u}{\partial t} - a\Delta u = - uv^2 + av \\ &\frac{\partial v}{\partial t} - b\Delta v = uv^2 - (a+1)v + h \\ &\frac{\partial u}{\partial n} = \frac{\partial v}{\partial n} = 0 \\ &u(0,\cdot) = u_0,\ \ v(0,\cdot) = v_0 \end{aligned}\right\} \tag{2}$$

where $u_0, v_0 \in L^1(\Omega)$ and $h \in L^1((0,T) \times \Omega)$. This system has been studied in several places in the literature with mostly u_0, v_0 h bounded except in [6]. Of interest for applications is the case when h is itself the solution of

$$\left.\begin{aligned} &\frac{\partial h}{\partial t} - c\Delta h = 0,\ \ \frac{\partial h}{\partial n} = 0 \\ &h(0,\cdot) = \delta \text{ Dirac mass in } \Omega. \end{aligned}\right\}$$

Thus, h has a singularity at $t = 0$.

II. The main result

The ideas used in the result below follow from discussions with P. Baras and Ph. Benilan for one part and from the paper by Hollis-Martin-Pierre [3] for the other part.

ASSUMPTION: Let $F, G : \mathbb{R}^2 \mapsto \mathbb{R}$ be locally Lipschitz continuous with

$$F(0,v) \geq 0,\ \ G(u,0) \geq 0 \quad \forall u,v \in [0,\infty[\tag{3}$$

$$\exists K > 0,\ \ F^+(u,v) \leq K(u+v+1) \quad \forall u,v \in [0,\infty[\tag{4}$$

where $F^+ = \max(0,F)$.

$$\left.\begin{array}{l}\exists K > 0, \ \exists\ \alpha,\beta > 0 \\ \alpha F(u,v) + \beta G(u,v) \leqq K(u+v+1) \quad \forall u,v \in [0,\infty[. \end{array}\right\} \quad (5)$$

We are also given Ω a bounded open subset of $\mathbb{R}^N$ with regular boundary and

$$a,b > 0 \quad (6)$$

$$u_0,v_0 \in L^1(\Omega),\ u_0,v_0 \geqq 0,\ g,h \in L^1(Q),\ \ g,h \geqq 0. \quad (7)$$

where $Q = (0,T) \times \Omega$.

We consider the problem

$$\left.\begin{array}{ll}\frac{\partial u}{\partial t} - a\Delta u = F(u,v) + g \text{ on } Q & \\ \frac{\partial v}{\partial t} - b\Delta v = G(u,v) + h \text{ on } Q & \\ \frac{\partial u}{\partial n} = \frac{\partial v}{\partial n} = 0 & \text{on } (0,T) \times \partial\Omega \\ u(0,\cdot) = u_0,\ \ v(0,\cdot) = v_0 & \end{array}\right\} \quad (8)$$

<u>Some remarks about the assumptions</u>

The condition (3) is natural to ensure the nonnegativity of the solutions.

Assumption (4) is rather restrictive: it ensures that the first equation of (8) is a "good" one and that *a priori* estimates can easily be obtained on u. Note that it is satisfied in the examples (1), (2) above.

For (1), $F(u,v) = -vf(v) \leqq 0$

For (2), $F(u,v) = av - uv^2 \leqq av$.

Assumption (5) is more natural and often appears in applications: for instance, in case of conservation of total mass we will have $F(u,v) + G(u,v) = 0$ as in (1). The more general condition (5) will be sufficient for our purpose. Note that it is also satisfied in (2) with $\alpha = \beta = 1$.

Finally, it is not essential to use Neumann conditions in (8); Dirichlet -

or Fourier - type conditions would work as well.

THEOREM: Under assumptions (3)-(7), there exists (u,v) solution of

$$\left.\begin{aligned}
&u,v \in C([0,T]; L^1(\Omega)) \\
&F(u,v),\ G(u,v) \in L^1(Q) \\
&u(t) = S_a(t)u_0 + \int_0^t S_a(t-s)[F(u(s),v(s)) + g(s)]ds \\
&v(t) = S_b(t)v_0 + \int_0^t S_b(t-s)[G(u(s),v(s)) + h(s)]ds
\end{aligned}\right\} \tag{9}$$

where $S_a(t)$, $S_b(t)$ are the semi-groups generated in $L^1(\Omega)$ by $-a\Delta$ and $-b\Delta$ with homogeneous Neumann boundary conditions.

REMARKS: Problem (9) is a weak formulation for (8). As a consequence serious difficulties occur for the uniqueness.

If g,h belong to $L^1_{loc}([0,\infty[\ ; L^\infty(\Omega))$ and if $|F(u,v)|$ and $|G(u,v)|$ do not grow faster than powers of u and v, we can prove a regularizing effect from $L^1(\Omega)$ into $L^\infty(\Omega)$ using in particular ideas from [3]. This will be published later in a more detailed version.

III. Outline of the proof of the theorem

If the data u_0, v_0, g, h are bounded, local existence of nonnegative solutions for (8) is classical. However, even in that case, global existence is not obvious. For this, we could use the results in [3]. We will give here instead a proof independent of [3] by claiming global existence for a simpler perturbed problem.

Let θ: $[0,\infty[\ \rightarrow [0,\infty[$ regular with

$$\lim_{r\to\infty} \theta(r) = +\infty\ ,\quad \theta(0) = 0. \tag{10}$$

For $\varepsilon > 0$, we set

$$F_\varepsilon(u,v) = F(u,v) - \varepsilon\theta(u)\ \ |F(u,v)| \tag{11}$$

$$G_\varepsilon(u,v) = G(u,v) - \varepsilon\theta(v)\,|G(u,v)| \tag{12}$$

$$u_0^\varepsilon = \min(u_0,\varepsilon^{-1})\ ,\ v_0^\varepsilon = \min(v_0,\varepsilon^{-1}) \tag{13}$$

$$g^\varepsilon = \min(g,\varepsilon^{-1}),\quad h^\varepsilon = \min(h,\varepsilon^{-1}) \tag{14}$$

LEMMA 1: There exists a global "regular" nonnegative solution on $(0,T) \times \Omega$ of

$$\left.\begin{aligned} &\frac{\partial u^\varepsilon}{\partial t} - a\Delta u^\varepsilon = F_\varepsilon(u_\varepsilon,v_\varepsilon) + g_\varepsilon \\ &\frac{\partial v^\varepsilon}{\partial t} - b\Delta v^\varepsilon = G_\varepsilon(u_\varepsilon,v_\varepsilon) + h_\varepsilon \\ &\frac{\partial u^\varepsilon}{\partial n} = \frac{\partial v^\varepsilon}{\partial t} = 0 \text{ on } (0,T) \times \partial\Omega \\ &u^\varepsilon(0,\cdot) = u_0^\varepsilon,\quad v^\varepsilon(0,\cdot) = v_0^\varepsilon. \end{aligned}\right\} \tag{15}$$

REMARK: By "regular" we mean here that u_ε, v_ε are uniformly bounded and as a consequence of the regularity theory for the heat operator,

$$\frac{\partial u}{\partial t} \in L^p_{loc}(]0,T];\ L^p(\Omega)),\ u \in L^p_{loc}(]0,T];\ W^{2,p}(\Omega)) \quad \forall 1 \leq p \leq \infty.$$

The proof of the lemma is straightforward:

- Local existence of nonnegative solutions follows from the Lipschitz continuous property of F_ε, G_ε and (3), (10).

- The particular structure of F_ε and G_ε and the choice of θ satisfying (10) yield an *a priori* uniform L^∞-estimate on $(0,T) \times \Omega$ for this solution.

- Then global existence follows by standard arguments (see for instance [2], [6]).

We now make several estimates on u^ε, v^ε.

LEMMA 2: (L^1-estimates) There exists M depending only on $\int_\Omega u_0$, $\int_\Omega v_0$, $\int_Q g$, $\int_Q h$, α, β, a, b, T such that for all $0 \leq t \leq T$

$$\int_\Omega u^\varepsilon(t),\ \int_\Omega v^\varepsilon(t),\ \int_\Omega (1+\varepsilon\theta(u^\varepsilon))|F(u^\varepsilon,v^\varepsilon)|,\ \int_\Omega (1+\varepsilon\theta(v^\varepsilon))|G(u^\varepsilon,v^\varepsilon)| \leq M. \tag{16}$$

<u>PROOF</u>: Integrating over Ω the equality

$$\frac{\partial}{\partial t}(\alpha u^\varepsilon + \beta v^\varepsilon) - \Delta(\alpha a u^\varepsilon + \beta b v^\varepsilon) = \alpha F_\varepsilon + \beta G_\varepsilon + \alpha g_\varepsilon + \beta h_\varepsilon \tag{17}$$

gives by also using (5)

$$\frac{\partial}{\partial t}\int_\Omega \alpha u^\varepsilon + \beta v^\varepsilon \leq K\int_\Omega u^\varepsilon + v^\varepsilon + 1 + \int_Q \alpha g + \beta h.$$

Since $\alpha, \beta > 0$, by integration of this differential inequality we obtain

$$\forall t \in [0,T] \int u^\varepsilon(t)+v^\varepsilon(t) \leq C(\int_\Omega u_0+v_0 + \int_Q g+h+1),\ C = C(\alpha,\beta,K,T). \tag{18}$$

On the other hand, integrating both equations in u^ε, v^ε gives

$$-\int_Q F_\varepsilon(u^\varepsilon,v^\varepsilon) \leq \int_\Omega u_0 + \int_Q g \tag{19}$$

$$-\int_Q G_\varepsilon(u^\varepsilon,v^\varepsilon) \leq \int_\Omega u_0 + \int_Q h. \tag{20}$$

We will now denote by M any expression depending only on the quantities mentioned in Lemma 2. Inequality (19) may be rewritten as

$$\int_Q \varepsilon\theta(u^\varepsilon)|F(u^\varepsilon,v^\varepsilon)| + F^-(u^\varepsilon,v^\varepsilon) \leq \int_\Omega u_0 + \int_Q g + \int_Q F^+(u^\varepsilon,v^\varepsilon) \leq M,$$

the last inequality coming from (4) and (18). We deduce

$$\int_Q (1 + \varepsilon\theta(u^\varepsilon))\ |F(u^\varepsilon,v^\varepsilon)| \leq M. \tag{21}$$

Similarly, we rewrite (20) as

$$\int_Q \varepsilon\theta(v^\varepsilon)|G(u^\varepsilon,v^\varepsilon)| + G^-(u^\varepsilon,v^\varepsilon) \leq \int_\Omega u_0 + \int_Q g + \int_Q G^+(u_\varepsilon,v_\varepsilon). \tag{22}$$

Now, from (5)

$$\beta G^+(u^\varepsilon,v^\varepsilon) \leq K(u^\varepsilon + v^\varepsilon + 1) + \alpha|F(u^\varepsilon,v^\varepsilon)|. \tag{23}$$

From (21)-(23) we deduce

$$\int_Q (1 + \varepsilon\theta(v^\varepsilon))|G(u^\varepsilon,v^\varepsilon)| \leq M. \tag{24}$$

LEMMA 3: Denote by $S_c(t)$ the semi-group generated by $-c\Delta$ $(c > 0)$ with homogeneous Neumann boundary conditions and consider the mapping

$$(w_0,f) \in L^1(\Omega) \times L^1(Q) \to w \in L^1(Q)$$

where

$$w(t) = S_c(t)w_0 + \int_0^t S_c(t-s)\, f(s)\, ds.$$

Then, this mapping is compact from $L^1(\Omega) \times L^1(Q)$ into $L^1(Q)$ and continuous from $L^1(\Omega) \times L^1(Q)$ into $C([0,T]; L^1(\Omega))$.

For a proof see [1].

Proof of the theorem

As a consequence of Lemma 3 and L^1-estimates of Lemma 2, there exists a subsequence $(u^{\varepsilon_n}, v^{\varepsilon_n})$ (which we denote by (u_n,v_n)) of $(u^\varepsilon,v^\varepsilon)$ and $(u,v) \in L^1(Q) \times L^1(Q)$ such that

(u_n,v_n) converges in $L^1(Q) \times L^1(Q)$ and almost everywhere to (u,v) as $n \to \infty$. (26)

But

$$u_n(t) = S_a(t)u_0^n + \int_0^t S_a(t-s)[F_n(u_n,v_n) + g_n](s)\, ds$$

and a similar expression is valid for $v_n(t)$. Assume that

$$\begin{aligned} &F_n(u_n,v_n) \text{ tends to } F(u,v) \text{ in } L^1(Q)\\ &G_n(u_n,v_n) \text{ tends to } G(u,v) \text{ in } L^1(Q). \end{aligned} \tag{27}$$

Then thanks to the continuity property in $C([0,T]; L^1(\Omega))$ stated in Lemma 2, the limit (u,v) is a solution of (9) and the proof is complete.

Therefore, the main point consists of demonstrating (27). According to (26), it is sufficient to prove

$$F_n(u_n,v_n),\ G_n(u_n,v_n) \text{ are uniformly integrable in } L^1(Q). \qquad (28)$$

The fact that (28) and (26) imply (27) is a consequence of the following lemma.

LEMMA 4: Let σ_n be a sequence in $L^1(Q)$ and σ in $L^1(Q)$ such that

(i) $\sigma_n \to \sigma$ a.e. on Q

(ii) σ_n is uniformly integrable in $L^1(Q)$.

Then σ_n converges to σ in $L^1(Q)$.

Indeed, from (i) we know that for all $\delta > 0$, there exists K_δ measurable with meas $K_\delta < \delta$ and σ_n converges uniformly to σ on $Q \setminus K_\gamma$. Moreover, by (ii), if $\varepsilon > 0$ is given, for δ small enough

$$\int_{K_\delta} |\sigma_n| < \varepsilon \quad \forall n \text{ (and } \int_{K_\delta} |\sigma| < \varepsilon\text{)}.$$

Now

$$\int_Q |\sigma_n - \sigma| \leq \int_{K_\delta} |\sigma_n| + |\sigma| + \int_{Q\setminus K_\delta} |\sigma_n - \sigma|$$

$$\Rightarrow \limsup_{n\to\infty} \int_Q |\sigma_n - \sigma| \leq 2\varepsilon \quad \forall\, \varepsilon > 0.$$

PROOF OF (28): $F_n(u_n,v_n)$, $G_n(u_n,v_n)$ are uniformly integrable in $L^1(Q)$.

By (4) and (11)

$$F_n^+(u_n,v_n) \leq F^+(u_n,v_n) \leq K(u_n + v_n + 1).$$

Since u_n, v_n converge in $L^1(Q)$, this implies uniform integrability of $F_n^+(u_n,v_n)$.

By (5) and (12)

$$\beta G_n^+(u_n,v_n) \leq \beta G^+(u_n,v_n) \leq K(u_n+v_n+1) + \alpha F^-(u_n,v_n).$$

Since $F^- \leq F_n^-$, we are reduced to prove that

$$F_n^-(u_n,v_n),\ G_n^-(u_n,v_n) \text{ are uniformly integrable in } L^1(Q). \qquad (29)$$

For this, we use the next lemma characterizing uniform integrability.

LEMMA 5: Let σ_n be a sequence of functions in $L^1(U)$ with U bounded open set in $\mathbb{R}^d$. Then the following statements are equivalent:

(i) σ_n is uniformly integrable in $L^1(U)$

(ii) There exists $J :]0,\infty[\to]0,\infty[$ with $J(0^+) = 0$ and

(a) J is convex, J' is concave, $J' \geq 0$

(b) $\lim_{r\to\infty} \frac{J(r)}{r} = +\infty$

(c) $\sup_n \int_U J(|\sigma_n|) < \infty$.

This lemma is also classical except perhaps for the possibility of choosing J' concave (see [4] for a proof).

Now fix such a J so that

$$\sup_n \int_Q J(u_n + v_n + g_n + h_n + 1) < \infty$$

$$\sup_n \int_\Omega J((\alpha+1)u_0^n + \beta v_0^n) < \infty \quad (\alpha,\beta \text{ as in (5)})$$

and set

$$j(r) = \int_0^r \min\{J'(s), (J^*)^{-1}(s)\}\, ds,$$

where J^* is the conjugate function of J. We easily check that j satisfies (a), (b) and

$$r \geq 0 \quad j(r) \leq J(r), \quad J^*(j'(r)) \leq r.$$

Multiplying the equation in u_n $(= u^{\varepsilon_n})$ by $j'(u_n)$ gives

$$\int_\Omega j(u_n)(T) + a \int_Q j''(u_n)\,|\nabla u_n|^2 + \int_Q j'(u_n)F_n^- =$$

$$\int_Q j'(u_n)\,(F_n^+ + g_n) + \int_\Omega j(u^n)$$

where we denote $F_n^+ = \max(F_n(u_n,v_n),0)$ and $F_n^- = -\min(F_n(u_n,v_n),0)$. We use (4) and the property

$$j'(r)\cdot s \leqq J(s) + J^*(j'(r)) \leqq J(s) + r \tag{31}$$

to bound the last integral in Q above:

$$\int_Q j'(u_n)(F_n^+ + g_n) \leqq \int_Q K\, j'(u_n)(u_n + v_n + 1) + j'(u_n)g_n \leqq \tag{32}$$

$$\leqq \int_Q K\, J(u_n + v_n + 1) + K\, u_n + u_n + J(g_n).$$

We retain from (30) and (32) that

$$\int_Q j''(u_n)\,|\nabla u_n|^2 \leqq C \text{ independent of } n. \tag{33}$$

Now, we set $w_n = (\alpha+1)u_n + \beta v_n$ where α,β are defined in (5). We multiply by $j'(w_n)$ the equation

$$\frac{\partial}{\partial t} w_n - \Delta((\alpha+1)au_n + \beta bv_n) = (\alpha+1)F_n + \beta G_n + (\alpha+1)g_n + \beta h_n.$$

We obtain after integration

$$\int_\Omega j(w_n)(T) + \int_Q j''(w_n)b|\nabla w_n|^2 + \int_Q j'(w_n)(F_n^- + \beta G_n^-) \leqq \int_\Omega j(w_n)(0) + \dots$$

$$\dots \int_Q j'(w_n)(F_n^+ + \alpha F_n + \beta G_n^+ + (\alpha+1)g_n + \beta h_n) + \int_Q (\alpha+1)(b-a)j''(w_n)\nabla w_n \nabla u_n. \tag{34}$$

We bound the last integral on Q using

$$(\alpha+1)(b-a)\nabla w_n \nabla u_n \leq b\ |\nabla w_n|^2 + C\ |\nabla u_n|^2,\ C = C(\alpha,b,a). \tag{35}$$

Since j' is concave, j" is decreasing so that

$$j''(w_n) = j''((\alpha+1)u_n + \beta v_n) \leq j''(u_n). \tag{36}$$

Using (33), (35), (36), we see that in (34) integrals on gradient terms are bounded from above independently of n. We now use (4), (5) as follows

$$F_n^+ \leq K(u_n+v_n+1),\ \alpha F_n + \beta G_n^+ \leq K(\alpha+1)\ (u_n+v_n+1). \tag{37}$$

Using again (31) and the assumption on j, we deduce from (34), (37)

$$\int_Q j'(w_n)\ (F_n^- + \beta G_n^-) \leq C \text{ independent of } n.$$

But, this last inequality implies the uniform integrability of $F_n^-(u_n,v_n)$ and $G_n^-(u_n,v_n)$ as required in (29). Indeed, let K be a measurable subset of Q.

$$\int_K F_n^- + G_n^- = \int_{K\cap[w_n>M]} F_n^- + G_n^- + \int_{K\cap[w_n<M]} F_n^- + G_n^-$$

$$\int_{[w_n>M]} F_n^- + G_n^- \leq \frac{1}{j'(M)} \int_Q j'(w_n)\ (F_n^- + G_n^-) \leq \frac{C}{j'(M)}$$

By construction, j'(M) tends to ∞ as M tends to ∞. We choose M so that

$$C/j'(M) < \varepsilon/2.$$

Now

$$\int_{K\cap[w_n<M]} F_n^- + G_n^- \leq \int_K \max_{0\leq\xi,\eta\leq a_M} \{F_n^-(\xi,\eta) + G_n^-(\xi,\eta)\}$$

where $a_M = \max\{M/(\alpha+1)a,\ M/\beta b\}$. But since

$$F_n^-(\xi,\eta) \leq |F(\xi,\eta)|\ (1+\theta(\xi)),\quad G_n^-(\xi,\eta) \leq |G(\xi,\eta)|\ (1+\theta(\eta)),$$

then the last integral can be made less than $\varepsilon/2$ by choosing meas K small enough independently of n. The uniform integrability follows.

References

1. P. Baras, J.C. Hassan, L. Véron, Compacité de l'opérateur définissant la solution d'une équation non homogène, C.R. Acad. Sc. Paris, t. 284 (1977), 799-802.
2. D. Henry, Geometric theory of semilinear parabolic equations, Lecture notes in Math. 840, Springer-Verlag, New-York (1981).
3. S.L. Hollis, R.M. Martin, M. Pierre, Global existence and boundedness in reaction-diffusion systems, to appear in SIAM J. on Math. Ana.
4. Lê-Châu-Hoàn, Etude de la classe des opérateurs m-accrétifs de $L^1(\Omega)$ et accrétifs dans $L^\infty(\Omega)$, Thèse de 3ème Cycle, Université de Paris VI (1977).
5. K. Masuda, On the global existence and asymptotic behaviour of solutions of reaction-diffusion equations, Hokkaido Math. J., XII (1983), 360-370.
6. F. Rothe, Global solutions of reaction-diffusion systems, Lecture notes in Math. 1072, Springer-Verlag, Berlin (1984).

M. Pierre
Département de Mathématiques
Université de Nancy I
B.P. 239
5406 - Vandoeuvre-les-Nancy
France.

J SABINA & J FRAILE

Directional wave fronts in reaction–diffusion systems: existence and asymptotic behaviour

Section 1. Introduction

The purpose of this work is to introduce new information about a kind of undulatory solutions to semilinear parabolic equations of the type

$$\frac{\partial u}{\partial t} = \Delta u + \sum_{i=1}^{n} b_i \frac{\partial u}{\partial x_i} - f(u), \tag{1}$$

where $u : \Omega \times \mathbb{R} \to \mathbb{R}$, $(x,t) \to u(x,t)$, Ω is a bounded domain of $\mathbb{R}^n$, $\vec{b} \doteq (b_1,\dots,b_n)$ is a constant vector in $\mathbb{R}^n$ and $f = f(u)$ a smooth function.

For equation (1), a well-known kind of undulatory solution is the plane wave front (PWF), i.e. solutions with the form

$$u(x,t) = v(kx - ct),$$

where $k \in \mathbb{R}^n$, $c \in \mathbb{R}$.

In this work, we are concerned with a more general type of wave-like solution to (1), namely, the solutions that we term directional wave fronts (DWFs), i.e.

$$u(x,t) = v(Kx - \vec{c}t) = v(\zeta), \tag{2}$$

where K is a diagonal matrix, $K = \operatorname{diag}(k_1,\dots,k_n)$ and $\vec{c} \in \mathbb{R}^n$. Hence $\zeta \in \mathbb{R}^n$ when $x \in \mathbb{R}^n$, $t \in \mathbb{R}$. From its own structure, the propagative nature of such a solution is clear. In the one-dimensional case ($x \in \mathbb{R}$), DWFs and PWFs are the same kind of solution. However, a DWF preserves the dimensionality of the spatial variable x through the phase ζ. This fact enables us to impose a certain kind of boundary condition on DWFs. That is not possible in the case of PWFs. On the other hand, there are two essential properties for a "good" wave-like solution to (1). The first one is boundedness and existence for every $t \in \mathbb{R}$ ("permanence" following Fife [2]). The second one is to exhibit "nice" asymptotic behaviour when t goes to $\pm\infty$. Testing these properties in the case of PWFs leads to a corresponding problem for an ODE which is more or less complicated (and generally very hard if (1) is not scalar). The objective of the present work is to describe the behaviour of of DWFs with respect to those properties. Because of the nature of the approach used, an implicit restriction throughout the work is the small amplitude character of our DWFs.

The paper is organized as follows. In Section 2 the Neumann Homogeneous Problem for DWFs is introduced. There we reformulate it as a second order semilinear evolution problem. Section 3 contains the hypothesis on the reaction term structure. This permits us to interpret the problem as one of bifurcation from the solution $u = 0$. In two previous works ([3], [6]) results of existence and asymptotic behaviour were deduced under an essential requirement: the non-vanishing of the drift term in (1), i.e., $\vec{b} \neq 0$. We summarize these results in Section 4. New results are presented in Section 5. There we establish the existence of DWFs with heteroclinic character in the case $\vec{b} = 0$, i.e., for the reaction-diffusion equation

$$\frac{\partial u}{\partial t} = \Delta u - f(u). \tag{3}$$

As for the results in Section 4, those of Section 5 are obtained by performing a reduction in the dimension of the problem. In both cases, the existence of a finite-dimensional centre manifold gives such a reduction. Following Kirchgässner's work ([4]), we establish in Section 6 the existence of the centre manifold used in Section 5. Kirchgässner's result is directly applicable in the case $\vec{b} = 0$. Finally, Section 7 describes a way of extending the study of DWFs to hyperbolic semilinear equations, following the techniques in Sections 5, 6.

Section 2. The homogeneous Neumann problem for DWFs

The equation for DWFs from (1) is

$$\sum_{i,j=1}^{n} k_i^2 v_{\zeta_i \zeta_j} + \sum_{i=1}^{n} (k_i b_i + c_i) v_{\zeta_i} - f(u) = 0, \tag{4}$$

the solution $v(\zeta)$ being defined in the unbounded domain

$$W = \bigcup_{t \in \mathbb{R}} (K\,\Omega - \vec{c}t).$$

Because of the cylindrical shape of W we can transform the phase ζ in such a way that (4) can be observed as an evolution problem with respect to the unbounded variable along the generator of W. A straightforward calculation ([3]) shows the existence of a linear invertible transformation

$$\Theta : \mathbf{R}^n \to \mathbf{R}^n \quad \eta \to \zeta = \Theta\eta$$

such that (4) is transformated to the equation

$$\Delta_\eta v + \vec{b}_0 \nabla_\eta v - f(v) = 0, \qquad (5)$$

where $\vec{b}_0 = \Theta^{-1}(K\vec{b} + \vec{c})$. Moreover, Θ can be chosen in order to get

$$\Theta^{-1}\vec{c} = b\,\vec{e}_n = (0,\ldots,b),\ b > 0.$$

Writing $\eta = (\xi,\tau) \in \mathbf{R}^{n-1} \times \mathbf{R}$ there exists a bounded domain $D \subset \mathbf{R}^{n-1}$, such that $\zeta \in W$ iff $(\xi,\tau) \in D \times \mathbf{R}$. On the other hand, there exists $\Gamma \subset \partial\Omega$ such that

"$(x,t) \in \Gamma \times \mathbf{R}$ iff $(\xi,\tau) \in \partial D \times \mathbf{R}$".

Assuming that D is a smooth domain in $\mathbf{R}^{n-1}$ it is possible to define a smooth vector field $\nu : \Gamma \subset \partial\Omega \to \mathbf{R}^n$, $x \to \nu(x)$ such that for every DWF $u = u(x,t)$ the following relation is satisfied

$$\frac{\partial u}{\partial \nu}(x,t) = 0,\ x \in \Gamma,\ t \in \mathbf{R} \quad \text{iff} \quad \frac{\partial v}{\partial n} = 0,\ \ \xi \in D,\ \tau \in \mathbf{R}, \qquad (6)$$

where $n = n(\xi)$ designs the outer unitary normal at $\xi \in \partial D$.

Thus, the Neumann homogeneous problem (NHP) for DWFs is defined to be the search for solutions to (1) with the form (2), supplemented with the boundary condition

$$\frac{\partial u}{\partial \nu}(x,t) = 0 \qquad \forall\, x \in \Gamma,\ \ t \in \mathbf{R}.$$

In (ξ,τ) coordinates the NHP can be written in the evolutionary form

$$\text{(P)}\qquad \begin{aligned} & v'' + b_n^o v' + \Delta_\xi v + \sum_{i=1}^{n-1} b_i^o v_{\xi_i} - f(v) = 0 \\ & \frac{\partial v}{\partial n} = 0 \quad \text{on } \partial D. \end{aligned}$$

Let us study now the role of $\vec{b}$ in the structure of (P). If $\vec{b} \neq 0$, it is possible to choose K and $\vec{c}$ in (3) such that $\vec{b}_0 = 0$. This suffices if

$K\vec{b} + \vec{c} = 0$ above. In this case the equation in (P) has the form

$$v'' + \Delta_\xi v - f(v) = 0. \tag{7}$$

If $\vec{b} = 0$ in (1) that equation is

$$v'' + b v' + \Delta\xi v - f(v) = 0, \quad b \in \mathbf{R} - \{0\}. \tag{8}$$

In the analysis of (P) the reversible nature in τ of equation (7) furnishes good symmetry properties which are fully exploited in the study of small amplitude solutions (see Section 5). In (8) it is not possible to remove the term in v' without losing the autonomous character of the equation. Because of this fact we label the situations $\vec{b} \neq 0$ and $\vec{b} = 0$ in (1) as the reversible and non-reversible cases, respectively. The terminology is imported from [4].

REMARK: In the same way, it is possible to define the Homogeneous Dirichlet Problem for DWFs. In fact, setting aside the obvious adaptations, a major part of the conclusions developed in this work for the NHP stand for the DHP.

Section 3. Hypothesis on the reaction term

We shall suppose that f in (1) depends on a real parameter λ, $|\lambda| \leqq \varepsilon$, and satisfies the following smallness conditions

(i) $f = f(\lambda,u)$ is a C^{r+1} function ($r \geqq 4$) and $f(\lambda,0) = 0$.

(ii) $f(\lambda,u) = a(\lambda)u + a_k(\lambda)\, u^k + r(\lambda,u)$, $k \leqq r-1$ and $r(\lambda,u) = O(|u|^{k+1})$ uniformly where $u \to 0$.

(iii) $a(0) = 0$, $a'(0) < 0$ and $a_k(0) \neq 0$.

(iv) f together with its derivatives up to the order r+1 are polynomially bounded, uniformly with respect to λ. Also, we shall suppose that $\Omega \subset \mathbf{R}^3$.

REMARK: (i) - (iii) imply a bifurcation of small amplitude constant solutions for (1) at $(\lambda,u) = (0,0)$. This bifurcation is accompanied with a stability transition between $u = 0$ and the bifurcated branches at $\lambda = 0$, with respect to the kinetic equation $u' = -f(\lambda,u)$. Those branches have the form:

$(\lambda(u),u)$, where $\lambda(u) = -(k-1)!\ (a_k(0)/a'(0))\ u^{k-1} + O(|u|^k)$. As we shall see later, these solutions are strongly related to the small DFWs solutions to (P).

Section 4. Results for the reversible case

In [3], we established the existence of a wide class of small amplitude DWFs for (P) under the hypothesis above for the reaction term f. We summarize the results in the following theorem:

THEOREM 1: Under the hypothesis (i) ... (iv) on f and $|\lambda|$ being small enough, the NHP for DWFs admits solutions $u = u(x,t)$ with the following properties

(1) If k is even (respectively, k odd, $\lambda a_k(\lambda) > 0$, $\lambda > 0$) there exists $u = u(\lambda)$ - a constant solution to (1) - and a DWF $u = u(\lambda,x,t)$ satisfying

$$|u(\lambda,\ .\ ,t) - u(\lambda)|_{C^0(\bar{\Omega})} \leq C_0 \exp(\beta_0 |t|) \quad C_0, \beta_0 > 0.$$

in $\lambda > 0$, $u(\lambda) = 0$. For $\lambda > 0$, $u(\lambda)$ is the non-trivial constant solution to (1), bifurcated from $u = 0$ at $\lambda = 0$ for k even.

(2) If k is odd, $f(\lambda,u)$ is odd in u, $\lambda a_k(\lambda) > 0$ and $\lambda > 0$, there exists a pair of constant solutions to (1) $u_+(\lambda)$, $u_-(\lambda)$, bifurcated from $u = 0$ at $\lambda = 0$ and there exists a DWF $u = u(\lambda,x,t)$ such that

$$|u(\lambda,\ .\ ,t) - u_+(\lambda)|_{C^0(\bar{\Omega})} \leq C_1 \exp(-\beta_1\ t) \qquad t \to \infty$$

$$|u(\ ,\ .\ ,t) - u_-(\lambda)|_{C^0(\bar{\Omega})} \leq C_2 \exp(\beta_2\ t) \qquad t \to -\infty,$$

C_i and β_i being positive constants, $i = 1,2$.

(3) Moreover, in both the cases (1), (2), for each λ small enough there exists a one-parameter family of t-periodic DWFs which are of small amplitude in the $C^0\ (\bar{\Omega})$-norm.

REMARKS:

(i) Point (1) asserts the existence of a one-parameter family of homoclinic DWFs. Point (2) gives the existence of a one-parameter family of heteroclinic DWFs. Heteroclinic and homoclinic character are two well-known types of behaviour for PWFs that are solutions to equation (1), particularly in the one-dimensional case ($x \in \mathbb{R}$).

(ii) In [6], the same conclusions have been stated for systems of equations as in (1) when the nonlinear term $f = f(\lambda,u)$ has an uncoupled linearization around $u = 0$. Again, an essential requirement on f is the presence of a constant solutions bifurcation at $(\lambda,u) = (0,0)$.

In the next section we shall outline the proof of Theorem 1.

Section 5. The results in the non-reversible case

5 - 1. Analysis of the linearization of (P) around v = 0.

The NHP (P) can be reformulated in abstract form as follows

$$v'' + b\,v' - T(\lambda)v = g(\lambda,v) \quad \left(' = \frac{d}{d\tau}\right), \qquad (9)$$

where $T(\lambda) = -\Delta_\xi + a(\lambda)$ is defined in $L^2 = L^2(D)$ with domain $D(T) = \{v \in H^2(D) \,/\, \frac{\partial v}{\partial n} = 0 \text{ on } \partial D\}$. In (9) we put $g \doteq a_k(\lambda)\,v^k + r(\lambda,v)$. For $\psi \in L^2$ let us define the projector P_o as

$$P_o\psi = \frac{1}{|D|}\int_D \psi(\xi)\,d\xi,$$

where $|D|$ defines the two-dimensional Lebesgue measure of D. If we define $P_1 = 1 - P_o$, $T_i(\lambda) = P_i\,T(\lambda)$ and $D(T_1) = \{v \in H^1(D) \,/\, P_o v = 0\}$, we can rewrite (9) in the form

$$\begin{aligned} v_o'' + b\,v_o' - a(\lambda)\,v_o &= g_o(\lambda, v_o, v_1) \\ v_1'' + b\,v_1' - T_1(\lambda)v_1 &= g_1(\lambda, v_o, v_1)\,. \end{aligned} \qquad (10)$$

In (10) $v_i \doteq P_i\,v$, $g_i \doteq P_i g(\lambda, v_o + v_1)$, $i = 0,1$.

Let us study now the spectrum of the linearization of (10) around $v_o = 0$, $v_1 = 0$. For equation (10-1) the eigenvalues of the linear part are given by

$$r_i(\lambda) = \frac{-b \pm (b^2 + 4\,a(\lambda))^{1/2}}{2}, \quad i = 1,2.$$

While $r_2(\lambda)$ is bounded away from 0 ($r_2(0) = -b$), $r_1(\lambda) \to 0$ when $\lambda \to 0$. On the other hand, let us observe that $T_1(\lambda)$ - with domain $D(T_1)$ - is an m-acretive operator with minimum eigenvalue $\sigma_1(\lambda) = a(\lambda) + \sigma_1$, $\sigma_1 > 0$ being the second eigenvalue of Δ_ξ under Neumann homogeneous conditions on D. Thus, $S_1(\lambda) \doteq T_1(\lambda)^{1/2}$ is defined and has a strict positive lower bound α for λ small. Moreover $D(S_1) = \{v \in H^1(D) \ / \ P_o \ v = 0\}$. Hence, the linearization of (10-2) can be written as

$$\begin{aligned} v_1' &= S_1(\lambda)\ v_2 \\ v_2' &= S_1(\lambda)\ v_1 - b\ v_2. \end{aligned} \tag{11}$$

A straightforward calculation establishes that the spectrum Σ of (11) is

$$\Sigma = \{ \frac{-b \pm (b^2 + 4\ \sigma_n(\lambda))^{1/2}}{2} \ / \ \sigma_n(\lambda) \in \Sigma(T_1(\lambda)) \}.$$

The existence of $\alpha_+ > 0$, $\alpha_- < 0$ such that $\forall\ \sigma \in \Sigma$: $\sigma > \alpha_+$ or $\sigma < \alpha_-$, uniformly in $|\lambda|$ small is now clear. From the discussion above, the existence of a one-dimensional centre manifold is expected because $r_1(\lambda) \to 0$ when $\lambda \to 0$ and the other eigenvalues remain bounded away from 0.

REMARK 1: In the reversible case, $b = 0$. Hence, eigenvalues of (10-1) are now $r_i(\lambda) = \pm(a(\lambda))^{1/2}$ and those of (10-2) are given now by $\pm\ \Sigma(S_1(\lambda))$. Because $r_i(\lambda) \to 0$, $\lambda \to 0$, the existence of a two-dimensional centre manifold is expected. In fact, this is the case as it is proved in [4].

By introducing new variables (y,z) in (10-1): $v_o = y + z$, $v_o' = r_1 y + r_2 z$, equation (10) is transformed into

$$\begin{aligned} y' &= r_1(\lambda)\ y - (r_2 - r_1)^{-1}\ g_o(\lambda,y,z,v_1) \\ z' &= r_2(\lambda)\ z + (r_2 - r_1)^{-1}\ g_o(\lambda,y,z,v_1) \\ v_1'' &+ b\ v_1' - T_1(\lambda)\ v_1 = g_1(\lambda,y,z,v_1), \end{aligned} \tag{12}$$

where $g_i \doteq g_i(\lambda, y + z + v_1)$, $i = 0,1$.

Let us state now our Centre Manifold existence theorem

THEOREM 2: Under the hypothesis (i) ... (iv) on f there exists $\eta > 0$, $\varepsilon > 0$ and a C^{r-1} function:

$$h : (-\varepsilon,\varepsilon) \times \mathbb{R} \rightarrow \mathbb{R} \times D(S_1)$$

$$(\lambda,y) \rightarrow (z,v_1) = (h_0(\lambda,y),h_1(\lambda,y))$$

satisfying

(i) $h(\lambda,y) = O(|y|^k)$ uniformly in λ, $y \rightarrow 0$.

(ii) For $|\lambda| < \varepsilon$ and every solution to (12) $(y,z) \in C^1(\mathbb{R},\mathbb{R}^2)$, $v_1 \in C^2(\mathbb{R},L^2) \cap C^1(\mathbb{R},D(S_1)) \cap C(\mathbb{R}, D(T_1))$ such that

$$|y(\cdot)|_{\infty,\mathbb{R}} + |z(\cdot)|_{\infty,\mathbb{R}} + |v_1(\cdot)|_{\infty,H^1,\mathbb{R}} < \eta$$

the following invariance relation holds

$$z(\tau) = h_0(\lambda,y(\tau))$$

$$v_1(\tau) = h_1(\lambda,y(\tau)) \qquad \tau \in \mathbb{R}$$ #

REMARK 2: For the reversible case ($b = 0$), Kirchgässner's result [4] gives the existence of $h = h(\lambda,v_0,v_0')$, a C^{r-1} function, which satisfies an analogous invariance relation. Because of the reversibility of (10) the relation $h(\lambda, v_0, -v_0') = h(\lambda, v_0,v_0')$ also holds.

5 - 2. Analysis of the small amplitude solutions to (P)

From Theorem 2, it follows that small amplitude solutions to (P) are given by those of the scalar ODE

$$y' = r_1(\lambda)y - (r_2 - r_1)^{-1} g_0(\lambda,y,h_0,h_1). \qquad (13)$$

This equation can be written as

$$y' = r_1(\lambda)y + (a_k(\lambda)/(b^2+4a(\lambda))^{1/2})y^k + O(|y|^{k+1}). \qquad (14)$$

Equation (14) has $(\lambda,y) = (0,0)$ as a bifurcation point of stationary solutions. For $|\lambda| + |y|$ small, bifurcated branches of solutions can be expressed by $(\lambda,(y),y)$ where

$$\lambda(y) = (-(k-1)!\, a_k(0)/a'(0))\, y^{k-1} + O(|y|^k).$$

Because $r_1(\lambda) = (a'(0)/b)\lambda + O(\lambda^2)$ we can state

(i) If k is even, for every λ small enough there exist $y(\lambda) \neq 0$, a stationary solution of (13) and a solution $y = y(\lambda,\tau)$ such that

$$\lim_{\tau \to -\infty} y(\lambda,\tau) = y(\lambda) \quad \text{and} \quad \lim_{\tau \to +\infty} y(\lambda,\tau) = 0 \text{ for } \lambda < 0,$$

$$\lim_{\tau \to -\infty} y(\lambda,\tau) = 0 \quad \text{and} \quad \lim_{\tau \to +\infty} y(\lambda,\tau) = y(\lambda) \text{ for } \lambda > 0.$$

(ii) If k is odd, for every λ small enough such that $\lambda\, a_k(\lambda) > 0$ there exist non-trivial stationary solutions of (13) $y_+(\lambda)$, $y_-(\lambda)$ and a pair $y_+(\lambda,\cdot)$ $y_-(\lambda,\cdot)$ of solutions to (13) satisfying

$$\lim_{\tau \to -\infty} y_\pm(\lambda,\tau) = y_\pm(\lambda) \text{ and } \lim_{\tau \to +\infty} y_\pm(\lambda,\tau) = 0 \text{ for } \lambda > 0$$

$$\lim_{\tau \to -\infty} y_\pm(\lambda,\tau) = 0 \quad \text{and} \quad \lim_{\tau \to +\infty} y_\pm(\lambda,\tau) = y_\pm(\lambda) \text{ for } \lambda > 0.$$

<u>REMARKS</u>:

(a) If we put $u(\lambda) = h_0(\lambda,\tilde{y}(\lambda)) + \tilde{y}(\lambda)$, where $\tilde{y}(\lambda)$ takes the values $y(\lambda)$, $y_\pm(\lambda)$ introduced above, we shall get the bifurcated constant solutions to (1) considered in Section 3. It is a consequence of the invariant character of h.

(b) To obtain heteroclinic small DWFs solutions to (P) it suffices to consider

$$v(\tau) = \tilde{y}(\lambda,\tau) + h_0(\lambda,\tilde{y}(\lambda,\tau)) + h_1(\lambda,\tilde{y}(\lambda,\tau)),$$

$\tilde{y}(\lambda,\cdot)$ being the solutions $y(\lambda,\cdot)$ and $y_\pm(\lambda,\cdot)$ introduced in (i), (ii). Working a little more to deduce the corresponding $C^o(\bar{\Omega})$ estimates we can already state

<u>THEOREM 3</u>: Let us assume that f satisfies conditions (i) ... (iv). For λ small enough, there exist solutions $u = u(x,t)$ to the NHP for DWFs with the following properties

(1) If k is even, there exist a constant solution to (1) $u = u(\lambda)$ and a DWF $u = u(\lambda,x,t)$ such that

$$|u(\lambda\cdot,t) - u(\lambda)|_{C^0(\bar{\Omega})} \leq C_1 \exp(-\beta_1 t) \quad t \to +\infty,$$

$$|u(\lambda,\cdot,t)|_{C^0(\bar{\Omega})} \leq C_2 \exp(\beta_2 t) \quad t \to -\infty, \text{ for } \lambda < 0,$$

and

$$|u(\lambda,\cdot,t) - u(\lambda)|_{C^0(\bar{\Omega})} \leq C_3 \exp(\beta_3 t) \quad t \to -\infty,$$

$$|u(\lambda,\cdot,t)|_{C^0(\bar{\Omega})} \leq C_4 \exp(-\beta_4 t) \quad t \to +\infty, \text{ for } \lambda > 0.$$

(2) If k is odd, $\lambda\, a_k(\lambda) > 0$, there exist a pair of constant solutions to (1) $u_\pm(\lambda)$ and a pair of DWFs $u_\pm = u_\pm(\lambda,x,t)$ such that

$$|u_\pm(\lambda,\cdot,t) - u_\pm(\lambda)|_{C^0(\bar{\Omega})} \leq C_\pm \exp(-\beta_\pm t) \quad t \to +\infty,$$

$$|u_\pm(\lambda,\cdot,t)|_{C^0(\bar{\Omega})} \leq \tilde{C}_\pm \exp(\tilde{\beta}_\pm t) \quad t \to -\infty, \text{ for } \lambda < 0,$$

and

$$|u_\pm(\lambda,\cdot,t) - u_\pm(\lambda)|_{C^0(\bar{\Omega})} \leq C_{0\pm} \exp(\beta_{0\pm} t) \quad t \to -\infty,$$

$$|u_\pm(\lambda,\cdot,t)|_{C^0(\bar{\Omega})} \leq \tilde{C}_{0\pm} \exp(-\beta_{0\pm} t) \quad t \to +\infty, \text{ for } \lambda > 0.$$

The C's and β's are all positive constants. #

REMARK 3: In the reversible case, the equation equivalent to (13) is

$$v_0'' - a(\lambda)\, v_0 = a_k(\lambda)\, v_0^k + r(\lambda,v_0,v_0'), \tag{15}$$

where $r(\lambda,v_o,v_o') = O(|v_o|^{k+1} + |v_0'|^2 |v_0|^{k-1})$ and $r(\lambda,v_0,-v_0') = r(\lambda,v_0,v_0')$. The two-dimensional character of (15) and its reversible nature illustrate the existence of a wider class of small amplitude DWFs to the NHP.

Section 6. A Centre Manifold existence theorem

To prove Theorem 2 we shall follow the programme developed in [4] for the existence of a centre manifold in the reversible case. The proof of those results that are somewhat standard will be omitted.

6 - 1. Preliminary results

In the sequel it is supposed that b in (9) is a fixed positive constant. For the rest of Section 6, $C_b^r(X,Y)$ will define the space of r-times continuous differentiable functions $h : X \to Y$ with bounded derivatives up to the order r; X, Y being Banach spaces. For $C_\kappa^r(X,Y)$ we mean the subspace of those h's satisfying $|D^{(k)}h|_{\infty,X} \leq \kappa$, $k \leq r$.

Let us begin by establishing the non critical nature of equations for (z,v_1) in (12) with respect to bounded solutions.

LEMMA 1: For $g_0 \in C_b^1(\mathbb{R},\mathbb{R})$, $g_1 \in C_b^1(\mathbb{R}, L^2)$ and λ small enough, there exists a unique solution $(z,v_1) = (z(\tau), v_1(\tau))$ to equation

$$z' = r_2(\lambda)\, z + (r_2 - r_1)^{-1} g_o$$
$$v_1'' + b\, v_1' - T_1(\lambda)\, v_1 = g_1, \tag{16}$$

which satisfies $z \in C_b^1(\mathbb{R},\mathbb{R})$, $v_1 \in C_b^2(\mathbb{R},L^2) \cap C_b^1(\mathbb{R},D(S_1) \cap C_b(\mathbb{R},D(T_1))$.

PROOF: Let us choose λ so small that $|r_2(\lambda)| \geq \rho > 0$ and $\sigma > \alpha_+ > 0$ or $\sigma < \alpha_- < 0$ $\forall\, \sigma \in \Sigma$ (see Section 5 - 1).

Then

$$z(\tau) = \int_{-\infty}^{\tau} e^{r_2(\lambda)(\tau-s)} (r_2 - r_1)^{-1} g_0(s)\, ds.$$

The analysis of equation (16-2) is a little more complicated. Firstly, it is easily seen that $v_1 \in C_b^2(\mathbb{R},L^2) \cap C_b^1(\mathbb{R},D(S_1)) \cap C_b(\mathbb{R},D(T_1))$ is a solution to (16-2) iff $(v_1,v_2) = (v_1(\tau),v_2(\tau))$ is a $C_b^1(\mathbb{R},D(S_1) \times D(S_1))$ solution to the equation

$$v_1' = S_1(\lambda)\, v_2$$
$$v_2' = S_1(\lambda)\, v_1 - b\, v_2 + S_1^{-1}(\lambda) g_1, \tag{18}$$

provided that $g_1 \in C_b^1(\mathbb{R},L^2)$.

Let us define

$$R_i(\lambda) = \frac{1}{2}\,(-b1 \pm (b^2 + 4T_1(\lambda))^{1/2}) \qquad i = 1,2.$$

Operators $R_1(\lambda)$ and $-R_2(\lambda)$ are m-accretive with domain $D(S_1)$, defined in L^2. Its lower bounds are α_+ and $-\alpha_-$, respectively. Because of this,

$$B(\lambda) = (R_2(\lambda) - R_1(\lambda))^{-1} = -\,(b^2 1 + 4T_1(\lambda))^{-1/2}$$

is a bounded operator from L^2 into $D(S_1)$. Observe that $D(S_1)$ with the S_1-graph norm is $D(S_1) = \{v \in H^1(D) \;/\; P_0(v) = 0\}$ with the $H^1(D)$-norm. Let us introduce now the equation

$$\begin{aligned} w_1' &= R_1\, w_1 - B\, g_1 \\ w_2' &= R_2\, w_2 + B\, g_1. \end{aligned} \tag{19}$$

For a g_1 as above, (19) has the unique solution $w_i \in C_b^1(\mathbb{R}, D(S_1))$ $i = 1,2$, given by

$$w_1(\tau) = \int_\tau^\infty e^{-R_1(s-\tau)}\, B\, g_1(s)ds$$

$$w_2(\tau) = \int_{-\infty}^\tau e^{R_2(\tau-s)}\, B\, g_1(s)ds.$$

By performing the transformation $v_1 = w_1 + w_2$, $v_2 = S_1^{-1}\,(R_1w_1 + R_2w_2)$ we obtain the required solution to equation (16-2). It has the form

$$v_i(\tau) = \int_{-\infty}^\infty K_i(\tau,s)\, g_1(s)\, ds \quad i = 1,2,$$

where $K_1(\tau,s) = e^{R_2(\tau-s)}\, B$ for $\tau \geqq s$ and $e^{-R_1(s-\tau)}\, B$ for $\tau < s$. $K_2(\tau,s) = e^{R_2(\tau-s)}\, S_1^{-1} R_2 B$ for $\tau \geqq s$ and $e^{-R_1(s-\tau)}\, S_1^{-1} R_1\, B$ for $\tau < s$.

<u>REMARK</u>: Let us put $\beta_1 = \min\,\{\rho, \alpha_+, -\alpha_-\}$. If $0 < \beta < \beta_1$ and $|g_0|_\beta \doteq \sup_{\mathbb{R}} e^{\beta|\tau|}\; |g_0(\tau)|$, $|g_1|_{\beta,L^2} \doteq \sup_{\mathbb{R}}\; e^{\beta|\tau|}\; |g_1(\tau)|_{L^2}$ are both finite, the same holds for $|z(\cdot)|_\beta$ and for $|v_1(\cdot)|_{\beta,D(S_1)}$. Using this β-norm the map $(g_0, g_1) \to (z, v_1)$ is continuous.

6 - 2. Redefinition of the non linear term

From condition (iv) in Section 3 it follows that $g \in C^r((-\varepsilon,\varepsilon) \times H^1(D), L^2)$, g being considered as a substitution operator. Let $\phi_\delta \in C^\infty(H^1(D), \mathbb{R})$, $0 \leqq \phi_\delta \leqq 1$ such that $\phi_\delta = 1$ at $|v| \leqq \delta$, $\phi_\delta = 0$ at $|v| \geqq 2\delta$. Taking $\phi_\delta\, g$ instead of g we can always assume that $g \in C_b^r \cap C_\eta^1((-\varepsilon,\varepsilon) \times H^1(D), L^2)$.

In the sequel, smallness of η will be at our disposal. Also we shall assume that

$$0 < \beta_0 \doteq r \max_{|\lambda| \leqq \varepsilon} \{r_1(\lambda)\} < \beta_1 \quad \text{(see the remark to Lemma 1).}$$

We are going to look for our centre manifold h in the set of functions $Y_\kappa^r \doteq C_b^r \cap C_\kappa^1((-\varepsilon,\varepsilon) \times \mathbb{R}, \mathbb{R} \times D(S_1))$ which additionally satisfy condition (i) in Theorem 2.

The following result is an immediate consequence of Theorem 1.1. Chapter 3 in [1].

LEMMA 2: Let us take $\beta > \beta' > \beta_0$ and $\eta \leqq \eta_0$, $\kappa \leqq \kappa_0$, η_0 and κ_0 being small enough. For $h \in Y_\kappa^r$ let us design $y = y(\cdot,\xi,\eta,h)$ the solution to the initial value problem

$$y' = r_1(\eta)\, y - (r_2 - r_1)^{-1}\, g_0(\lambda, y, h_0(\lambda,y), h_1(\lambda,y))$$

$$y(0) = \xi.$$

Then

$$y \in C_b^r(\mathbb{R} \times (-\varepsilon,\varepsilon) \times Y^r, C^1(I,\mathbb{R})),$$

I being a bounded interval in $\mathbb{R}$. Moreover, there exists $C_0 = C_0(\beta,\beta',\kappa,\eta)$ such that

$$|y(\tau;\xi,\lambda,h)| \leqq C_0(\eta + |\xi|)\, e^{\beta'|t|} \qquad t \in I$$

$$\left|\frac{\partial^\gamma y}{\partial\xi^{\gamma_1}\partial^{\gamma_2}\partial\lambda^{\gamma_3}}(\tau;\xi,\lambda,h)\right| \leqq C_0\, e^{\gamma\beta|t|} \qquad t \in I,$$

where $\gamma = \gamma_1 + \gamma_2 + \gamma_3$. #

The next result is essential to ensure the invariant property of the centre manifold h.

LEMMA 3: Let us take $\beta_0 < \beta' < \beta < \beta_1$, η_2 small and let $v(\tau) = (y(\tau), z(\tau), v_1(\tau))$ and $\tilde{v}(\tau) = (\tilde{y}(\tau), \tilde{z}(\tau), \tilde{v}_1(\tau))$ be two bounded solutions to equation (12) which satisfy $y(0) = \tilde{y}(0)$. Then $v(\tau) = \tilde{v}(\tau)$, $\forall\ \tau \in \mathbb{R}$.

PROOF: Let us define $y_0 = y - \tilde{y}$, $z_0 = z - \tilde{z}$ and $v_{10} = v_1 - \tilde{v}_1$. Then it is established that

$$|y_0|_{\infty,\beta} \leqq A\phi,\ |z_0|_{\infty,\beta} \leqq B\phi,\ |v_{10}|_{D(S_1),\infty,\beta} \leqq C\phi,$$

where $|f|_{X,\infty,\beta} = \sup\limits_{\mathbb{R}} e^{-\beta|t|}|f(\tau)|$ and

$$A = \frac{\eta}{|r_2 - r_1|(\beta - \beta')}$$

$$B = \frac{-2\eta\ r_2(\lambda)}{|r_2 - r_1|(r_2^2 - \beta^2)}$$

$$C = \{\frac{C(\alpha_+)}{\beta-\alpha_-} - \frac{2\ C(\alpha_-)\alpha_-}{\alpha_-^2 - \beta^2}\}\ |S_1\ B|\ \eta$$

$$\phi = |y_0|_{\infty,\beta} + |z_0|_{\infty,\beta} + |v_{10}|_{D(S_1),\infty,\beta}.$$

The $C(\alpha_+)$, $C(\alpha_-)$ above are those constants such that

$$|e^{(-1)^i R_i(\lambda)\ t}| \leqq C(\alpha_\pm)\ e^{\alpha_\pm t} \qquad t > 0,\ i = 1,2.$$

It suffices to take η_2 so small that $\max\{A,B,C\} < 1$. #

6 - 3. Existence of the centre manifold

For $h \in Y_\kappa^r$ define $T\,h = (T_0 h,\ T_1 h) = (\hat{h}_1,\ \hat{h}_2)$ as

$$\hat{h}_0(\xi) = \int_{-\infty}^{0} (r_2 - r_1)^{-1}(\lambda)\, e^{-r_2(\lambda)\, s}\, g_0(\lambda, y(s), h_0, h_1)\, ds$$

$$\hat{h}_1(\xi) = \int_{-\infty}^{\infty} K_1(0,s)\; g_1(\lambda, y(s), h_0, h_1)\, ds,$$

where $y(\tau) = y(\tau;\xi,\lambda,h)$ designs the solution to the initial value problem in Lemma 2.

If η_3, κ_3 are chosen sufficiently small and $\eta \leqq \eta_3$, $\kappa \leqq \kappa_3$, it is proved - via Lemma 2 - that T maps Y^r_κ into itself.

On the other hand, some standard calculations show that

$$|T h_1 - T h_2|_{\infty, \mathbb{R}} \leq C\, |h_1 - h_2|_{\infty,\mathbb{R}} \qquad h_i \in Y^r_\kappa,$$

where $C = O(\eta)$. Hence, by taking η small, T defines a contractive mapping in the supremum norm. If we put h_* as the initial data and carry out the iterations $h_n = T^n h_*$ ($h_* \in Y^r$); $h_n \in Y^r_\kappa$ and h_n converges to $h \in C^o_b((-\varepsilon,\varepsilon) \times \mathbb{R}, \mathbb{R} \times D(S_1))$. However, for every n, h_n is a C^r function whose derivatives up to the order r are all Lipschitzian with a uniform Lipschitz constant. Hence, $h \in C^{r-1}$ with Lipschitzian derivatives (see Lemma 2.5, page 39 in [5]). From this fact it is shown that h satisfies (i) in Theorem 2.

To finish the proof it is established that $(y(\tau), h_0(\lambda,y(\tau)), h_1(\lambda,y(\tau)))$ is a solution to equation (12) provided that $y(\tau)$ is just a solution to equation (13). This fact and Lemma 3 yield the invariant property of h.#

Section 7. Semilinear hyperbolic equations

Let us consider now the hyperbolic semilinear equation

$$u_{tt} = \Delta u - f(\lambda,u), \tag{20}$$

for $u : \Omega \times \mathbb{R} \to \mathbb{R}$, $(x,t) \to u(x,t)$, Ω a bounded domain in $\mathbb{R}^n$ and f as in Section 3.

The equation for DWFs (20) has the form

$$\sum_{i,j=1}^{n} (\delta_{ij}\, k_i k_j - c_i c_j)\, v_{\zeta_i \zeta_j} - f(\lambda,v) = 0 \tag{21}$$

The matrix $A_n = (\delta_{ij} k_i k_j - c_i c_j)_{i,j}$ is symmetric. On the other hand - from a theorem of Gershgorin - the eigenvalues of A_n are included into

$$\bigcup_{i=1}^{n} \{z \in \mathbb{C} \ / \ |z-z_i| \leq r_i\} ,$$

where $z_i = k_i^2 - c_i^2$, $r_i = \sum_{j > 1} |c_i c_j|$. Thus, it is clear that, for a fixed matrix $K = \text{diag}(k_1,\ldots,k_n)$, many choices of the vector $\vec{c}$ are possible in order to get A_n to be positive definite.

As in Section 2, a suitable choice of the linear map $\Theta : \mathbf{R}^n \to \mathbf{R}^n$, $\eta \to \zeta = \Theta = \Theta(\xi,\tau)$ transforms (21) into the equation

$$v'' + \Delta_\xi v - f(\lambda,v) = 0. \tag{22}$$

In the case of a damped semilinear hyperbolic equation

$$u_{tt} - a\, u_t = \Delta u - f(\lambda,u), \tag{23}$$

the equation for DWFs is

$$v'' + a\, b\, v' + \Delta_\xi v - f(\lambda,v) = 0. \tag{24}$$

In arriving at equation (24), we have assumed that $\vec{c}$ and K have been chosen in order that A_n be positive definite and - as in Section 2 - under the additional requirement $\Theta^{-1} \vec{c} = (0,\ldots,b)$, $b > 0$.

The study of existence and asymptotic behaviour of small DWFs in accordance with the homogeneous Dirichlet and Neumann problems for hyperbolic semilinear equations is then contained in the discussion developed in the present work. The case of damped semilinear hyperbolic equations is an additional application of the results established in Sections 5 - 6 for the non reversible case.

References

1. Chow, S.N., Hale, J.K. (1982). "Methods of bifurcation theory". Springer, Berlin.
2. Fife, P.C. (1980). "Mathematical aspects of reacting and diffusing systems". Lec. Notes in Biomathematics no. 28, Springer, Berlin.
3. Fraile, J., Sabina, J. (1984). Boundary value conditions for wave fronts in reaction-diffusion systems. The Proc. of the Royal Soc. of Edinburgh, 99 A, 127-136.

4. Kirchgässner, K. (1983). Homoclinic bifurcations of perturbed reversible systems. Lec. Notes in Mathematics no. 1017, Springer, Berlin.
5. Marsden, J.E., MacKracken, M. (1976). "The Hopf bifurcation and its applications". Springer, Berlin.
6. Sabina, J. (1986). Directional Wave Fronts in reaction-diffusion systems. Bull. of the Australian Math. Soc. 33, 1-20.
7. Fischer, G. (1984). Zentrumsmannigfaltigkeiten bei elliptischen differential-gleichungen. Math. Nachr. 115, 137-157.

NOTE: After completing this manuscript we became acquainted with Fischer's work [7]. It contains a general centre manifold theorem for 2m-order elliptic equations on strip-like domains. Theorem 2 in our work is in the scope of Fischer's centre manifold theorem. However, the inclusion of the proof of Theorem 2 remains interesting for the applications to Directional Wave Fronts given here.

J. Sabina
ETSI de Montes
Un. Politécnica de Madrid
28040 Madrid
Spain.

J. Fraile
Departamento de Matematica Aplicada
Universidad Complutense de Madrid
28040 Madrid
Spain.

JUAN L VAZQUEZ

Convexity properties of the solutions of nonlinear heat equations

1. Introduction

In this paper we report on some recent progress in the qualitative theory of nonlinear heat equations, based on the study of the concavity properties of their solutions.

One of the most widely studied examples of a nonlinear heat equation is

$$u_t = \Delta(u^m). \tag{1.1}$$

When $m > 1$ it is called the porous medium equation (PME), and it is relevant both from the point of view of its mathematical properties and for its applications. Among the latter, we count the flow of isentropic gases in porous media [M], where the label 'porous medium equation' originates, thermal conduction at high temperatures [ZR], spread of crowd-avoiding populations [GM] and others (see [P] for a general reference). In all of these applications the solutions are nonnegative.

From the mathematical point of view, the interest of the PME lies in the fact that it is of degenerate parabolic type: it is parabolic at the points where $u > 0$ but degenerates where u vanishes, in accordance with the fact that the diffusion coefficient is given by $D(u) = mu^{m-1}$. This implies a series of properties that strongly depart from the linear heat equation ($m = 1$). In particular, if the initial data $u(x,0)$ have compact support, then the same happens with $u(\cdot,t)$ for every $t > 0$ [OKC] and there appears a free boundary that separates the regions $[u > 0]$ and $[u = 0]$. Across this free boundary u^{m-1} is not a C^1 function. In the case of one space dimension, the behaviour of the free boundary and of the solution near it is now well understood and can be explained in the form of jump discontinuities for the quantity $(u^{m-1})_x$ and Rankine-Hugoniot conditions, cf. [V2] for a reference to this topic.

The lack of regularity together with the nonlinearity of the PME explain the need of new techniques in order to establish a mathematical theory for it. This theory is now, to a large extent, complete, at least in the case of the one-dimensional Cauchy problem and the techniques introduced for it may prove to be useful in studying more general classes of diffusion equations.

In this sense, the extent to which the theory of the Cauchy problem for equation (1.1) in $(x,t) \in Q_T = \mathbb{R}^N \times (0,T)$, with initial conditions

$$u(x,0) = u_o(x) \text{ for } x \in \mathbb{R}^N, \tag{1.2}$$

and $m > 1$, $N \geq 1$, $u_o \geq 0$, relies on a particular *a priori* estimate which is valid for all continuous, nonnegative solutions of (1.1), (1.2) defined in a strip Q_T, $0 < T \leq \infty$, independently of the initial conditions is very remarkable. If we define the pressure variable v as

$$v = \frac{m}{m-1} u^{m-1}, \tag{1.3}$$

then the Aronson-Bénilan estimate [AB] reads

$$v \geq -\frac{k}{t} \text{ in } \mathcal{D}'(Q_T)$$

where $k = (m-1 + \frac{2}{N})^{-1}$. A function v satisfying an estimate like (1.4) is sometimes called semisubharmonic (with respect to the spatial variables). If $N = 1$ this coincides with the concept of semiconvexity.

Let us point out that (1.4) holds for the nonnegative solutions of (1.1) in a strip not only in the case $m > 1$ but also for $m \leq 1$. For $m = 1$, we have the linear heat equation $u_t = \Delta u$ and v is defined as $v = \log(u)$. Apart from (1.4) other results about log-convexity of solutions are well-known, cf. [BL]. Equation (1.1) for $m < 1$ is known as the fast-diffusion equation because $D(u) \to \infty$ as $u \to 0$. In [AB], it is proved that (1.4) holds as long as k is positive, i.e. for $m > (N-2)_+/N$. For this equation, the nonnegative solutions are C^∞, strictly positive functions and no free boundaries appear.

Estimate (1.4) plays a fundamental role in the existence theory for (1.1), (1.2) as developed in [BCP]. In view of its usefulness, one may be interested in

(i) Finding upper estimates corresponding to (1.4),

(ii) Extending these results to other equations.

We will address both questions in the following sections in the context of nonnegative solutions for initial-value problems in one space dimension.

Thus, Section 2 presents an account of known theory for the PME. Among other things, we explain in some detail the recent concavity result of Bénilan and Vazquez [BV] and its consequences, as an answer to question (i).

In Section 3 we present the extension of the semi convexity estimate (1.4) to the equation of nonlinear heat diffusion with absorption

$$u_t = (u^m)_{xx} - \lambda u^p; \; m,p > 1, \tag{1.5}$$

as done in Herrero-Vazquez [HV].

The rest of the paper is devoted to introducing and improving a series of concavity results for equation (1.5). Contrary to the first two sections, the results are new and we give detailed proofs of them. Thus, in Section 4 we prove that the property of "concavity inside the postivity set" is also preserved by (1.5) and we obtain upper estimates for v_{xx} in $[u > 0]$ (Theorem 9). An interesting consequence is that for very large absorption exponents, namely if $p \geqq 2m+3$, we recover the same asymptotic convexity established in [BV] for the nonabsorption case (Theorem 10).

Finally, Section 5 contains an application of this concavity result to establish the behaviour of the free boundary. The results give new information for the range of exponents $1 < p < m$.

This paper reflects work done by a number of people with whom I had the opportunity of discussing the subject. I am especially indebted to Philippe Bénilan, whose contribution has been fundamental in this topic.

2. Convexity and concavity for the PME

We will consider here the Cauchy problem (1.1), (1.2) with $m > 1$ in the case of one space dimension $N = 1$. As we said above, the lower estimate on v_{xx} (1.4) plays an important role in the theory. This estimate is valid for all nonnegative solutions of (1.1) in Q_T. If we can control v_{xx} at $t = 0$ then a more precise estimate follows:

THEOREM 1 [BV]: Let u_0 be a continuous, nonnegative real function, let $v_0 = (m/(m-1))u_0^{m-1}$ and assume that

$$v_{0xx} \geqq -C \text{ in } \mathcal{D}'(\mathbb{R}) \tag{2.1}$$

Then, if u is the solution of (1.1), (1.2) and v is the pressure given by (1.3) we have

$$v_{xx} \geqq - \frac{C}{1 + C(m+1)t}. \tag{2.2}$$

The proof of this result can be done as in [AB]. We first assume that u is smooth and conveniently bounded as $|x| \to \infty$. Since $p = v_{xx}$ satisfies the equation

$$p_t = (m-1)vp_{xx} + 2mv_x p_x + (m+1)p^2 \tag{2.3}$$

and the right-hand term of (2.2) is also a solution of (2.3), the estimate follows from the maximum principle after comparing the values at $t = 0$. For general u we can use approximation by smooth solutions.

The bound (1.4) implies that the function $x \mapsto v_x(x,t) + x/((m+1)t)$ is nondecreasing for every fixed $t > 0$. Hence there exists the limit

$$v_x(s(t)-,t) = \lim_{\substack{x \to s(t) \\ x < s(t)}} v_x(s(t),t). \tag{2.4}$$

This enabled Knerr [Kn] (see also [A]) to prove that the (right-hand) free boundary $x = s(t)$ of a solution u which is compactly supported in the space variable, is a Lipschitz-continuous curve and s satisfies the equation

$$D^+s(t) = -v_x(s(t)-,t). \tag{2.5}$$

The proof is based in a local comparison argument near $(s(t),t)$. Caffarelli and Friedman [CF] exploited (1.3) and (2.5) to prove that after a certain <u>waiting time</u> t^* $(0 \leqq t^* < \infty)$ the free boundary $x = s(t)$ not only moves but is also smooth: $s \in C^1(t^*,\infty)$, and never stops again: $s'(t) > 0$ for $t > t^*$. In fact this is due to a certain semiconvexity of s. Subsequently, Vazquez showed that this semiconvexity is an exact replica of that of v. The result is

<u>THEOREM 2 [V1]</u>: With the above notations we have

$$s''(t) + \frac{m}{(m+1)t} s'(t) \geq 0 \tag{2.6}$$

in the sense of measures in $[0,\infty)$.

For a very easy proof of this fact see [ACV]. Estimate (2.6) has deep consequences for the asymptotic behaviour of both u and s as $t \to \infty$ as we explain next. In fact it can be reformulated as: $s'(t)t^{m/(m+1)}$ is non-decreasing in $(0,\infty)$. Therefore the following limit exists

$$\lim_{t\to\infty} s'(t)t^{m/(m+1)} = K \tag{2.7}$$

It is not difficult to prove that $s(t) = ct^{1/(m+1)} + O(1)$ as $t \to \infty$ with a constant c that depends only on m and $\|u_0\|_1$, the L^1-norm of u_0, also called the initial mass, cf. [V1, p. 514]. Therefore, we conclude that $K = c/(m+1)$. If we call $r(t) = ct^{1/(m+1)}$, then (2.7) means that $s'(t) \leq r'(t)$. Therefore we get a new limit

$$x_0 = \lim_{t\to\infty} (s(t)-r(t)). \tag{2.8}$$

By the estimate on s(t), x_0 is finite. In fact it can be proved that x_0 coincides with the <u>centre of mass</u> of the initial distribution.

$$x_0 = M^{-1} \int_{\mathbb{R}} u_0(x)x\,dx \tag{2.9}$$

where $M = \|u_0\|_1$. In this way we get

<u>THEOREM 3 [V1]</u>: As $t \to \infty$ we get

$$s(t) = c_m(M^{m-1}t)^{1/(m+1)} + x_0 + o(1) \tag{2.10a}$$

$$s'(t) = \frac{c_m}{m+1} (M^{m-1}t^{-m})^{1/(m+1)} + o(1/t) \tag{2.10b}$$

From these formulae, we obtain estimates that relate u,v and v_x to the respective quantities $\bar{u}$, $\bar{v}$, $\bar{v}_x$ belonging to the solution of (1.2) which has as initial data $\bar{u}_0 = M\delta(x - x_0)$, where δ is Dirac's mass. This solution is explicit and known as the Barenblatt solution. Cf. details and proofs in [V1].

Recently, Bénilan and Vázquez [BV] have studied the existence of upper estimates for v_{xx}. Here the difficulty stems from the fact that across a point (x_0,t_0) which belongs to a moving free boundary, i.e. $x_0 = s(t_0)$ and $s'(t_0) > 0$, the first derivative v_x has a jump of value $|s'(t_0)|$, hence v_{xx} is a Dirac mass at this point. Because of this, we estimate v_{xx} from above only inside the positivity set

$$P[u] = \{(x,t) \in Q: u(x,t) > 0\}. \tag{2.11}$$

In case v has compact support I = [a,b] and is concave inside I, i.e.

$$v''_0 \leq -c < 0 \text{ in } \mathcal{D}'(I), \tag{2.12}$$

the positivity set can be described as $\{(x,t) \in Q: s_-(t) < x < s_+(t)\}$, where $s_-(t)$, $s_+(t)$ are the left-hand and right-hand free boundaries resp., and the following result holds.

THEOREM 4 [BV]: Under the above assumptions

$$v_{xx} \leq - \frac{c}{1 + (m+1)ct} \text{ in } P . \tag{2.13}$$

A direct proof of this result via the equation satisfied by v_{xx} faces the problem of the bad behaviour of v_{xx} at $x = s_\pm(t)$. Instead, in [BV] we study the two Cauchy problems

$$v_t = (m-1)vv_{xx}, \qquad v(x,0) = v_0(x), \tag{2.14}$$

$$v_t = v_x^2, \qquad v(x,0) = v_0(x). \tag{2.15}$$

The first problem is also degenerate parabolic, but it has the nice property that the support is stationary. Using this fact and an approximation process we get for its solutions the estimate

$$v_{xx} \leq - \frac{c}{1 + (m-1)ct} \text{ in } P, \tag{2.16}$$

if the initial data are given by a continuous, nonnegative function with compact support and satisfying (2.12). On the other hand $v_t = v_x^2$ is a first-order equation which can be studied along characteristics if the initial data are as above. We obtain in this case

$$v_{xx} \leqq - \frac{c}{1 + 2ct} \quad \text{in } P. \tag{2.17}$$

Both equations are then combined through a Trotter-Kato formula to obtain a solution of (1.2) that satisfies the estimate (2.13). Observe the simple arithmetics behind the passage from formulas (2.16), (2.17) to (2.13): $(m-1) + 2 = (m+1)$!

We remark that, since there are no free boundaries when $0 < m < 1$, we can give a direct proof of Theorem 4 in that case, based on the Maximum Principle as in Theorem 1.

The upper bound (2.13) has consequences on the behaviour of the interface or on the asymptotics as $t \to \infty$. Thus we have

THEOREM 5 [BV]: On the assumptions of Theorem 4 we have

$$s_+''(t) + \frac{mc}{1 + (m+1)ct} s_+'(t) \leqq 0 \text{ in } \mathcal{D}'(\mathbb{R}^+). \tag{2.18}$$

In particular s is a concave function of t. Similarly for $s_-(t)$.

THEOREM 6 [BV]: As $t \to \infty$ we have

$$v_{xx} = - \frac{1}{(m+1)t} + O(\frac{1}{t^2}) \tag{2.19}$$

uniformly in x for $(x,t) \in P$ and

$$s_+''(t) = r''(t)(1 + O(\frac{1}{t})), \ r(t) = c_m(\|u_0\|_1^{m-1} t)^{1/(m+1)}. \tag{2.20}$$

Explicit examples show that the convergence rates in these asymptotic expressions are optimal. Therefore we have obtained very detailed information on the behaviour of solutions with concave data.

Recently Aronson and Vazquez have improved these results in the following sense.

THEOREM 7 [AB]: Every solution u whose initial data u_0 is a continuous, nonnegative function with compact support has an eventually concave pressure, namely there exists a time $T > 0$ such that for every $t > T$ and $(x,t) \in P[u]$ we have

$$v_{xx} \leqq -c(t) < 0. \tag{2.21}$$

In this way the asymptotic results obtained for concave solutions extend to every nonnegative solution of (1.2) with compact support.

Let us point out to end this section that in the case of problem (1.1), (1.2) in several space dimensions, no results similar to our Theorem 4 are known and, in particular, an upper estimate on Δv of the form c/t seems not to be true under conditions on v_0 similar to (2.12). A main difference with respect to the one-dimensional case lies in the fact that there are several different concepts that translate the one-dimensional condition: $v_0'' \leqq 0$, e.g. concavity or superharmonicity, and none of them seems to be preserved by the equation $v_t = (m-1)v\Delta v + |\nabla v|^2$.

3. Heat diffusion with absorption: semiconvexity

We now consider the extension of the above results to the one-dimensional nonlinear heat equation with absorption

$$u_t = (u^m)_{xx} - \lambda u^p, \ (x,t) \in Q \tag{3.1}$$

where λ, m, p are positive constants, $m,p > 1$, with initial data

$$u(x,0) = u_0(x), \ x \in \mathbb{R}. \tag{3.2}$$

Here u_0 is a continuous, nonnegative and bounded function. The existence and uniqueness of weak solutions to problem (3.1), (3.2) as well as the main qualitative properties of them have been studied by a number of authors, beginning with Kalashnikov, Kersner and Knerr, cf. [HV] and its references. In this latter work Herrero and Vazquez studied the existence of lower bounds for v_{xx}, v being defined as in the PME by (1.3). Their main result is

THEOREM 8 [HV]: Let u be the solution to problem (3.1), (3.2) and let $p \geq m$. Then

$$v_{xx} \geqq -\frac{k}{t} \quad \text{with} \quad k = k(m,p) > \frac{1}{m+1}\,. \tag{3.3}$$

The proof is based on the study of the PDE satisfied by v_{xx}. Since v satisfies

$$v_t = (m-1)vv_{xx} + v_x^2 - \mu v^\beta, \tag{3.4}$$

with $\beta = (p+m-2)/(m-1)$ and $\mu = \mu(\lambda,m) > 0$, differentiating twice with respect to x gives for $v_{xx} = w$ the following PDE with coefficients and second member depending on v:

$$w_t - (m-1)vw_{xx} - 2mv_x w_x - (m+1)w^2 + \mu\beta\, v^{\beta-1} w = \tag{3.5}$$

$$= \mu\beta(\beta-1)v^{\beta-2}{v_x}^2.$$

As in the case of the PME we compare $w = v_{xx}$ with a subsolution of (3.5) of the form $\bar{w}(x,t) = -k/t$. In order to do that we have to estimate the last two terms. Since $\beta > 1$, $\bar{w} < 0$ and $v \geqq 0$ the former term, $-c\,v^{\beta-1}\bar{w}$, is non-negative, which is good for our purpose. As for the latter, we use the following decay estimates on v and v_x.

LEMMA 3.1 [HV]: For every solution of (3.1), (3.2) we have

$$u(x,t) \leq (\lambda(p-1)t)^{-1/(p-1)} \tag{3.6}$$

and

$$|v_x(x,t)|^2 \leq \frac{2}{m}\,\|v_0\|_\infty\, t^{-1}. \tag{3.7}$$

Both estimates combine to give an estimate for the term in question of the form

$$\mu\beta(\beta-1)v^{\beta-2}{v_x}^2 \leq Ct^{-2}, \qquad C = C(m,p). \tag{3.8}$$

Therefore, there is a value of K such that $\bar{w} = -k/t$ satisfies (3.3) with $\leq$ instead of =. Since $v_{xx} \geq \bar{w}$ at t = 0, by the maximum principle we conclude that $w \geq \bar{w}$ in Q and this ends the proof. The value of K is estimated in [HV] as

$$K \leq \frac{1}{m+1} + \left(\frac{2^{\sigma}(m+p-2)}{m(m^2-1)}\right)^{1/2}, \sigma = 2 + \frac{m-1}{p-1} . \tag{3.9}$$

Two remarks are in order regarding Theorem 8. Firstly, the theorem does not apply to the range of parameters <u>$1 < p < m$</u>. It is to be noted that the case p = 1 can be easily reduced to the nonabsorption case $\lambda = 0$ (see [HV, formula (2.13)]), hence a lower estimate for v_{xx} follows in this case. It is probable that no such estimate holds for $1 < p < m$ since the result is false if we eliminate the diffusion terms, i.e. if we consider the equation

$$v_t = - \mu v^{\beta} \text{ with } 1 < \beta < 2. \tag{3.10}$$

The second remark concerns the case of <u>large</u> absorption exponents: <u>$p > m+2$</u>. It is well known that in this case the solution of (3.1), (3.2) corresponding to initial data $u_0 \in L^1(\mathbb{R})$ has the same decay rates for $t \to \infty$ as the non-absorption equation (the PME), i.e. $u(x,t) = O(t^{-1/(m+1)})$, uniformly in $x \in \mathbb{R}$, [HV] (this does not happen for $m \leq p < m+2$). Now, if we combine this estimate for u with (3.7) we obtain the bound for the last term of (3.5)

$$v^{\beta-2}{v_x}^2 \leq Ct^{-2-\theta}, \theta = \frac{p-m+2}{m+1} , \tag{3.11}$$

where C depends on u_0, m and p. In this way, the argument of Theorem 8 can be improved to yield a bound that is asymptotically the one of the PME:

<u>COROLLARY 3.2</u>: Let u be the solution to problem (3.1), (3.2) with $u_0 \in L^1(\mathbb{R})$. If $p > m+2$ we have

$$v_{xx} \geq - \frac{1}{(m+1)t} - \frac{k}{t^{1+\theta}}$$

where $\theta = (p-m+2)/(m+1)$ as above and $k = C/(1+\theta)$.

Again the proof consists of checking that the second member of (3.12) is a subsolution for equation (3.5).

In much the same way as was done for the PME, these bounds on v_{xx} can be used to get regularity of solutions and interfaces as well as asymptotic behaviour. We refer to [HV] for details about such results.

4. Heat diffusion with absorption: concavity

In this section, we establish some new concavity results for the solutions of the Cauchy problem (3.1), (3.2). We will assume that the initial data are given by a continuous, nonnegative and bounded function u_0 supported in a bounded interval $I_0 \subset \mathbb{R}$ and such that its pressure v_0 satisfies

$$v_0'' \leq -c_0 < 0 \text{ in } \mathcal{D}'(I_0). \tag{4.1}$$

Here is the basic concavity result.

<u>THEOREM 9</u>: Let u be a solution of (3.1), (3.2) with $m,p > 1$ and u_0 as above. Then for every $t > 0$ the support of the function $u(\cdot,t)$ is an interval $I(t)$,$v(\cdot,t)$ is concave in this interval and moreover

$$v_{xx}(x,t) \leq -c(t) \text{ in } I(t), \tag{4.2}$$

where $c(t)$ is determined as follows: $c(0) = c_0$ and $y(t) = 1/c(t)$ satisfies

$$y'(t) = \mu\beta M(t)y(t) + (m+1), \tag{4.3}$$

with $M(t) = \|v(\cdot,t)\|_\infty^\gamma$ and $\gamma = (p-1)/(m-1)$.

In case v_0 is only assumed to be concave in its support we have

<u>COROLLARY 4.1</u>: Let u be a solution of (3.1), (3.2) under the above conditions but with $c_0 = 0$. Then for every $t > 0$ $v(\cdot,t)$ is a concave function in the interval $I(t)$ where it is positive.

The proof of this corollary follows from Theorem 9 by simply approximating v_0 by strictly concave functions $v_{0,\varepsilon}$ satisfying (4.1).

PROOF OF THEOREM 9: Following the ideas of [BV], we will reduce the study of problem (3.1), (3.2) to the study of the two partial problems

$$u_t = (u^m)_{xx},\ u(x,0) = u_0(x) \tag{4.4}$$

and

$$u_t = -\lambda u^p,\ \ u(x,0) = u_0(x). \tag{4.5}$$

Both Cauchy problems generate contraction semigroups in $L^1(\mathbb{R})$, cf. [C]. It follows from the explanation in Section 2 that Theorem 10 holds for the solutions of (4.4) with a bound

$$c_1(t) = \frac{c_0}{1 + (m+1)c_0 t}. \tag{4.6}$$

A direct integration gives for the solution of (4.5) - i.e. of $v_t = -\mu v^\beta$, $v(x,0) = v_0(x)$ - the expression

$$v(x,t) = v_0(x)(1 + \ \mu\gamma v_0(x)^\gamma t)^{-1/\gamma} \tag{4.7}$$

with $\gamma = \beta - 1 = (p-1)/(m-1)$. From (4.7) it easily follows that whenever v_0 satisfies (4.1) then $v(x,t) > 0$ if and only if $x \in I_0$ and then

$$v_{xx}(x,t) \leqq v_0''(x)(1 + \mu\gamma v_0(x)^\gamma t)^{-\beta/(\beta-1)}.$$

In fact this calculation can be done at any time $t_0 > 0$ so that we actually get for every t_0, $h > 0$

$$v_{xx}(x,t_0+h) \leqq v_{xx}(x,t_0)(1 + \mu\gamma v^\gamma(x,t_0)t_0)^{-\beta/(\beta-1)}. \tag{4.8b}$$

Since $v(x,t) \leqq \|v_0\|_\infty$, it follows that $v_{xx} \leqq -c_2(t)$, where $c_2(t)$ satisfies

$$c_2(t) = c_0(1 + \ \mu\gamma\|v_0\|_\infty^\gamma\ t)^{-\beta/(\beta-1)}. \tag{4.9}$$

This means that $c_2'(0) = \ \mu\beta\|v_0\|_\infty^\gamma c(0)$. Again the same calculation can be performed starting at any time $t_0 > 0$ so that this differential equation is valid for all times. We sum up this result for later use as follows:

LEMMA 4.1: Under the above hypotheses on m,p and u_0 the solution to problem (4.4) satisfies in I_0 the estimate (4.2) with a bound $c_2(t)$ whose inverse $y_2(t)$ is determined by

$$y_2'(t) = \mu\beta \| v(\cdot,t)\|_\infty^\gamma \, y_2(t), \; y_2(0) = 1/c_0. \tag{4.10}$$

The result (4.3) is now obtained from the estimates (4.6), (4.10) by using a Trotter-Kato formula. Let us denote by $S_1(t)$ the semigroup generated by (4.4) (i.e. $u(x,t) = (S(t)u_0)(x)$ is the solution of (4.4) with initial data u_0), $S_2(t)$ the semigroup generated by (4.5) and $S(t)$ the one generated by (3.1), (3.2). We may obtain S from S_1 and S_2 through the formula

$$S(t)u_0 = \lim_{n\to\infty} (S_1(t/n)S_2(t/n))^n(u_0) \tag{4.11}$$

The fact that (4.11) holds follows from properties of semigroups of contractions in a Banach space, cf. [BI], but a direct proof under our hypotheses is possible by repeating the arguments of [BV] for the PME (explained in Section 2). The limit in (4.11) is uniform on compact subsets of $\mathbb{R} \times [0,\infty)$. To calculate an upper bound for v_{xx} at a time $t > 0$ if v is the pressure of the solution of (3.1), (3.2), we look first for an estimate on v_{xx}^n where u^n is given as

$$u^n(x) = \{[S_1(t/n)S_2(t/n)]^n(u_0)\}\,(x). \tag{4.12}$$

Let us introduce some notation: Let $u_j^{\,n} = [S_1(t/n)S_2(t/n)]^j(u_0)$ for $j = 0,1,\ldots,n$, and let $v_j^{\,n} = \{(m/(m-1)\}(u_j^{\,n})^{m-1}$. Combining (4.6) and (4.10) we can obtain bounds, y_j^n, for $1/(-\sup(v_j^{\,n})_{xx})^{(\dagger)}$. We have $y_0^{\,n} = 1/c_0$ and for $j \geqq 1$

$$y_j^{\,n} = (y_{j-1}^n + (m+1)t/n)) + \mu\beta \, \|v_0\|_\infty^\gamma \, (t/n) y_j^{\,n}, \tag{4.13}$$

where we have used the fact that $\|v_j^{\,n}\|_\infty \leq \|v_0\|_\infty$. Since $\{y_j^{\,n}\}$ is increasing in j we may replace $y_j^{\,n}$ by $y_n^{\,n}$ in the last term of (4.13) and work out the recurrence to obtain, with $y^n = y_n^{\,n}$,

---

(†) All the sup are taken in the corresponding positivity sets.

$$y^n \leqq (1/c_0) + (m+1)t + \mu\beta\|v_0\|_\infty^\gamma t y^n. \tag{4.14}$$

For t small enough we may express y_n in the form

$$y^n \leqq (1/c_0 + (m+1)t)/(1-\mu\beta\|v_0\|_\infty^\gamma t) = \tag{4.15}$$

$$= 1/c_0(1 + \mu\beta\|v_0\|_\infty^\gamma t) + (m+1)t + o(t),$$

where the last term is small uniformly in n. Letting $n \to \infty$ we obtain, in this way, an estimate on $\bar{y}(t) = (-\sup v_{xx})^{-1}$ at time t given by $\bar{y}(t) \leqq$ second member of (4.15). Clearly we have as $t \to 0$

$$\frac{1}{t}|\bar{y}(t) - \bar{y}(0)| \leqq y(0)\mu\beta\|v_0\|_\infty^\gamma + (m+1) + o(t). \tag{4.16}$$

This estimate on the right-hand derivative of y(t) at t = 0 has obviously a similar counterpart at every $t > 0$. Moreover replacing the $\leqq$ sign by an equals sign in (4.16) means passing from the bound $\bar{y}(t)$ to a possibly larger bound y(t) which satisfies (4.3). This ends the proof.

In order to get an explicit value for y(t) we need an estimate on $\|v(\cdot,t)\|_\infty$. This can be done by remarking that if $N = \|v_0\|_\infty$, the function

$$\bar{v} = (N^{-\gamma} + \mu\gamma t)^{-1/\gamma}, \quad \gamma = \beta - 1,$$

is a solution of equation (3.4) with initial data $\bar{v}(x,0) = M \geqq v(x,0)$. By the Maximum Principle (cf. [HV, Theorem 0]) we have $\bar{v}(x,t) \geqq v(x,t)$ in Q, hence

$$\|v(\cdot,t)\|_\infty \leqq (N^{-\gamma} + \mu\gamma t)^{-1/\gamma} \tag{4.18}$$

Substituting (4.18) into (4.3) and integrating gives the following result

<u>COROLLARY 4.2</u>: With the notation of Theorem 10 we have

$$y(t) \leqq (1/c_0)(1 + \mu\gamma N^\gamma t)^{\beta/(\beta-1)} + \tag{4.19}$$
$$+ \frac{m+1}{N}[(1 + \mu\gamma N^\gamma t)^{\beta/(\beta-1)} - (1 + \mu\gamma N^\gamma t)].$$

Therefore $c(t) \geqq O(t^{-\beta/(\beta-1)})$ for $t >> 0$.

Remark that, when either $\beta \to 1$ or $\mu \to 0$, we get in the limit the bound $c(t) = 1/((m+1)t)$ of the PME.

In the case of large diffusion exponent, $p > m+2$, this result can be considerably improved. As we said above for that range of exponents we get an estimate $\|u(\cdot,t)\|_\infty \sim t^{-1/(m+1)}$ for initial data u_0 in $L^1(\mathbb{R})$, which is the case under our assumptions. Using this estimate in (4.3) and integrating the resulting ODE we get

COROLLARY 4.3: Under the above hypotheses if moreover $p > m+2$ we have

$$c(t) \geqq \frac{1}{(m+1)t} - K\left(\frac{1}{t^2} + \frac{1}{t^{1+\theta}}\right) \tag{4.20}$$

where K depends on u_0, m, p and $\theta = (p-m-2)/(m+1)$.

Corollaries 3.2 and 4.3 can be summarized in the following result.

THEOREM 10: Let u be a solution of (3.1), (3.2) whose initial pressure satisfies the concavity assumption (4.1). If $p > m+2$ we have as $t \to \infty$

$$v_{xx} = -\frac{1}{(m+1)t} + O(t^{-1-\eta}),\ \eta = \min\left(1, \frac{p-m-2}{m+1}\right), \tag{4.21}$$

uniformly in the positivity set of the solution.

This result shows how close these solutions are for large t to the explicit Barenblatt solutions of the PME, cf. [AB] or [V1]. In particular for $p \geqq 2m+3$ we again obtain estimate (2.19).

5. An application

As we have explained in Sections 2 and 3, the existence of a lower bound of the form $v_{xx} \geqq -C(\tau)$ valid for $t \geqq \tau > 0$ is the basic tool in establishing that the free boundary $x = s(t)$ is a Lipschitz-continuous curve governed by equation (2.5) and, moreover, that s is a strictly increasing C^1-function for $t > t^*$. This programme failed in the case of the Cauchy problem (3.1), (3.2) with $1 < p < m$ because of the lack of such an estimate. Nevertheless, if v_0 is concave in its support, the results of Section 4 imply the estimate $v_{xx} \leqq 0$ in P[u]. Such an upper estimate is also sufficient to derive the above mentioned properties for the free boundary as can be checked by repeating the arguments in [Kn], [CF], [ACK]. We get

THEOREM 11: Let $u(x,t)$ be a solution of (3.1), (3.2) with $m,p > 1$, let v_0 satisfy $v_0'' \leqq -c_0 \leqq 0$ in its support and let $x = s(t)$ be the right-hand free boundary of u. Then s is a C^1 concave function in $(0,\infty)$ and

$$s'(t) = -v_x(s(t),t) > 0 \text{ for every } t > 0. \tag{5.1}$$

PROOF: Formula (5.1) is established as explained in Section 2. Remark that a concave function $v(\cdot,t)$, that does not vanish identically, necessarily has a negative derivative at the right end of its interval of definition. Therefore, $s'(t) > 0$. This positivity, together with the concavity of $v(\cdot,t)$ allows us to repeat the arguments in [CF], [ACK] to establish C^1 smoothness.

A precise concavity result for the PME is established in [BV]. The C^1 smoothness and (5.1) are new in the absorption case if $1 < p < m$.

References

[A] D.G. Aronson, Regularity properties of flows through porous media: the interface, Arch. Rat. Mech. Anal., 37 (1970), 1-10.

[AB] D.G. Aronson, Ph. Bénilan, Régularité de l'équation des milieux poreux dans $\mathbb{R}^N$, C.R. Acad. Sci. Paris, A-B 288 (1979), 103-105.

[AV] D.G. Aronson, J.L. Vazquez, Eventual C^∞-regularity and concavity for the solutions of the one-dimensional porous medium equation, preprint, 1985.

[BCP] Ph. Bénilan, M.G. Crandall, M. Pierre, Solutions of the porous medium equation in $\mathbb{R}^N$ under optimal conditions on initial values, Indiana Univ. Math. J., 33 (1984), 51-87.

[BI] Ph. Bénilan, S. Ismail, Générateur des semi-groupes non linéaires et la formule de Lie-Trotter, Annales Fac. Sci. Toulouse, 7 (1985), 151-160.

[BV] Ph. Bénilan, J.L. Vazquez, Concavity of solutions of the porous medium equation, Trans. Amer. Math. Soc., to appear.

[BL] H.J. Brascamp, E.H. Lieb, On extension of the Brunn-Minkowski and Prékopa-Leindler Theorems. Including Inequalities for Log Concave

Functions and with an Application to the Diffusion Equation, J. Funct. Anal. 22, 366-389 (1976).

[CF] L.A. Caffarelli, A. Friedman, Regularity of the free boundary for the one-dimensional flow of gas in a porous medium, Amer. J. Math., 101 (1979), 1191-1218.

[C] M.G. Crandall, An introduction to evolution governed by accretive operators, in Dynamical Systems, An International Symposium (L. Cesari et al. eds.), Academic Press, New York, 1976, pp. 131-165.

[GM] M.E. Gurtin, R.C. McCamy, On the diffusion of biological populations, Math. Biosci., 33 (1977), 35-49.

[HV] M.A. Herrero, J.L. Vazquez, The one-dimensional nonlinear heat equation with absorption: regularity of solutions and interfaces, SIAM J. Math. Analysis, to appear.

[K] B.F. Knerr, The porous medium equation in one dimension, Trans. Amer. Math. Soc., 234 (1977), 381-415.

[M] M.Muskat, The flow of homogeneous fluids through porous media, McGraw-Hill, New York, 1937.

[OKC] O.A. Oleinik, A.S. Kalashinikov, Y.L. Czhou, The Cauchy problem and boundary problems for equations of the type of nonstationary filtration, Izv. Akad. Nauk SSSR Ser. Mat. 22 (1958), 667-704 (Russian).

[P] L.A. Peletier, The porous media equation, in Applications of Nonlinear Analysis in the Physical Sciences (H. Ammamm et al. eds.), Pitman, London, 1981, pp. 229-241.

[VI] J.L. Vazquez, Asymptotic behaviour and propagation properties of the one-dimensional flow of gas in a porous medium, Trans. Amer. Math. Soc., 277 (1983), 507-527.

[V2] J.L. Vazquez, Hyperbolic aspects in the theory of the porous medium equation, Proceedings of the Workshop on Metastability and PDE's, Minnesota, 1985, to appear in Springer Lecture Notes in Mathematics.

[ZR] Y.B. Zel'dovich, Y.P. Raizer, Physics of shock-waves and high-temperature hydrodynamic phenomena, Vol. 2, Academic Press, New York, 1966.

The author was supported by CAICYT 2805-83 and Programme de Coopération Hispano Français 1985.

Juan L. Vazquez
Departamento Matemáticas
Universidad Autónoma
28049 Madrid
Spain.

L VÁZQUEZ

Some open computational problems in quantum lattice theory

I. Introduction

In the last years we have witnessed a rapidly growing interest in the application of the finite difference method for solving the operator equations of motion in quantum mechanics and in quantum field theory. This line of work was started in 1983 by Bender and his group [1,2] and in this framework they are getting promising results [3,4,5]. At the same time other contributions in this line of research have been made by Moncrief [6] and Vázquez [7-13].

In this contribution, we describe general features which must satisfy the discretizations of the Heisenberg equations of motion, and at the same time such schemes are given. In Section II the numerical approach to the one-dimensional equation of motion for classical and quantum particles is summarized. In Section III we analyse the properties of the finite difference method in quantum field theory with application to the scalar field. In both sections we describe the open questions that arise in the finite difference approach to quantum theory.

II. One-dimensional motion of a particle in a potential

Let us consider the application of the finite difference method to solve the classical and quantum equations of motion of a particle in a potential.

A. Newtonian mechanics

The Newton equation

$$m \frac{d^2 q}{dt^2} = F(q)$$

$$q(0) = q_0,\ q_t(0) = V_0 \tag{1}$$

describes the displacement q(t) of a particle of mass m under the conservative force F(q). The physics of the system is contained in the following properties:

(1) The conservation of the energy E

$$\frac{dE}{dt} = 0 \tag{2}$$

where

$$E = \frac{1}{2} m \left(\frac{dq}{dt}\right)^2 + U(q) \tag{3}$$

and

$$F(q) = - dU/dq.$$

(2) The system is time-inversion symmetry-preserving.

(3) The behaviour of the potential energy U(q).

A possible numerical approach to the dynamical equation (1) consists of approximating the ordinary differential equation by a finite difference equation. On the other hand the corresponding numerical scheme must preserve the conservation law and symmetry of the underlying continuous equation, otherwise the numerical solutions do not show the physics of the dynamical model. A conservative and time reversible scheme [14] which approaches the equation (1) is

$$m \frac{q^{n+2}-2q^{n+1}+q^n}{(\Delta t)^2} = - \frac{U(q^{n+2})-U(q^n)}{q^{n+2}-q^n} \tag{4}$$

where q^n is the position at time $t = n\Delta t$. If we multiply (4) by $\frac{q^{n+2}-q^n}{2\Delta t}$ and we rearrange the terms, we get

$$\frac{E^{n+1}-E^n}{\Delta t} = 0, \quad \forall n \tag{5}$$

where

$$E^n = \frac{1}{2} \left(\frac{q^{n+1}-q^n}{\Delta t}\right)^2 + \frac{1}{2} \left(U(q^{n+1}) + U(q^n)\right) \tag{6}$$

which is a consistent discretization of (2). On the other hand the scheme (4) is invariant under time-inversion. In this way the conservation law and symmetry of the continuous equation (1) are shown by the scheme. These properties are very useful in studying numerically the long time behaviour of the solutions as well as the stability and convergence of the numerical scheme. It is clear that there are many consistent discretizations of the ordinary differential equation (1), but to our knowledge the scheme (4) together with the one proposed by Greenspan [15] are the only ones which show a discrete conserved energy for a general potential.

B. Quantum mechanics

Now let us consider the motion of a quantum particle. This motion can be studied either in the Schrödinger picture or in the Heisenberg picture:

B1. Schrödinger picture

In the nonrelativistic approach we have to solve the Schrödinger equation which is a partial differential equation

$$i \frac{\partial \psi}{\partial t} = - \frac{1}{2m} \frac{\partial^2 \psi}{\partial x^2} + U(x)\psi \tag{7}$$

$$\psi(x,0) = f(x).$$

This equation has two relevant conserved quantities:

The charge

$$Q = \int dx \quad \psi\psi^* \tag{8}$$

The energy

$$E = \frac{1}{2m} \int dx \left(\frac{\partial \psi}{\partial x} \frac{\partial \psi^*}{\partial x} + U(x) \psi\psi^*\right) \tag{9}$$

where ψ^* is the complex conjugate of ψ.

Among the numerical methods for solving the Schrödinger equation there are finite difference schemes [11, 16] that show discrete analogies of the conservation laws (8) - (9), which are the cornerstone of the theory. These numerical schemes allow us to get reliable numerical solutions that show the physics of the underlying continuous equation. In this framework the implicit scheme by Delfour et al [16] admits conserved discrete charge and energy:

$$i \frac{\psi_\ell^{n+1} - \psi_\ell^n}{\Delta t} = -\frac{1}{4m} \left(\frac{\psi_\ell^{n+1} - 2\psi_\ell^{n+1} + \psi_{\ell-1}^{n+1}}{(\Delta x)^2} + \frac{\psi_{\ell+1}^n - 2\psi_\ell^n + \psi_{\ell-1}^n}{(\Delta x)^2} \right) \tag{10}$$

$$+ \frac{1}{2} U_\ell \left(\psi_\ell^{n+1} + \psi_\ell^n \right)$$

$$Q^n = \sum_\ell \Delta x \, \psi_\ell^{*n} \psi_\ell^n \, , \; Q^n = Q^0 \tag{11}$$

$$E^n = \frac{1}{2m} \sum_\ell \Delta x \left[\left(\frac{\psi_{\ell+1}^{*n} - \psi_\ell^{*n}}{\Delta x} \right) \left(\frac{\psi_{\ell+1} - \psi_\ell^n}{\Delta x} \right) + U_\ell \, \psi_\ell^{*n} \psi_\ell^n \right] \tag{12}$$

$$E^n = E^0$$

where ψ_ℓ^n denotes the value of the function ψ at the point $(n\Delta t, \; \ell\Delta x)$.

It is obvious that the problem of convergence and stability of the numerical shceme (10) for the Schrödinger equation fits into the general theory of the numerical approach to partial differential equations.

B2. Heisenberg picture

As is well-known, the motion of a quantum particle can also be studied by the Heisenberg picture. While in the Schrödinger picture we have to solve a partial differential equation, now we have to deal with ordinary differential equations of operators.

For the one-dimensional quantum system the Hamiltonian is

$$H = P^2/2m + U(q) \tag{13}$$

and the Heisenberg equations of motion are

$$\frac{dq(t)}{dt} = \frac{p(t)}{m} \; ; \quad \frac{dp(t)}{dt} = F(q(t)) \tag{14}$$

where $F(q) = -dU/dq$, and $p(t)$ and $q(t)$ are operators which must satisfy the equal-time commutation relation

$$[q(t), p(t)] \equiv q(t)\,p(t) - p(t)q(t) = i \tag{15}$$

The quantum mechanical problem consists of solving the ordinary differential equations (14) for the operators $q(t)$ and $p(t)$ with the constraint (15).

By applying the finite-difference approach to solve (14), the Heisenberg equations become operator difference equations, such that the equal-time commutation relation is satisfied in each step:

$$[q^n, p^n] = i \quad \forall n \tag{16}$$

where q^n and p^n are the operators q and p at time $t = n\Delta t$. We must stress that (16) is similar to the existence of a discrete conserved energy in the classical case. (Eq. (5)). The above condition is a necessary property of the discretization because the physical properties of the discrete solution must be the same as those of the underlying continuous equation. As an example [10] of a discretization which preserves the equal-time commutation relation, we have:

$$\frac{q^{n+1}-q^n}{\Delta t} = \frac{1}{m} p^n$$
$$\frac{p^{n+1}-p^n}{\Delta t} = F(q^{n+1}) \tag{17}$$

From the point of physical applications it is important to remark that from the sequence of operators

$$\{(q^0, p^0), (q^1, p^1), \ldots, (q^n, p^n)\}$$

we can obtain useful information about the system: the spectrum, the estimation of unequal-time commutators and the computation of S-matrix elements.

In the study of the finite difference schemes of operators as (17), several mathematical questions arise, questions that can be summarized as follows:

(1) Conditions which guarantee the convergence of the solution to a proper continuum limit as $\Delta t \to 0$.

(2) Stability conditions of the finite-difference scheme.

(3) Relation between the classical finite-difference schemes and the quantum finite-difference schemes.

III. The quantum field theory

A quantum field equation is simply a partial differential equation of operators. An example is the nonlinear quantum scalar field

$$\frac{\partial^2\phi}{\partial t^2} - \frac{\partial^2\phi}{\partial x^2} + \sigma\phi + \lambda\phi^3 = 0 \tag{18}$$

where $\phi(x,t)$ is an operator. When the equation is linear ($\lambda = 0$) the properties of the system are well-known [17].

In the case of nonlinear quantum fields we have three methods of study:

A. Rigorous mathematical approach

This line of work has been developed, mainly, by the school created by J. Glimm and A. Jaffee [18].

B. Numerical approach

This method allows us to make nonperturbative computations in quantum field theory and it is based on the numerical estimation of functional integrals using a Monte Carlo algorithm. The different aspects of this method are described in the book by Creutz [19].

C. Minkowski lattice approach

This method was introduced by Bender and his group in 1983, and consists simply of using the finite difference method to solve the partial difference equations of operators which appear in quantum field theory. As an example let us consider system (18). More precisely the quantum problem consists of solving the equations

$$\begin{aligned} \frac{\partial\phi}{\partial t} &= \Pi \\ \frac{\partial\Pi}{\partial t} &= \frac{\partial^2\phi}{\partial x^2} - \sigma\phi - \lambda\phi^3 \end{aligned} \tag{19}$$

for the operators $\phi(x,t)$ and $\Pi(x,t)$ such that the following equal-time commutation relations are satisfied:

$$[\phi(x,t),\ \phi(y,t)] = 0$$

$$[\Pi(x,t),\ \Pi(y,t)] = 0 \qquad (20)$$

$$[\phi(x,t),\ \Pi(y,t)] = i\delta\ (x-y)$$

An appropriate consistent discretization of the equations (19) must show discrete equal-time commutation relations analogous to (20), otherwise the mathematical and physical properties of the finite-difference equations are not related to those of the underlying continuous equation. An example of such discretization is the following [10]

$$\frac{\phi_\ell^{n+1}-\phi_\ell^n}{\Delta t} = \Pi_\ell^{n+1}$$

$$\frac{\Pi_\ell^{n+1}-\Pi_\ell^n}{\Delta t} = \frac{\phi_\ell^{n+1}-2\phi_\ell^n+\phi_{\ell-1}^n}{(\Delta x)^2} - \sigma\phi_\ell^n - \lambda(\phi_\ell^n)^3 \qquad (21)$$

where ϕ_ℓ^n and Π_ℓ^n are the values of the field operators ϕ and Π at the point $(t = n\Delta t,\ x = \ell\Delta x)$. This scheme is explicit and preserves the equal-time commutation relations

$$[\phi_\ell^n,\ \phi_k^n] = 0,\quad [\Pi_\ell^n,\ \Pi_k^n] = 0,\quad [\phi_\ell^n,\ \Pi_k^n] = \frac{i}{\Delta x}\ \delta_{\ell k} \qquad (22)$$

With the Minkowski lattice approach it is possible to obtain spectrum estimations, S-matrix elements and unequal-time commutators. To date the results obtained on the lattice are consistent with those known for the underlying continuous system. In particular for the scalar field [8,9] we have:

(a) $\phi_{tt} - \phi_{xx} + \sigma\phi + \lambda\phi^3 = 0$

When $\lambda = 0$ (linear theory) all the levels of the spectrum mass are equally separated and the mass gap is $\sigma^{1/2}$. This means that the Minkowski lattice approach shows the same spectrum as the underlying continuous theory.

If $\lambda \neq 0$ two-particle bound states are absent from the lattice, in correspondence with the rigorous mathematical results for the continuous system.

(b) The quantum Sine-Gordon field

$$\phi_{tt} - \phi_{xx} + \frac{m^3}{\sqrt{\lambda}} \sin\left(\frac{\sqrt{\lambda}}{m}\phi\right) = 0. \tag{23}$$

The spectrum energy on the lattice is consistent with the results of Dashen-Hasslacher-Neveu for the underlying continuous Sine-Gordon field.

The results on the Minkowski lattice concerning the quantum spinor fields are contained in references [2], [3] and [7].

As in the case of the Heisenberg equations of a particle, the following open questions appear about the finite-difference equations of operators obtained from the quantum field equations:

(1) The study of the relation between the results on the Minkowski lattice and the properties of the underlying continuous system.

(2) Convergence of the theory based on the lattice when Δx, $\Delta t \to 0$.

(3) Stability of the finite-differentiating approximations.

(4) The study of the relation between the theory of the classical finite-difference equations and the theory of the quantum finite-difference equations.

References

1. C.M. Bender and D.H. Sharp: "Solution of operator field equations by the method of finite elements" Phys. Rev. Lett. 50, 1535 (1983).
2. C.M. Bender, K.A. Milton and D.H. Sharp: "Consistent formulation of fermions on a Minkowski lattice", Phys. Rev. Lett. 51, 1815 (1983).
3. C.M. Bender, K.A. Milton and D.H. Sharp: "Gauge invariance and the finite-element solution of the Schwinger model", Phys. Rev. D31, 383 (1985).
4. C.M. Bender, F. Cooper, J.E. O'Dell and L.M. Simmons Jr. "Quantum tunneling using discrete-time operator difference equations", Phys. Rev. Lett. 55, 901 (1985).
5. C.M. Bender, K.A. Milton, D.H. Sharp, L.M. Simmons Jr. and R. Stong: "Discrete time quantum mechanics", Phys. Rev. D32, 1476 (1985).

6. V. Moncrief: "Finite-difference approach to solving operator equations of motion in quantum theory", Phys. Rev. D28, 2485 (1983).
7. L. Vázquez: "Dirac field on a Minkowski lattice", Phys. Rev. D32, 2066 (1985).
8. L. Vázquez: "The two dimensional quantum field theory $\Box\phi + \lambda\phi + \lambda\phi^3 = 0$ on a Minkowski lattice", Phys. Rev. (to be published).
9. L. Vázquez: "Particle spectrum estimations for the quantum field theory $\phi_{tt} - \phi_{xx} + \frac{m^3}{\sqrt{\lambda}} \sin\left(\frac{\sqrt{\lambda}}{m}\phi\right) = 0$ on a Minkowski lattice", BiBoS 95/85 (Research Center Bielefeld-Bochum-Stochastics, F.R. Germany).
10. L. Vázquez: "On the discretization of certain operator field equations". Zeitschrift für Naturforschung 41a, 788 (1986).
11. L. Vázquez: "Explicit schemes to solve the Schroedinger field on a Galileo lattice", Phys. Rev. D.34, 3253 (1986).
12. L. Vázquez: "A more accurate explicit scheme to solve certain quantum operator equations of motion". Submitted to Zeitschrift für Naturforschung.
13. L. Vázquez: "About the unitary discretizations of Heisenberg equations of motion". BiBoS 233/86
14. L. Vázquez: "Long time behvior in numerical solutions of certain dynamical systems". To appear in Anales de Fisica.
15. D. Greenspan: "Conservative numerical methods for $\ddot{x} = f(x)$", J. Comp. Phys. 56, 28 (1984).
16. M. Delfour, M. Fortin and G. Payre: "Finite-Difference solutions of a nonlinear Schroedinger equation", J. Comp. Phys. 44, 277 (1981).
17. S.S. Schweber: "An Introduction to Relativistic Quantum Field Theory" (Harper & Row, London, 1966).
18. J. Glimm and A. Jaffee: "Quantum Physics: A Functional Integral Point of View" (Springer-Verlag, N.Y. Inc. 1981).
19. M. Creutz: "Quarks, Gluons and Lattices" (Cambridge University Press, 1983).

L. Vázquez
Departamento de Física Teórica
Facultad de Ciencias Físicas
Universidad Complutense
28040 Madrid
Spain.

J M VEGAS

Irregular variations of the domain in elliptic problems with Neumann boundary conditions

1. Introduction

We consider the scalar semilinear parabolic equation

$$u_t = \Delta u + f(x,u) \text{ in } \Omega$$
$$\frac{\delta u}{\delta n} = 0 \qquad \text{on } \delta\Omega \tag{1}$$

where $f:\mathbb{R}^n \times \mathbb{R} \to \mathbb{R}$ is a smooth function such that $f(x,u)u \leq 0$ for $|u| \geq M$ and Ω is a bounded domain in $\mathbb{R}^n$ with smooth boundary $\delta\Omega$; $\delta u/\delta n$ denotes the normal derivative of u. (1) defines a dissipative semigroup on $H^1(\Omega)$ which has a gradient structure given by the energy functional

$$J(u) = \int_\Omega (\tfrac{1}{2}|\nabla u|^2 - F(x,u))dx, \quad u \in H^1(\Omega) \tag{2}$$

where $F(x,u) = \int^u f(x,u)dv$. Due to this fact, the asymptotic behaviour of the semigroup depends essentially on the set $E(\Omega)$ of its equilibrium points, which is *a priori* bounded in $L^\infty(\Omega)$, by our hypotheses on f. Thus, we are led to concentrate on the stationary problem

$$-\Delta u = f(x,u) \text{ in } \Omega$$
$$\frac{\delta u}{\delta n} = 0 \qquad \text{on } \delta\Omega \tag{3}$$

Our main interest is to study the variations in the general behaviour of (1) as a function of Ω, and, in particular, the structure of $E(\Omega)$ as dependent on Ω.

In order to do this, we imbed Ω in a "continuous" family of domains Ω_ε, where $\Omega \equiv \Omega_0$, and Ω_ε converges to Ω_0 in some sense. The results concerning the possible "continuation" of the structure of $E(\Omega_0)$ to $E(\Omega_\varepsilon)$ will vary according to the type of convergence considered.

Thus, for very regular perturbations (that is, convergence of $\delta\Omega_\varepsilon$ to $\delta\Omega_0$ in C^1), one can obtain asymptotic expansions of the associated Green's functions by writing $\delta\Omega_0$ as the zero set of a smooth, 1-1 function $h(x,\varepsilon)$ for $\varepsilon = 0$ (Hadamard) (see Fujiwara and Ozawa [1]). By defining Ω_ε as the image of Ω_0 under a C^k-small perturbation of the identity, Mignot, Murat and Puel [5] study the nonlinear eigenvalue problem $-\Delta u = \lambda f(u)$ in Ω with homogeneous Dirichlet boundary conditions (though their method can easily be adapted to the Neumann case), obtaining asymptotic formulas for some critical values of λ in terms of "derivatives with respect to the domain".

Hale and Vegas [2], Vegas [8,9], Matano and Mimura [4] and others study the case of some families Ω_ε that can be referred to as semi-regular perturbations, since, without the above C^k-convergence of $\delta\Omega_\varepsilon$, they assume that $|\Omega_\varepsilon \setminus \Omega_0| \to 0$ as $\varepsilon \to 0$ ($|\cdot|$ denotes the Lebesgue measure in $\mathbb{R}^n$), plus a certain type of spectral convergence of the first few eigenvalues of the Laplacian on Ω_ε (with Neumann boundary conditions) to the ones corresponding to Ω_0, together with the associated spectral projections. Here Ω_0 is the union of two connected domains. In these hypotheses, Hale and Vegas (2) show that the method of Liapunov-Schmidt can be applied, and the assumed spectral properties ensure the continuity of the resulting bifurcation functions. Then, under some hypotheses of f, any locally constant hyperbolic solution u_0 can be uniquely "continued" to a "curve" u_ε of equilibrium solutions which share the stability properties of u_0.

In this paper we drop any restriction on the convergence of Ω_ε to Ω_0, except for the basic one of convergence in measure: $|\Omega_\varepsilon \setminus \Omega_0| \to 0$ as $\varepsilon \to 0$. In this case, Lobo and Sánchez-Palencia [3] show that the spectral family associated to $-\Delta$ on Ω_ε converges to the spectral family associated to $-\Delta$ on Ω_0; however, this does not imply convergence of eigenvalues and spectral projections, so, in general, the Liapunov-Schmidt method gives rise to bifurcation functions which are discontinuous in ε; thus, the results in Hale and Vegas [2] cannot be generalized. Here, by applying a variational method, we show that, in spite of such discontinuities, there is an underlying continuous character in the simple convergence in measure, in the sense that a hyperbolic stationary solution of (1) on Ω_0 cannot "disappear" when we perturb Ω_0 to Ω_ε as long as $|\Omega_\varepsilon \setminus \Omega_0|$ is sufficiently small. Nothing is stated, though, about the number of solutions into which it "continuates" or "bifurcates". The result is based on the following theorem of Castro, in the version quoted by Nirenberg [6]:

THEOREM (Castrol) [6]: Let f be a C^1 real function defined on a Banach space X satisfying the Palais-Smale condition. Assume that X has a direct sum decomposition $X = X_1 \oplus X_2$, with $\dim X_1 < \infty$. Let S_1 (B_1) be the unit sphere (ball) in X_1, and S_2 a sphere $\|x\| = R$ in X_2. Assume for some constants $\beta < \alpha$; $c_0 < c_1$:

$$f(x) \leqq c_0 \text{ for } x \in X_1, \ \|x\| < 1,$$

$$f(x) \leqq \beta \text{ for } x \in X_1, \ \|x\| = 1.$$

$$f(x) \geqq \alpha \text{ for } x \in X_2, \ \|x\| \leqq R$$

$$f(x) \geqq c_1 \text{ for } x \in X_2, \ \|x\| = R.$$

Then $c = \inf\{\max(f(x) : x \in \Sigma) : \Sigma$ is a k-dimensional surface in X spanning S_1 and homotopic to B_1 in $X \setminus S_2\}$ is a critical value of f.

2. Statement and proof of the results

The first thing we have to make sure is that no solution may arise "out of nothing", that is, any branch of solutions u_ε of (3) on Ω_ε, as well as any branch of critical values of the associated energy functional J given by (2) must have a "starting point":

THEOREM 1: Let $f : \mathbb{R}^n \times \mathbb{R} \to \mathbb{R}$ be smooth, globally Lipschitz in u and satisfy $f(x,u)u < 0$ for $|u| \geqq M$ for some M. Let Ω_ε be a family of domains containing Ω_0 and such that $|\Omega_\varepsilon \setminus \Omega_0| \to 0$ as $\varepsilon \to 0$. Then, if $\varepsilon_n \to 0$ as $n \to \infty$, u_{ε_n} is a solution of (1) on Ω_{ε_n} for each n, and u_{ε_n} converges to u_0 weakly in $H^1(\Omega_0)$, then u_0 is a solution of (3) on Ω_0. Furthermore, if J_ε denotes the functional

$$J_\varepsilon(u) = \int_{\Omega_\varepsilon} (\tfrac{1}{2}|\nabla u|^2 - F(x,u))dx, \quad u \in H^1(\Omega_\varepsilon), \quad \varepsilon \geqq 0,$$

then $J_{\varepsilon_n}(u_{\varepsilon_n}) \to J_0(u_0)$ as $n \to \infty$.

PROOF: Let $u = \lim u_{\varepsilon_n}$ in $H^1(\Omega_0)$ weak. Then, $u_{\varepsilon_n} \to n$ strongly in $L^2(\Omega_0)$, and then for all $v \in H^1(\Omega_0)$.

$$\int_{\Omega_0} (\nabla u_{\varepsilon_n} . \nabla v + f(x,u_{\varepsilon_n})v)dx \to \int_{\Omega_0} (\nabla u . \nabla v + f(x,u)v)dx \text{ as } n \to \infty.$$

On the other hand, if we denote $\Omega_\varepsilon \setminus \Omega_0$ by R_ε,

$$\int_{R_{\varepsilon_n}} \nabla u_{\varepsilon_n} . \nabla v \, dx \quad 0, \quad \int_{R_\varepsilon} f(x,u_{\varepsilon_n}) \, dx \to 0 \text{ as } n \to \infty.$$

This implies

$$\int_{\Omega_0} (\nabla u . \nabla v + f(x,u)v)dx = 0 \text{ for all } v \in H^1(\Omega_0),$$

hence u is a weak solution of (1) on Ω_0.

To prove the second part of the theorem, we observe that

$$\int_{\Omega_{\varepsilon_n}} |\nabla u_{\varepsilon_n}|^2 \, dx = \int_{\Omega_{\varepsilon_n}} f(x,u_{\varepsilon_n})u_{\varepsilon_n} \, dx.$$

This implies

$$J_{\varepsilon_n}(u_{\varepsilon_n}) = \int_{\Omega_{\varepsilon_u}} (\tfrac{1}{2}|\nabla u_{\varepsilon_n}|^2 - F(x,u_{\varepsilon_n}))dx$$

$$= \int_{\Omega_{\varepsilon_n}} (\tfrac{1}{2}f(x,u_{\varepsilon_n})u_{\varepsilon_n} - F(x,u_{\varepsilon_n}))dx.$$

All we have to do now is observe that

$$\int_{R_{\varepsilon_n}} f(x,u_{\varepsilon_n})u_{\varepsilon_n} \, dx \to 0 \text{ as } n \to \infty,$$

since $|R_{\varepsilon_n}| \to 0$ and there is an *a priori* bound in $L^\infty(\Omega_\varepsilon)$ for the solutions of (3) which is independent of ε. This finishes the proof.

THEOREM 2: Assume Ω_ε and f satisfy the hypotheses of Theorem 1. Let u_0 be either a hyperbolic stationary solution of (1) on Ω_0 or a strict minimum of the energy functional J_0 on Ω_0, and let $\tilde{u}_0$ be any extension of u_0 in $H^1(\mathbb{R}^N)$. Then there exists $\varepsilon_0 > 0$ and $r > 0$ such that for every $0 \leqq \varepsilon \leqq \varepsilon_0$ there exists at least one solution $u_\varepsilon \in H^1(\Omega_\varepsilon)$ of (1) on Ω_ε such that

$$\|\nabla(u_\varepsilon-\tilde{u}_0)\|^2_{L^2(\Omega_\varepsilon)} + \|u_\varepsilon - u_0\|^2_{L^2(\Omega_0)} \leqq r^2.$$

PROOF: Define in $H^1(\Omega_\varepsilon)$ the equivalent norm

$$\|\nabla u\|^2_{L^2(\Omega_\varepsilon)} + \|u\|^2_{L^2(\Omega_0)} \equiv \|u\|^2_\varepsilon$$

Let $\lambda_1, \lambda_2, \ldots$ and $w_1, w_2, \ldots$ be the eigenvalues and their corresponding eigenfunctions of the linearized operator $-\Delta - f_u(x,u_0)$ on Ω_ε with Neumann boundary conditions. f_u stands for the partial derivative of f with respect to u.

1. u_0 is hyperbolic

Assume that $\lambda_1 < \lambda_2 < \ldots < \lambda_N < 0 < \lambda_{N+1} \ldots$. Let $U_0 = \text{span}\,(w_1,\ldots,w_N)$, $S_0 = (u \in H^1(\Omega_0) \int_{\Omega_0} uw_k dx = 0 \text{ for } k = 1,\ldots,N)$ denote the unstable and stable manifolds, respectively, associated with $-\Delta - f_u(x,u_0)$.

Let $J_0(u) = \int_{\Omega_0} (\frac{1}{2}|\nabla u|^2 - F(x,u))dx$; then, denoting the second differential of J_0 by J_0'', for $v \in S_0$ we have

$$J_0''(u_0)(v,v) = \int_{\Omega_0} (\tfrac{1}{2}|\nabla v|^2 - f_u(x,u_0)v^2)dx \geqq \lambda_{N+1} \|v\|^2_{L^2(\Omega_0)},$$

and there exists $\mu > 0$, which can be assumed to be less than 1, such that $J_0''(u_0)(v,v) \geqq \mu \|v\|^2_{H^1(\Omega_0)}$, whereas, if $v \in U_0$, then

$$J_0''(u_0)(v,v) \leqq -\mu \|v\|^2_{H^1(\Omega_0)}.$$

Let us consider arbitrary extensions of $w_1, w_2, \ldots, w_N$ in $H^1(\mathbb{R}^n)$, which will be denoted by $\tilde{w}_1, \tilde{w}_2, \ldots, \tilde{w}_N$, and let us define the following subspaces of $H^1(\Omega_\varepsilon)$:

$$S_\varepsilon = \{u \in H^1(\Omega_\varepsilon) \quad \int_{\Omega_0} u w_k \, dx = 0 \text{ for } k = 1,\ldots,N\}$$

$$U_\varepsilon = \text{span } (\tilde{w}_1,\ldots,\tilde{w}_N),$$

Let us now consider $J_\varepsilon(u) = \int_{\Omega_\varepsilon} (\frac{1}{2}|\nabla u|^2 - F(x,u))dx$ on $H^1(\Omega_\varepsilon)$. Let $v \in S_\varepsilon$. Then, if we still denote $\Omega_\varepsilon \setminus \Omega_0$ by R_ε,

$$J_\varepsilon(\tilde{u}_0+v) = J_0(u_0+v) + \int_{R_\varepsilon} (\tfrac{1}{2}|\nabla(\tilde{u}_0+v)|^2 - F(x,\tilde{u}_0+v))dx.$$

Now, since $J_0'(u_0)(v) = 0$ for all $v \in H^1(\Omega_0)$, we have

$$J_0(u_0+v) = J_0(u_0) + \tfrac{1}{2}J_0''(u_0)(v,v) + o(\|v\|_0^2).$$

Let $v \in S_\varepsilon$, that is, assume that $v|\Omega_0 \in S_0$. Then

$$J_0''(u_0)(v,v) \geq \mu \|v\|_0^2,$$

which implies

$$J_\varepsilon(\tilde{u}_0+v) \geq J_0(u_0) + \mu\|v\|_0^2 + o(\|v\|_0^2) + \int_{R_\varepsilon} (\tfrac{1}{2}|\nabla(\tilde{u}_0+v)|^2 -$$

$$- F(x,\tilde{u}_0+v))dx \geq J_0(u_0) + (\mu/2)\|v\|_0^2 + O(|R_\varepsilon|^{\frac{1}{2}}),$$

for $\|v\|_0$ sufficiently small.

Let now $v \in U_\varepsilon$, that is, $v = \Sigma\alpha_k\tilde{w}_k$ for some $\alpha = (\alpha_1,\ldots,\alpha_N)$. Then

$$J_0(u_0+v) = J_0(u_0) + \tfrac{1}{2}\Sigma\alpha_k^2\lambda_k + (|\alpha|^2), \text{ and}$$

$$\int_{R_\varepsilon} (\tfrac{1}{2}|\nabla(\tilde{u}_0+v)|^2 - F(x,\tilde{u}_0+v))dx = \tfrac{1}{2}(\int_{R_\varepsilon} |\nabla u_0|^2 + |\nabla v|^2 - 2\nabla\tilde{u}_0,\nabla v)dx +$$

$$+ O(|R_\varepsilon|^{\frac{1}{2}}),$$

uniformly in α for α in compact sets. Therefore, for $v \in U_\varepsilon$,

$$J_\varepsilon(\tilde{u}_0+v) = J_0(u_0) + \tfrac{1}{2}\Sigma\lambda_k\alpha_k^2 + o(|\alpha|^2) + o(|R_\varepsilon|^{\frac{1}{2}}).$$

Let $v \in S_\varepsilon$ be such that $\|v\|_\varepsilon = r$. The, v satisfies

$$J_\varepsilon(\tilde{u}_0+v) \geq J_0(u_0) + (\mu/2)(r^2 - \int_{R_\varepsilon} |\nabla v|^2 dx) + \tfrac{1}{2}\int_{R_\varepsilon} |\nabla v|^2 + O(|R_\varepsilon|^{\frac{1}{2}}) \geq$$

$$\geq J_0(u_0) + (\mu/2)r^2 + \tfrac{1}{2}(1-\mu)\int_{R_\varepsilon} |\nabla v|^2 dx + O(|R_\varepsilon|^{\frac{1}{2}}) \geq$$

$$\geq J_0(u_0) + (\mu/2)r^2 + O(|R_\varepsilon|^{\frac{1}{2}}).$$

On the other hand, for $v \in U_\varepsilon$, it is clear that

$$\|v\|_\varepsilon^2 = \|\Sigma a_k w_k\|^2_{L^2(\Omega_0)} + \|\Sigma\alpha_k\tilde{w}_k\|^2_{H^1(\Omega_\varepsilon)} = |\alpha|^2(1+O(|R_\varepsilon|)),$$

uniformly in α for α in compact sets. Thus, if $\|v\|_\varepsilon = r$,

$$J_\varepsilon(\tilde{u}_0+v) = J_0(u_0) + \tfrac{1}{2}\Sigma\lambda_k\alpha_k^2 + o(|\alpha|^2) + o(|R_\varepsilon|^{\frac{1}{2}}) =$$

$$= J_0(u_0) + \tfrac{1}{2}\Sigma\lambda_k\alpha_k^2 + o(r^2) + o(|R_\varepsilon|^{\frac{1}{2}}).$$

We now have all the elements necessary to apply Castro's Theorem:

(1) $H^1(\Omega_\varepsilon) = S_\varepsilon \oplus U_\varepsilon$.

Indeed, if $u \in S_\varepsilon \cap U_\varepsilon$, then $u = \Sigma\alpha_k\tilde{w}_k$ and $u|\Omega_0 \in S_\varepsilon$, which implies that $\alpha = 0$. On the other hand, given $u \in H^1(\Omega_\varepsilon)$, we may write $u|\Omega_0 = \Sigma\alpha_k w_k$ and $v = u - \Sigma\alpha_k\tilde{w}_k$, and then $v \in S_\varepsilon$, since $(u - \Sigma\alpha_k w_k)|\Omega_0$ belongs to S_0.

(2) Let $v \in S_\varepsilon$, $w \in U_\varepsilon$ be such that $\|v\|_\varepsilon = r$, $\|w\|_\varepsilon \leq r$. Then if $w = \Sigma\alpha_k\tilde{w}_k$, we have

$$J_\varepsilon(\tilde{u}_0+v) - J_\varepsilon(\tilde{u}_0+w) \geq (\mu/2)r^2 - \Sigma\lambda_k\alpha_k^2 + o(|\alpha|^2) + O(|R_\varepsilon|^{\frac{1}{2}}) \geq$$

$$\geq (\mu/2)r^2 + o(r^2) + O(|R_\varepsilon|^{\frac{1}{2}}),$$

which is strictly positive if r is fixed and $|R_\varepsilon|$ is sufficiently small.

(3) Let $v \in S_\varepsilon$, $w = \Sigma\alpha_k w_k \in U_\varepsilon$ be such that $\|v\|_\varepsilon \leq r$, $\|w\|_\varepsilon = r$.

Then,

$$J_\varepsilon(\tilde{u}_0+v) - J_\varepsilon(\tilde{u}_0+v) \geq (\mu/2)\|v\|_\varepsilon^2 - \Sigma\lambda_k\alpha_k^2 + o(|\alpha|^2) + O(|R_\varepsilon|^{\frac{1}{2}}) \geq$$

$$\geq |\lambda_N||\alpha|^2 + o(|\alpha|^2) + O(|R_\varepsilon|^{\frac{1}{2}}) =$$

$$= |\lambda_N|r^2 + o(r^2) + O(|R_\varepsilon|^{\frac{1}{2}}),$$

which is strictly positive if r is fixed and $|R_\varepsilon|$ is sufficiently small.

Thus, all the hypotheses are satisfied, and we can conclude the existence of a critical point of J_ε. Now we have to prove that it lies in the specified neighbourhood of $\tilde{u}_0$.

From the value of the critical value c given by Castro's Theorem, we see that if we consider the part of the "extended" unstable manifold U_ε in that neighbourhood, i.e., $(\tilde{u}_0 + w\colon w \in U_\varepsilon, \|w\|_\varepsilon = r)$ as spanning surface Σ, and take into account that

$$J_\varepsilon(u) = J_0(u) + \int_{R_\varepsilon} (\tfrac{1}{2}|\nabla u|^2 - F(x,u))dx \geq J_0(u) + \tfrac{1}{2}\int_{R_\varepsilon} |\nabla u|^2\, dx +$$

$$+ O(|R_\varepsilon|^{\frac{1}{2}})$$

for all $u \in H^1(\Omega_\varepsilon)$, we find that, for $u = \tilde{u}_0 + w$, $w \in U_\varepsilon$, $J_0(\tilde{u}_0+w)$ reaches a maximum $J_0(u_0) + \frac{1}{2}r^2 + O(|R_\varepsilon|^{\frac{1}{2}})$ for $w = 0$, and this means that

$$c \leq J_0(u_0) + \tfrac{1}{2}r^2 + O(|R_\varepsilon|^{\frac{1}{2}}).$$

On the other hand, if $u = \tilde{u}_0 + v$, $v \in S$, $\|v\|_\varepsilon \leq r$, we have

$$J_\varepsilon(\tilde{u}_0+v) \geq J_0(u_0) + O(|R_\varepsilon|^{\frac{1}{2}}).$$

Hence,

$$c \geq J_0(u_0) + O(|R_\varepsilon|^{\frac{1}{2}}).$$

These inequalities imply that, if r is sufficiently small and $|R_\varepsilon|$ is taken accordingly, the critical value corresponding to the given solution lies as close to $J_0(u_0)$ as desired. Assume now that u_0 is the only critical point corresponding to the critical value $J_0(u_0)$, and assume, by contradiction, the existence of a sequence u_{ε_n} of critical points of J_{ε_n} lying outside the given neighbourhood of $\tilde{u}_0$, for some sequence $\varepsilon_n \to 0$. The restrictions $u_{\varepsilon_n}|\Omega_0$ are bounded in L^∞, and hence in H^1, because $J_{\varepsilon_n}(u_{\varepsilon_n})$ is bounded. This means that $u_{\varepsilon_n}|\Omega_0$ can be assumed to converge weakly in H^1 to some v_0, which, by Theorem 1, must be a critical point of J_0, and satisfy $J_{\varepsilon_n}(u_{\varepsilon_n}) \to J_0(v_0)$. On the other hand, $J_{\varepsilon_n}(u_{\varepsilon_n}) \to J_0(u_0)$, which implies $u_0 = v_0$, by our assumption that $J_0(u_0)$ is a critical value corresponding only to u_0. But this easily implies that

$$\int_{\Omega_{\varepsilon_n}} |\nabla u_{\varepsilon_n}|^2 \, dx \to \int_{\Omega_0} |\nabla u_0|^2 \, dx,$$

which contradicts the assumption that $\|u_{\varepsilon_n} - \tilde{u}_0\|_{\varepsilon_n} \geq r$. This solves the problem under this extra hypothesis. Finally, a standard perturbation argument enables us to reduce the general case to the one we have just discussed.

2. $\underline{u_0 \text{ is a strict, non-hyperbolic minimum of } J_0}$

In this case, the proof is based on the following lemma:

LEMMA: Let u_0 be a strict minimum of J on a bounded domain Ω. Then, for r sufficiently small, $m(r) \equiv \inf \{J(u) : \|u-u_0\|_{H^1(\Omega)} = r\}$ is strictly positive.

PROOF OF THE LEMMA: Let us assume that the linearized operator $-\Delta - f_u(x,u_0)$ has an N-dimensional null space U (N may be zero if u_0 is hyperbolic, but must be finite, since Ω is bounded). Its orthogonal S corresponds to the stable manifold of the associated gradient flow, and the second differential $J''(u_0)$ is positive definite on S.

Suppose there exist sequences of functions $v_n \in S$, $w_n \in U$ such that $\|v_n + w_n\| = r$ and $J(u_0 + v_n + w_n) \to 0$ as $n \to \infty$, where $\|\cdot\|$ stands for the H^1 norm. Without loss of generality we may assume that v_n and w_n converge weakly in H^1 to some functions $v \in S$, $w \in U$, and, since U is finite-dimensional, the convergence $w_n \to w$ is actually strong.

By the weak continuity of J, $J(u_0+v+w) \leq \lim\inf J(u_0+v_n+w_n) = 0$, which implies that $v = 0$, $w = 0$ if r is sufficiently small, since u_0 is a strict minimum. But, since w_n converges strongly to zero and $\|v_n + w_n\| = r$, $\|v_n\|$ must converge to r. But then:

$$J(u_0 + v_n + w_n) = J(u_0) + \tfrac{1}{2}J''(u_0)(v_n,v_n) + o(r),$$

which implies that

$$0 = \lim \|v_n\| - \tfrac{1}{2}(J(u_0 + v_n + w_n) - J(u_0)) = \tfrac{1}{2}J''(u_0)(v/r,v/r) + o(1),$$

which is impossible, since $J''(u_0)$ is positive definite on S.

<u>END OF THE PROOF OF THE THEOREM</u>: Let m(r) be the strictly positive function given by the previous lemma. Let $v \in H^1(\Omega_\varepsilon)$ be such that $\|v\|_\varepsilon = r$. Then

$$J_\varepsilon(\tilde{u}_0 + v) = J_0(u_0 + v) + \int_{R_\varepsilon} (\tfrac{1}{2}|\nabla\tilde{u}_0+v|^2 - F(x,\tilde{u}_0+v))dx \geq$$

$$\geq m(r^2 - [\int_{R_\varepsilon}|\nabla v|^2dx])^{\frac{1}{2}}) + \tfrac{1}{2}\int_{R_\varepsilon}|\nabla v|^2dx + \int_{R_\varepsilon} \nabla v.\nabla u_0 dx +$$

$$+ \tfrac{1}{2}\int_{R_\varepsilon}|\nabla\tilde{u}_0|^2dx - \int_{R_\varepsilon} F(x,\tilde{u}_0+v)dx.$$

Let $n(r) = \min\{(m((r^2-\delta) + \tfrac{1}{2}\delta^2)^{\frac{1}{2}}) : 0 \leq \delta \leq r\}$. Clearly, $n(r) > 0$ for r small. With the help of this function, we have

$$J_\varepsilon(\tilde{u}_0+v) \geq n(r) + \int_{R_\varepsilon} \nabla v.\nabla\tilde{u}_0 dx + \tfrac{1}{2}\int_{R_\varepsilon} |\nabla\tilde{v}_0|^2dx - \int_{R_\varepsilon} F(x,\tilde{u}_0+v)dx =$$

$$= n(r) + O(|R_\varepsilon|^{\frac{1}{2}}).$$

Therefore, if r is fixed and $|R_\varepsilon|$ is sufficiently small, the minimum of J_ε in the given ball (which must exist, since J_ε is lower semicontinuous in $H^1(\Omega_\varepsilon)$) cannot be attained at the boundary of the ball, and thus corresponds to a critical point of J_ε. This finishes the proof.

3. Final remarks

1. The results obtained in Hale and Vegas [2] contain additional information:

the uniqueness of solutions which may arise out of this perturbation process, and the "continuation" of their stability properties. Under our present general hypotheses, this is no longer the case; indeed, it can be easily shown that, if there is no convergence of eigenvalues and eigenfunctions in the sense discussed in the introduction, we can have a hyperbolic minimum u_0 such that, for a given extension $\tilde{u}_0$, the linearized operator $-\Delta - f_u(x,\tilde{u}_0)$ on Ω_ε has a negative eigenvalue μ_ε bounded away from zero as $\varepsilon \to 0$. The general theory of dynamical systems suggests that, besides the minimum of J_ε which must lie in the r-ball of $\tilde{u}_0$ in $H^1(\Omega_\varepsilon)$, there may be at least another critical point of J_ε in that ball corresponding to a saddle point.

2. It would be interesting to know if similar results can be obtained for systems of parabolic equations, with continuation of more general hyperbolic or stable isolated invariant sets. In this case, the variational method should be substituted by either the consideration of two Liapunov functions (Yorke) (see Rybakowski [7]) or by degree theory, or the Conley index as generalized by Rybakowsky to noncompact phase spaces [7].

References

1. Fujiwara, D. and Ozawa, S.: The Hadamard variational formula for the Green functions on some normal elliptic boundary value problems. Proc. Japan Acad., 54 ser. A, 215-220 (1976).
2. Hale, J. and Vegas, J.: A nonlinear parabolic equation with varying domain. Arch. Rat. Mech. Anal, 86 (2) (1984).
3. Lobo Hidalgo, M. and Sánchez-Palencia, H.: Sur certaines propriétés spectrales des perturbations du domaine dans les problèmes aux limites, Comm. in Partial Differential Equations, 4 (10), 1085-1098 (1979).
4. Matano, H. and Mimura, M.: Pattern formation in competition-diffusion system in nonconvex domains. Publ. RIMS, Kyoto Univ., 19 (3), 1049-1079 (1983).
5. Mignot, F., Murat, F. and Puel, J.P.: Variation d'un point de retourement par rapport au domaine. Comm. in Partial Differential Equations, 4 (11), 1236-1297 (1979).
6. Nirenberg, L.: Variational and topological methods in nonlinear problems. Bull. A.M.S., 4 (3), 267-302 (1981).

7. Rybakowski, Ch.: On the homotopy index for infinite-dimensional semi-flows, Trans. A.M.S., 269, 351-383 (1982).
8. Vegas, J.M.: Perturbations caused by perturbing the domain in an elliptic equation, J. Differential Equations, 48 (2), 189-226 (1983).
9. Vegas, J.M.: A Neumann elliptic problem with variable domain, in Contributions to nonlinear partial differential equations, C. Bardos, A. Damlamian, J.I. Diaz and J. Hernández, eds. 264-273. Pitman, 1973.

This work has been supported by a grant from the Comisón Asesora para la Investigación Científica y Técnica CAICYT no. 3308/83.

J.M. Vegas
Departamento de Matematica Aplicada
Universidad Complutense
28040 Madrid
Spain.

L VERON

Singularities of some degenerate elliptic equations

Introduction

Let Ω be an open subset of $\mathbb{R}^N$, $N \geq 2$, containing 0, $\Omega' = \Omega - \{0\}$ and $p > 1$. A function $u \in C^1(\Omega')$ is a weak singular solution of the p-Laplace equation

$$\operatorname{div}(|\nabla u|^{p-2} \nabla u) = 0 \tag{0.1}$$

in Ω' (or u is p-harmonic in Ω') if

$$\int_{\Omega'} |\nabla u|^{p-2} \nabla u . \nabla \zeta \, dx = 0 \tag{0.2}$$

holds for any $\zeta \in C_o^1(\Omega')$. It is clear that (0.1) admits a simple radial solution (with ω_N = volume of $B_1(0)$)

$$\mu(x) = \mu(|x|) = \begin{cases} \frac{p-1}{N-p} (N\omega_N)^{-1/(p-1)} |x|^{(p-N)/(p-1)} & \text{if } p \neq N, \\ (N\omega_N)^{-1/(N-1)} \operatorname{Log}(1/|x|) & \text{if } p = N, \end{cases} \tag{0.3}$$

which is singular when $1 < p \leq N$ and satisfies

$$-\operatorname{div}(|\nabla \mu|^{p-2} \nabla \mu) = \delta_o \tag{0.4}$$

in $\mathcal{D}'(\Omega)$. When u is singular and nonnegative Serrin proved in [16] that when $1 < p \leq N$ there exist $C > 0$ and $K > 0$ such that

$$C^{-1} \leq u/\mu \leq C \tag{0.5}$$

holds near 0 and

$$-\operatorname{div}(|\nabla u|^{p-2} \nabla u) = K \delta_o \tag{0.6}$$

holds in $\mathcal{D}'(\Omega)$. In Section 1 we give some results which have been obtained with Kichenassamy [11] and precise (0.5) and (0.6) under a more general assumption.

Another interesting singularity problem is to study the behaviour near 0 of any $u \in C^1(\Omega')$ satisfying

$$\mathrm{div}(|\nabla u|^{p-2} \nabla u) = u|u|^{q-1}. \tag{0.7}$$

When $0 < q \leqq p-1$, (0.7) falls into the scope of Serrin's works [16] so we shall study the case $q > p-1 > 0$. If we first look for solutions of (0.7) under the radial form $u(x) = \alpha|x|^{\beta}$ we find $\beta = -p/(q+1-p)$ and

$$\alpha = \gamma_{N,p,q} = [(\frac{p}{q+1-p})^{p-1} (\frac{pq}{q+1-p} - N)]^{1/(q+1-p)} \tag{0.8}$$

and it is clear that $\gamma_{N,p,q}$ exists only if $p \geqq N$ or $1 < p < N$ and $p-1 < q < N(p-1)/(N-p)$. In fact if $1 < p < N$ and $q \geqq N(p-1)/(N-p)$ Vázquez and Veron [25] proved that all the isolated singularities of solutions of (0.7) were removable and so $u \in C^1(\Omega)$. This is not the case when $\gamma_{N,p,q}$ does exist and in Section 2 we shall give a classification theorem obtained with Friedman [10] which extends Veron's results in the semilinear case ($p = 2$) [27].

In Section 3 we look for solutions of (0.1) under the form

$$u(x) = |x|^{-\beta} \omega(x/|x|) \tag{0.9}$$

$\beta > 0$ and we obtain all of them when $N = 2$. In higher dimensions, we use a shooting method due to Tolksdorf [20] to prove the existence of the first anisotropic singular solution of (0.1) under the form (0.9). Anisotropic singularities of (.0.7) are given under the form

$$u(x) = |x|^{-p/(q+1-p)} \omega(x/|x|). \tag{0.10}$$

1. Singularities of the p-Laplace equation

The main result of this section is the following [11]:

THEOREM 1.1: Assume $1 < p \leqq N$ and $u \in C^1(\Omega')$ is p-harmonic in Ω' such that

$$u/\mu \in L^{\infty}(B_r(0)) \tag{1.1}$$

for $r > 0$ small enough, then there exists $\gamma \in \mathbb{R}$ such that

$$u - \gamma\mu \in L^{\infty}_{loc}(\Omega). \tag{1.2}$$

Moreover ∇u satisfies

$$\lim_{x\to 0} |x|^{(N-1)/(p-1)} \nabla(u-\gamma\mu)(x) = 0 \tag{1.3}$$

and the following equation holds in $\mathcal{D}'(\Omega)$

$$-\text{div}(|\nabla u|^{p-2} \nabla u) = \gamma|\gamma|^{p-2} \delta_0. \tag{1.4}$$

PROOF: Without any loss of generality we assume $\Omega \supset \bar{B}_1(0)$.

Step 1. A priori estimates. For $0 < a < 1/2$ and $0 < |x| < 1$ we set $x = ay/(1 + a)$ and $v(y) = u(ay/(1 + a))/\mu(a)$. The function v is p-harmonic and bounded in $\Gamma = \{y : 1/2 \leq |y| \leq 3\}$. So we deduce from the *a priori* estimates of [19] (see also [7], [14], [23]) that

$$\|\nabla v\|_{C^{\alpha}(\bar{B}_2(0)-B_1(0))} \leq C \tag{1.5}$$

where $\alpha \in (0,1)$, $C > 0$. This implies

$$|\nabla u(x)| \leqq C|x|^{-1} \mu(x), \tag{1.6}$$

$$|\nabla v(x) - \nabla x(x')| \leqq C|x-x'|^{\alpha} |x|^{-1-\alpha} \mu(x), \tag{1.7}$$

for $0 < |x| \leqq |x'| \leqq 1/2$.

Step 2. Strict comparison principles. The following result is due to Tolksdorf [20, Prop. 3.3.2]: If u_1 and u_2 are two p-harmonic functions in a connected open subset of $\mathbb{R}^N$ such that $u_1 \geqq u_2$ in G and if u_1 and u_2 are not identical and if ∇u_2 never vanishes; then $u_1 > u_2$ in G. As a consequence if G does not contain 0, if ϕ is p-harmonic in G and $v = \phi/\mu$ (or $v = \phi-\mu$) achieves its maximum in G, then v is constant.

Step 3. The case $1 < p < N$. We define

$$\gamma = \limsup_{x \to 0} u(x)/\mu(x) \tag{1.8}$$

and we may assume $\gamma > 0$. We can also assume $\sup_{|x|=1} u(x) = 0$ and let $\tilde{\gamma}$ be the function defined on (0,1] by

$$\tilde{\gamma}(r) = \sup_{r \leqq |x| \leqq 1} u(x)/\mu(x). \tag{1.9}$$

$\tilde{\gamma}$ is nonnegative and nonincreasing on (0,1) (from Step 2) and $\lim_{r \to 0} \tilde{\gamma}(r) = 0$. Moreover there exists x_r such that $|x_r| = r$ and $\tilde{\gamma}(r) = u(x_r)/\mu(r)$. We now define the function u_r on $A_r = \{\xi: 0 < |\xi| < 1/r\}$ by

$$u_r(\xi) = u(r\,\xi)/\mu(r). \tag{1.10}$$

From Step 1 we have

$$\begin{cases} |u_r(\xi)| \leqq C \quad \mu(\xi), \\ |\nabla u_r(\xi)| \leqq C\ |\xi|^{(1-N)/(p-1)}, \\ |\nabla u_r(\xi) - \nabla u_r(\xi)| \leqq C|\xi-\xi'|^{\alpha} \quad |\xi|^{(1-N)/(p-1)-\alpha}, \end{cases} \tag{1.11}$$

for $0 < |\xi| \leqq |\xi'| \leqq 1/r$, where C is independent of r. From the Ascoli's theorem there exists a p-harmonic function v and a sequence $\{r_n\} \to 0$ such that $\{u_{r_n}\}$ converges to v in the $C^1_{loc}(\mathbb{R}^N - \{0\})$ topology. Moreover we have

$$u_r(\xi)/\mu(\xi) = u(r\xi)/(\mu(r\xi)\ \mu(1)) \leqq \gamma/\mu(1); \tag{1.12}$$

and if we set $\xi_r = x_r/r$, then $u_r(\xi_r)/\mu(\xi_r) = \tilde{\gamma}(r)/\mu(1)$. We can also assume that $\{\xi_{r_n}\}$converges to some $\xi_0 \in S^{N-1}$ so we get

$$v(\xi)/\mu(\xi) \leqq \gamma/\mu(1) \quad \text{and} \quad w(\xi_0)/\mu(\xi_0) = \gamma/\mu(1). \tag{1.13}$$

Step 2 implies $v(\xi) = (\gamma/\mu(1))\ \mu(\xi) = \lim_{r \to 0} u_r(\xi)$. If we take in particular $|\xi| = 1$ we deduce

$$\lim_{x\to 0} u(x)/\mu(x) = \gamma \tag{1.14}$$

and (1.3).

In order to prove the boundedness of $u - \gamma\mu$ we consider for $\varepsilon > 0$ the following p-harmonic functions

$$v_\varepsilon^+(x) = (\gamma+\varepsilon)\,\mu(x) - (\gamma+\varepsilon)\,\mu(1) + \sup_{|x|=1} u(x), \tag{1.15}$$

$$v_\varepsilon^-(x) = (\gamma-\varepsilon)\,\mu(x) - (\gamma-\varepsilon)\,\mu(1) + \inf_{|x|=1} u(x). \tag{1.16}$$

From (1.4) we have $v_\varepsilon^-(x) \leqq u(x) \leqq v_\varepsilon^+(x)$ for $0 < |x| \leqq 1$, which implies (1.2). As for (1.4) it is a straightforward application of (1.3) and Green's formula.

<u>Step 4</u>. <u>The case p = N</u>. As in Step 3 we define γ and $\tilde{\gamma}$, $\gamma > 0$ and $\tilde{\gamma}(r) = \sup_{r\leqq|x|\leqq 1/2} u(x)/\mu(x)$, and $u_r(\xi) = u(r\xi)/\mu(r)$. From Step 1 we have the following estimates

$$\begin{cases} |u_r(\xi)| \leqq C(1 + |\mathrm{Log}\,|\xi\,||/\mathrm{Log}(1/r)), \\ |\nabla u_r(\xi)| \leqq C(1 + |\mathrm{Log}\,|\xi\,||/\mathrm{Log}(1/r))|\xi|^{-1}, \\ |\nabla u_r(\xi) - \nabla u_r{}'\xi')| \leqq C|\xi-\xi'|^\alpha\,(1 + |\mathrm{Log}\,|\xi||/\mathrm{Log}(1/r))|\xi|^{-1-\alpha}, \end{cases} \tag{1.17}$$

for $0 < |\xi| \leqq |\xi'| \leqq 1/(2r)$. So there exists a sequence $\{r_n\}$ and a p-harmonic function v defined in $\mathbb{R}^N$-{0} such that $\{u_{r_n}\}$ converges to v in the $C^1_{loc}(\mathbb{R}^N\text{-}\{0\})$ topology. From (1.17) v is bounded; hence, it can be extended to $\mathbb{R}^N$ as a N-harmonic function [16], say $\tilde{v}$, and $\tilde{v}$ is constant from [15]. As in Step 3 we have

$$\gamma = \tilde{v}(\xi) = \lim_{x\to 0} u(x)/\mu(x). \tag{1.18}$$

The proof of the boundedness of $u-\gamma\mu$ is as in Step 3. To prove (1.3) and (1.4) we define

$$v_r(\xi) = u(r\xi) - \gamma\mu(r); \tag{1.19}$$

v_r is N-harmonic in Λ_r and satisfies

$$|v_r(\xi)| \leq \gamma|\mu(\xi)| + C. \tag{1.20}$$

This implies as in Step 1

$$\begin{cases} |\nabla v_r(\xi)| \leq C|\xi|^{-1}, \\ |\nabla v_r(\xi) - \nabla v_r(\xi')| \leq C|\xi-\xi'|^{\alpha} \quad |\xi|^{-1-\alpha}, \end{cases} \tag{1.21}$$

for $0 < |\xi| \leq |\xi'| \leq 1/2r$. Moreover the same estimates hold with v_r replaced by u, ξ by x and ξ' by x', with $0 < |x| \leq |x'| \leq 1$. From Step 2 the supremum of $u(x) - \gamma\mu(x)$ in $B_{1/2}(0)$ is achieved either at 0 or for $|x| = 1/2$ (otherwise $u-\gamma\mu$ is constant and everything is done). If we assume that it is achieved at 0 then the function $\tilde{\lambda}$ defined by

$$\tilde{\lambda}(r) = \sup_{r \leq |x| \leq 1/2} (u(x) - \gamma\mu(x))$$

is nonincreasing and $\lim_{r \to 0} \tilde{\lambda}(r) = \lambda$. From (1.22) and (1.21) there exist a N-harmonic function v and a sequence $\{r_n\} \to 0$ such that $\{v_{r_n}\}$ converges to v in the $C^1_{loc}(\mathbb{R}^N - \{0\})$ topology and $\xi_0 \in S^{N-1}$ such that

$$v(\xi) \leq \gamma\mu(\xi) + \lambda, \quad v(\xi_0) = \gamma\mu(\xi_0) + \lambda. \tag{1.22}$$

Hence $v = \gamma\mu + \lambda$ which implies $\lim_{x \to 0} (u(x) - \gamma\mu(x)) = \lambda$ and (1.3). If the supremum of $u - \gamma\mu$ is achieved for $|x| = 1/2$ we can use the conformal invariance of the N-Laplace equation through the inversion $x \to x/|x|^2$ and we prove (1.3) as above. The proof of (1.4) is the same as in Step 3.

Thanks to Theorem 1.1 we can study the singular Dirichlet problem

$$\begin{cases} -\operatorname{div}(|\nabla u|^{p-2}\nabla u) = \gamma|\gamma|^{p-2}\delta_0 \text{ in } \mathcal{D}'(\Omega), \\ u = g \quad \text{on } \partial\Omega. \end{cases} \tag{1.23}$$

<u>THEOREM 1.2</u>: Assume $1 < p < N$, Ω is bounded with a regular boundary $\partial\Omega$, $\gamma \in R$ and $g \in L^{\infty}(\Omega) \cap W^{1,p}(\Omega)$. Then there exists a unique $u \in C^{1,\alpha}(\Omega')$ such that $|\nabla u|^{p-1} \in L^1(\Omega)$, $\nabla u \in L^p(\Omega - B_1(0))$ for $r > 0$ small enough and (1.1) satisfying (1.23). Moreover u satisfies (1.2) and (1.3).

REMARK 1.1: The proof of the existence of u is rather easy but the uniqueness of u is explicitly based on (1.2) and (1.3). A similar result has been obtained in [5] with a more general quasilinear operator but in a smaller class of uniqueness. The general problem of proving existence and uniqueness for

$$- \operatorname{div}(|\nabla u|^{p-2} \nabla u) = \nu \tag{1.24}$$

where ν is a bounded measure is still unsolved.

As a consequence of Theorem 1.1 we have a global result:

THEOREM 1.3: Assume $1 < p \leq N$ and u is p-harmonic in $\mathbb{R}^N - \{0\}$ such that $|u(x)| \leq a|\mu(x)| + b$ for any $x \neq 0$. Then there exist two constants γ and λ such that $u(x) = \gamma\mu(x) + \lambda$ for $x \neq 0$.

Another interesting consequence of Theorem 1.1 is the following

THEOREM 1.4: Any non-negative N-harmonic function in $\mathbb{R}^N$-C, where C is discrete, is a constant.

2. The classification theorem

The main result of this section which generalizes those of [25] is the following [10]:

THEOREM 2.1: Assume $0 < p-1 < q < N(p-1)/(N-p)$ if $1 < p < N$ or $q > p-1$ if $p = N$ and $u \in C^1(\Omega')$ is a weak nonnegative solution of (0.7) in Ω'. Then we have the following

(i) either $\lim_{x\to 0} |x|^{p/(q+1-p)} u(x) = \gamma_{N,p,q}$ which is defined by (0.8).

(ii) either there exists a real number γ which can take any positive value such that $\lim_{x\to 0} u(x)/\mu(x) = \gamma$ and u satisfies

$$- \operatorname{div} (|\nabla u|^{p-2} \nabla u) + u^q = \gamma|\gamma|^{p-2} \delta_0 \tag{2.1}$$

in $\mathcal{D}'(\Omega)$.

(iii) or $u(x)$ admits a finite limit when x tends to 0 and u can be extended as a $C^1(\Omega)$ function satisfying (0.7) in $\mathcal{D}'(\Omega)$.

The main ingredient for proving Theorem 2.1 is the following extension of Theorem 1.1

PROPOSITION 2.1: Assume $0 < p-1 < q < N(p-1)/(N-p)$ if $1 < p < N$ or $q > p-1$ if $p = N$ and $u \in C^1(\Omega')$ is a weak solution of (0.7) in Ω' such that u/μ remains bounded near 0. Then there exists a real number γ such that

$$\lim_{x\to 0} u(x)/\mu(x) = \gamma. \tag{2.2}$$

Moreover

$$\lim_{x\to 0} |x|^{(N-1)/(p-1)} \nabla(u-\gamma\mu)(x) = 0, \tag{2.3}$$

and (2.1) holds in $\mathcal{D}'(\Omega)$.

The proof of Proposition 2.1 is essentially based on the same ingredients as those of Theorem 1.1: estimates on ∇u, scaling methods. Moreover an extended strict comparison principle and Theorem 1.3 allow us to end the proof (see [10] for details). Moreover singularities of type (ii) do exist as we have the following result:

PROPOSITION 2.2: Assume $0 < p-1 < q < N(p-1)/(N-p)$ if $1 < p < N$ or $q > p-1$ if $p = N$ and let R be any positive number. Then for any real numbers α, γ there exists a unique function $\psi = \psi^{\alpha,\gamma}$ in $C^1(0,R]$ satisfying

$$\begin{cases} -(r^{N-1} |\psi_r|^{p-2} \psi_r)_r + r^{N-1}|\psi|^{q-1} \psi = 0 \text{ in } (0,R), \\ \lim_{r\to 0} \psi(r)/\mu(r) = \gamma,\ \psi(R) = \alpha. \end{cases} \tag{2.4}$$

Furthermore

$$\lim_{r\to 0} \psi_r(r)/\mu_r(r) = \gamma, \tag{2.5}$$

the function $(\alpha,\gamma) \to \psi^{\alpha,\gamma}$ is separately nondecreasing in α or in γ and for any α' we have

$$|\psi^{\alpha,\gamma}(r) - \psi^{\alpha',\gamma}(r)| \leq |\alpha - \alpha'| \tag{2.6}$$

on (0,R].

The proof is essentially based upon the fact that under the hypotheses on N, p, q, $|\mu|^q$ is integrable near 0. The equation (2.4) is replaced by

$$\begin{cases} -(r^{N-1}|\psi_r^\varepsilon|^{p-2}\psi_r^\varepsilon)_r + r^{N-1}|\psi^\varepsilon|^{q-1}\psi^\varepsilon = 0 \text{ in } (\varepsilon,R), \\ \psi^\varepsilon(R) = \alpha, \\ \psi_r^\varepsilon(\varepsilon) = \gamma\,\mu_r(\varepsilon), \end{cases} \tag{2.7}$$

for $0 < \varepsilon < R$, and it can be proved that $\lim_{\varepsilon\to 0} \psi^\varepsilon$ exists and satisfies (2.4) and (2.5). As a consequence of Theorem 2.1 and Proposition 2.2 we can solve the singular Dirichlet problem for (0.7)

$$\begin{cases} -\operatorname{div}(|\nabla u|^{p-2}\nabla u) + u|u|^{q-1} = \gamma|\gamma|^{p-2}\delta_0 \text{ in } \mathcal{D}'(\Omega), \\ u = g \quad \text{on } \partial\Omega . \end{cases} \tag{2.8}$$

<u>THEOREM 2.2</u>: Assume $0 < p-1 < q < N(p-1)/(N-p)$ if $1 < p < N$ or $q > p-1$ if $p = N$, Ω is bounded and $\partial\Omega$ regular. Then for any $g \in L^\infty(\Omega) \cap W^{1,p}(\Omega)$ there exists a unique function $u \in C^1(\Omega')$ such that $|\nabla u|^{p-1} \in L^1(\Omega)$, $|u|^q \in L^1(\Omega)$, $\nabla u \in L^p(\Omega - B_r(0))$ ($r > 0$ small enough) and $u/\mu \in L^\infty(B_r(0))$ satisfying (2.8). Moreover (2.2) and (2.3) are satisfied.

We can give now the sketch of the proof of Theorem 2.1.

<u>LEMMA 2.1</u>: There exists a constant $C > 0$ such that for any $\rho \in (0,1]$ ($B_1(0) \subset \Omega$) we have

$$\max_{|x|=\rho} u(x) \leq C \min_{|x|=\rho} u(x). \tag{2.9}$$

<u>PROOF</u>: From Vázquez's *a priori* estimate [24] (see also [3] and [25]) there exists $A = A(N, p, q) > 0$ such that

$$u(x) \leq A|x|^{-p/(q+1-p)} \tag{2.10}$$

for $0 < |x| \leq \frac{1}{2}$. Moreover it is clear that $u > 0$ in $B_{1/2}(0)$. Let

$y \in B_{1/2}(0)$, $\rho = |y|/2$ then $B_\rho(y) \subset B_1(0) - \{0\}$. We write (0.7) in $B_\rho(y)$ in the form

$$- \operatorname{div} (|\nabla u|^{p-2} \nabla u) + b^p u^{p-1} = 0 \tag{2.11}$$

with $b^p = u^{q+1-p}$. From a result of Trudinger [21], there exists a constant $C_1 = C_1(n, p, \rho \|b\|_{L^\infty(B_\rho(y))})$ such that

$$\max_{x \in B_{\rho/3}(y)} u(x) \leq \min_{x \in B_{\rho/3}(y)} u(x) \tag{2.12}$$

and by (2.10) $\rho \|b\|_{L^\infty(B_\rho(y))} \leq C_2$ independent of ρ. If x_1 and x_2 are such that $0 < |x_1| = |x_2| \leq 1/2$ we can join them by 10 connected balls of radius $\frac{1}{6} |x_1|$ and centre on $\partial B_{|x_1|}(0)$. Hence (2.9) follows with $C = C_1^{10}$.

<u>LEMMA 2.2</u>: Assume $0 < p-1 < q < N(p-1)/(N-p)$ if $1 < p < N$ or $q > p-1$ if $N = p$ and let $\psi \in C^1(0,1]$ be a nonnegative solution of

$$-(r^{N-1} |\psi_r|^{p-2} \psi_r)_r + r^{N-1} \psi^q = 0 \text{ in } (0,1] \tag{2.13}$$

such that $\psi(r)/\mu(r)$ is unbounded in $(0, 1/2]$. Then there exists a constant $C = C(N, p, q)$ such that

$$|r^{p/(q+1-p)} \psi(r) - \gamma_{N,p,q}| \leq C r^\tau \tag{2.14}$$

for $0 < r \leq 1/2$ where τ is the positive root of the equation

$$X^2 + (N - \frac{p(q+1)}{q+1-q}) X + \frac{p}{p-1} (N - \frac{pq}{q+1-p}) = 0.$$

<u>Sketch of the proof</u>. <u>Step 1</u>. There exists $\alpha > 0$ such that

$$\alpha^{-1} \leq r^{p/(q+1-p)} \psi(r) \leq \alpha , \tag{2.16}$$

on $(0, 1/2)$. We give the proof when $1 < p < N$ and set

$$s = r^{(p-N)/(p-1)}, \ \phi(s) = \psi(r). \tag{2.17}$$

Then

$$(|\phi_s|^{p-2}\ \phi_s)_s = (\frac{p-1}{N-p})^p\ s^{-p(N-1)/(N-p)}\ \phi^q \tag{2.18}$$

in $[1, +\infty)$. Hence ϕ_s is increasing and $\lim_{s \to +\infty} \phi_s(s) = +\infty$ from the hypothesis. From convexity $\phi(s) \leq s\ \phi_s(s)$ which implies

$$(\phi_s^{p-1})_s \leq (\frac{p-1}{N-p})^p\ s^{q-p(N-1)/(N-p)}\ \phi_s^q(s) \tag{2.19}$$

for s large enough. Integrating (2.19) yields

$$\liminf_{s \to +\infty} s^{[q-N(p-1)/(N-p)]/(q+1-p)}\ \phi_s(s) > 0, \tag{2.20}$$

which gives the left-hand side of (2.16). The right-hand side is just (2.10).

<u>Step 2</u>. $\lim_{r \to 0} r^{p/(q+1-p)}\ \psi(r) = \gamma_{N,p,q}$.

We consider $v(r) = r^{p/(q+1-p)}\ \psi(r)$; v satisfies

$$\begin{aligned} |r\ v_r - \frac{p}{q+1-p}\ v|^{p-2}\ [r^2(p-1)v_{rr} &+ r(N-1-2\ \frac{p(p-1)}{q+1-p})v_r \\ &+ p\ \frac{pq-N(q+1-p)}{(q+1-p)^2}\ v] = v^q, \end{aligned} \tag{2.21}$$

in (0,1) except at the points where $rv_r - \frac{p}{q+1-p}\ v = 0$.

But it is clear that v is monotone near 0. If we set $v_k(r) = v(r/k)$ then v_k is bounded and there exists $\{k_n\} \to +\infty$ and $\tilde{v}$ such that $\{v_{k_n}\}$ converges to $\tilde{v}$ in the $C^1_{loc}(0, +\infty)$ topology and $\tilde{v} = \lim_{r \to 0} v(r) = v_0$. Using the weak form of (2.21) we deduce

$$(\frac{p}{q+1-p}\ v_0)^{p-2}\ (p\ \frac{pq-N(q+1-p)}{(q+1-p)^2})\ v_0 = v_0^q. \tag{2.22}$$

As $v_0 \geq \frac{1}{\alpha}$ we deduce $v_0 = \gamma_{N,p,q}$.

<u>Step 3</u>. We set $\gamma = \gamma_{N,p,q}$, $\psi(r) = \gamma\ r^{-p/(q+1-p)}(1+z(r))$; from Step 2 $z(r)$, $r\ z_r(r)$ and $r^2 z_{rr}(r) \to 0$ as $r \to 0$ and

$$\gamma^{p-1}|rz_r - \frac{p}{q+1-p}(z+1)^{p-2}\{(p-1)r^2 z_{rr}$$

$$+ r(N-1-2\frac{p(p-1)}{q+1-p})z_r + p\frac{pq-N(q+1-p)}{(q+1-p)^2}(1+z)\} = \gamma^q(1+z)^q \tag{2.23}$$

which gives

$$r^2(p-1)z_{rr} + r[N(p-1)-1+2p - \frac{p^2q}{q+1-p}]z_r + p(N - \frac{pq}{q+1-p})z$$

$$= O(z^2 + r^2z_r^2 + r^4 z_{rr}^2) \tag{2.24}$$

and by standard elliptic estimates the right-hand side of (2.24) is $O(z^2)$. If we introduce now the linear equation

$$r^2(p-1)Z_{rr} + r[N(p-1)-1+2p - \frac{p^2q}{q+1-p}]Z_r + p(N - \frac{pq}{q+1-p})Z = 0 \tag{2.25}$$

whose solutions are $Z_1(r) = r^\tau$ and $Z_2(r) = r^{\tau'}$ where τ and τ' are respectively the positive and negative roots of (2.15). By a classical comparison argument we deduce $|z(r)| \leq C\, Z_1(r)$ which ends the proof.

PROOF OF THEOREM 2.1: We can assume that u/μ is not bounded near 0 otherwise we use Proposition 2.1. So there exists $x_n \to 0$ such that

$$\lim_{n\to+\infty} u(x_n)/\mu(x_n) = +\infty. \tag{2.26}$$

If we set $r_n = |x_n|$, then Lemma 2.1 implies

$$\lim_{n\to+\infty} \min_{|y|=r_n} u(y)/\mu(y) = +\infty. \tag{2.27}$$

We consider now ϕ_n (resp. ψ_n) the solution of (2.13) in $(r_n, 1]$ such that $\phi_n(1) = 0$ (resp. $\psi_n(1) = \max_{|x|=1} u(x)$ and $\phi_n(r_n) = \min_{|x|=r_n} u(x)$ (resp. $\psi_n(r_n) = \max_{|x|=r_n} u(x)$). From the maximum principle $\phi_n \leq u \leq \psi_n$ in $(r_n,1)$. From estimate (2.10) applied to ϕ_n and ψ_n there exists ϕ and ψ satisfying (2.13) in $(0,1]$ and $n_k \to +\infty$ such that $\phi_{n_k} \to \phi$ and $\phi_{n_k} \to \psi$ as $n_k \to +\infty$. Moreover, $\phi \leq u \leq \psi$. An easy verification shows that $\phi(r)/\mu(r)$ is not bounded near 0 which implies

$$||x|^{p/(q+1-p)} u(x) - \gamma_{N,p,q}| \leq C|x|^{\tau} \tag{2.28}$$

for $0 < |x| \leq 1/2$, from Lemma 2.2, which ends the proof.

REMARK 2.1: Under the same hypotheses on N, p and q we can show that if u is positive in $\mathbb{R}^N$-K (K compact) and satisfies (0.7) in $\mathbb{R}^N$-K, then

$$\limsup_{|x|\to+\infty} |x|^{-\tau'} \, ||x|^{p/(q+1-p)} u(x)-\gamma_{N,p,q}| < +\infty, \tag{2.29}$$

where τ' is the negative root of (2.15). As a consequence we have a global result

THEOREM 2.3: Assume $0 < p-1 < q < N(p-1)/N(N-p)$ if $1 < p < N$ or $q > p-1$ if $p = N$, $u \in C^1(\mathbb{R}^N - \{0\})$ is a nonnegative solution of (0.7) in $\mathbb{R}^N - \{0\}$. Then u is radial and we have the following alternative

(i) either $u(x) = \gamma_{N,p,q} \, |x|^{-p/(q+1-p)}$,

(ii) or there exists $\gamma \geq 0$ such that $\tilde{u}(|x|) = u(x)$ and $\tilde{u}$ is the unique $C^1(0, +\infty)$ function satisfying

$$\begin{cases} -(r^{N-1}|\tilde{u}_r|^{p-2}\tilde{u}_r)_r + r^{N-1}|\tilde{u}|^{q-1}\,\tilde{u} = 0 \text{ on } (0, +\infty), \\ \lim_{r\to 0} \tilde{u}(r)/\mu(r) = \gamma. \end{cases} \tag{2.30}$$

3. Antisotropic singularities

If we look for p-harmonic singular functions in the form

$$u(x) = |x|^{-\beta}\omega(x/|x|) = r^{-\beta}\omega(\sigma), \; \beta > 0, \tag{3.1}$$

where $(r,\sigma) \in (0, +\infty) \times S^{N-1}$ are the spherical coordinates in $\mathbb{R}^N - \{0\}$, a straightforward calculation shows that ω is C^1 and must satisfy on S^{N-1}

$$-\mathrm{div}_{S^{N-1}}((\beta^2\omega^2 + |\nabla_\tau\omega|^2)^{(p-2)/2} \nabla_\tau\omega) \tag{3.2}$$

$$+ \beta\,\gamma(\beta^2\omega^2 + |\nabla_\tau\omega|^2)^{(p-2)/2}\,\omega = 0$$

where $\gamma = N-1(\beta+1)(p-1)$, ∇_τ is the "tangential" gradient on S^{N-1} identified to the covariant derivative on S^{N-1} and $\mathrm{div}_{S^{N-1}}$ is the divergence operator acting on C^1 vector fields on S^{N-1}. Using the conformal invariance of the N-Laplace equation and Tolksdorf' results [20] we have

THEOREM 3.1: Assume $p = N$ and S is an open hemisphere of S^{N-1}, then there exists only one C^1 function ω_1 which is positive on S, negative on $S^{N-1}-\bar{S}$ with $\max_{\sigma\in S} \omega_1(\sigma) = 1$ such that

$$u(x) = r^{-1}\,\omega_1(\sigma) \tag{3.3}$$

is N-harmonic in $\mathbb{R}^N - \{0\}$. If $S = \{\sigma = (\sigma' \cos\theta, \sin\theta) : \sigma' \in S^{N-2}, \theta \in (-\frac{\pi}{2}, \frac{\pi}{2})\}$ then $\omega_1(\sigma) = \cos\theta$.

Using a shooting method we can get a similar result when $p \neq N$.

THEOREM 3.2: Assume $1 < p < N$ or $p > N$ and S is an open hemisphere of S^{N-1}. Then there exists only one (β_1, ω_1) with $\beta_1 > 0$, ω_1 C^1 and positive on S, negative on $S^{N-1} - \bar{S}$ and $\max_{\sigma\in S^{N-1}} \omega_1(\sigma) = 1$ such that the function u defined by (3.1) is p-harmonic in $\mathbb{R}^N - \{0\}$.

Sketch of the proof

Step 1. Existence. As in [20] we define a conical domain $K(R,R') = \{r\sigma : \sigma \in S, R < r < R'\}$, $R > 0$, and its conical boundary $K_S(R,R') = \{r\sigma : \sigma \in \partial S, R < r < R'\}$ and let g be the function defined by

$$g(x) = \begin{cases} 2 - |x| & \text{if } |x| \leqq 2, \\ 0 & \text{if } |x| \geqq 2, \end{cases} \tag{3.4}$$

and u_n be the solution of the Dirichlet problem (with $n \geqq 2$)

$$\begin{cases} -\operatorname{div}(|\nabla u_n|^{p-2}\nabla u_n) = 0 & \text{in } K(1,n), \\ u_n = g & \text{on } \partial K(1,n). \end{cases} \tag{3.5}$$

From the Hopf maximum principle [20], $u_n > 0$ in $K(1,n)$. When $n \to +\infty$, the sequence $\{u_n\}$ converges in $W^{1,p}_{loc}(\overline{K(1,\infty)})$ to some u which is the unique function in $L^p_{loc}(\overline{K(1,\infty)})$ such that $\nabla u \in L^p(K(1,\infty))$ satisfying

$$\begin{cases} -\operatorname{div}(|\nabla u|^{p-2}\nabla u) = 0 & \text{in } K(1,\infty), \\ u = g & \text{on } \bar{S} \cup K_S(1,\infty), \\ \lim\limits_{|x|\to+\infty} u(x) = 0. \end{cases} \tag{3.6}$$

Moreover $u > 0$ in $K(1,\infty)$ and $u \in C^{1,\alpha}(\overline{K(2+\varepsilon,+\infty)})$.

From the strict comparison principle the function

$$R \to C(R) = \sup_{x\in K(R,\infty)} u(x)$$

is decreasing and the supremum is achieved on $|x| = R$. Using Tolksdorf's equivalence principle [20 Lemma 2.1] we have

$$u(Rx) \leq (1-\varepsilon(R-1))\, u(x) \tag{3.7}$$

for some $\varepsilon > 0$ and any $R \in (1,2)$ and

$$C(R) \leq k\, C(2R) \tag{3.8}$$

for some positive constant k and $R > 3$. We now introduce

$$u_R(x) = u(Rx)/C(R). \tag{3.9}$$

From (3.7) and (3.8) we have as in Theorem 1.1

$$|\nabla u(x)| \leq M\, C(|x|)\, |x|^{-1} \tag{3.10}$$

which implies that $\int_{K(1,n)} |\nabla u_R|^p \, dx \leq C'_n$ independent of R. Hence there exist a sequence $\{R_\nu\}$ converging to $+\infty$ and a function $u^* \in C^1(K(1,\infty))$ which vanishes on $K_S(1,\infty)$ and is p-harmonic in $K(1,\infty)$ such that $\{u_{R_\nu}\}$ converges to u^* in the C^1_{loc} topology of $K(1,\infty) \cup K_S(1,\infty)$. Moreover $u^* > 0$ and u^* is C^∞ in $K(1,\infty) \cup K_S(1,\infty)$.

The main point now is to prove that there exists $\beta_1 > 0$ such that

$$u^*(r,\sigma) = r^{-\beta_1} u^*(1,\sigma). \tag{3.11}$$

We take $R > 1$ and set for $r \geqq 4$

$$\Sigma_{r,R} = \sup \{C : C\, u(x) \leqq u(Rx), \forall\, x \in K(r,\infty)\},$$

$$\Sigma_R = \sup \{\Sigma_{r,R} : r \geqq 4\}.$$

Moreover $r \mapsto \Sigma_{r,R}$ is increasing, hence $\Sigma_R < 1$ and $\Sigma_R\, u^*(x) \leqq u^*(Rx)$ in $K(1,\infty)$. If we assume now that the following relation

$$\Sigma_R\, u^*(x) = u^*(Rx) \tag{3.12}$$

does not hold in $K(1,\infty)$, then from the strict comparison principle we would have $\Sigma_R\, u^*(x) < u^*(Rx)$ in $K(1,\infty)$ and there would exist $\delta > 0$ such that $(\Sigma_R + 2\delta)\, u^*(x) \leqq u^*(Rx)$ on $\partial K(2,\infty)$ which would imply $(\Sigma_R + \delta)\, u(x) \leqq u(Rx)$ on $\partial K(r,\infty)$ for some $r > 0$ and $(\Sigma_R + \delta)\, u(x) < u(Rx)$ in $K(r,\infty)$ which is a contradiction. Hence (3.12) holds. In order to prove that for some $\beta_1 > 0$:

$$\Sigma_R = R^{-\beta_1}, \tag{3.13}$$

we first deduce from (3.12) that the function $R \to \Sigma_R$ is decreasing and C^1 on $(1, +\infty)$. Moreover we have for any $n \in \mathbb{N}^*$ $\Sigma_{R^n} u^*(x) = (\Sigma_R)^n u^*(x) = u^*(R^n x)$ so $\Sigma_{R^n} = (\Sigma_R)^n$. In the same way $\Sigma_{R^{n/q}} = (\Sigma_R)^{n/q}$ for $q \in \mathbb{N}^*$ and finally

$$(\Sigma_R)^\alpha = \Sigma_{R^\alpha} \tag{3.14}$$

for any $R > 1$ and $\alpha > 1$. An elementary exercise shows that (3.13) is true for some $\beta_1 > 0$ and then (3.11). We now define

$$\tilde{\omega}(\sigma) = u^*(1,\sigma) \tag{3.15}$$

then $\tilde{\omega}$ satisfies (3.2) in S and vanishes on ∂S. The function ω is now defined by reflection as

$$\omega(\sigma) = \begin{cases} \tilde{\omega}(\sigma) & \text{if } \sigma \in \bar{S}, \\ -\tilde{\omega}(-\sigma) & \text{if } \sigma \in S^{N-1} - \bar{S}. \end{cases} \tag{3.16}$$

$\omega \in C^\infty(S^{N-1})$ and satisfies (3.2) in S^{N-1}.

Step 2. Uniqueness. If (β',ω') is another solution then Tolksdorf's equivalence principle first implies that $\beta' = \beta$. An application of the strict comparison principle and Hopf maximum principle as in [20] shows that $\omega' = \omega$.

REMARK 3.1: It is not easy to compute β_1 except in some cases (p = 2, p = N or N = 2) but we can give an estimate in assuming that $S = \{(\sigma' \cos\theta, \sin\theta),\ \sigma' \in S^{N-2},\ \theta \in (-\frac{\pi}{2}, \frac{\pi}{2})\}$. In that case ω depends only on θ and $u(x) = u(r,\theta)$ satisfies ([12])

$$\sin^{n-2}\theta\{(u_r^2 + \frac{1}{r^2}u_\theta^2)^{(p-2)/2}\, r^{N-1}\, u_r\}_r$$
$$+ r^{N-3}\{(u_r^2 + \frac{1}{r^2}u_\theta^2)^{(p-2)/2}\, u_\theta \sin^{n-2}\theta\}_\theta = 0. \tag{3.17}$$

If we look for super or sub-solutions under the form $r^{-\alpha}\cos\theta$ we can prove the following result (see [28]) with the notations of Theorem 3.2:

THEOREM 3.3: Assume $p > \frac{3}{2}$ and $N > 2$ then

$$\text{Min}\,(\frac{N-1}{p-1}, \frac{N+p-3}{2p-3}) \leq \beta_1 \leq \text{Max}\,(\frac{N-1}{p-1}, \frac{N+p-3}{2p-3}). \tag{3.18}$$

When $1 < p \leq \frac{3}{2}$ we just have $\frac{N-1}{p-1} < \beta_1$.

When N = 2, equation (3.2) takes the particular form

$$-((\beta^2\omega^2+\omega_\theta^2)^{(p-2)/2}\,\omega_\theta)_\theta + (1-(\beta+1)(p-1)\beta(\beta^2\,{}^2+\omega_\theta^2)^{(p-2)/2} = 0. \tag{3.19}$$

Introducing the variable $Y = \omega_\theta/\omega$ we get the following result [11].

THEOREM 3.4: Assume N = 2 and p > 1. Then for each positive integer k there exist a $\beta_k > 0$ and $\omega_k : \omega_k : \mathbb{R} \to \mathbb{R}$ with least period $2\pi/k$ of class C^∞ such that

$$u_k(x) = |x|^{-\beta k} \omega(x/|x|) \tag{3.20}$$

is p-harmonic in $\mathbb{R}^2 - \{0\}$; β_k is the positive root of

$$(\beta + 1)^2 = (1 + \frac{1}{k})^2 (\beta^2 - \beta(2-p)/(p-1)). \tag{3.21}$$

(β_k, ω_k) is unique up to translation and homothety over ω_k.

REMARK 3.2: A similar result has been obtained by Krol and Mazja in the case of regular p-harmonic functions [13] (see also [6], [12]).

If we look now for solutions of (0.7) under the form (3.1) it is clear that $\beta = p/(q+1-p)$ for homogeneity reasons and ω must satisfy

$$-\mathrm{div}_{S^{N-1}}(\beta^2\omega^2 + |\nabla_\tau\omega|^2)^{(p-2)/2} \nabla_\tau\omega) + \omega|\omega|^{q-1} = \lambda(\beta^2\omega^2 + |\nabla_\tau\omega|^2)^{(p-2)/2} \omega, \tag{3.22}$$

on S^{N-1} with $\lambda = (\frac{p}{q+1-p})(\frac{pq}{q+1-p}) - N)$. As in the semilinear case (p = 2) we have an existence result as soon as $\beta_1 < p/(q+1-p)$.

THEOREM 3.5: Assume $0 < p-1 < q$, $p/(q+1-p) > \beta_1$ and let S be an open hemisphere of S^{N-1}. Then there exists at least one C^1 function which is positive on S, negative on $S^{N-1} - \bar{S}$ satisfying (3.22) in S^{N-1}.

The proof of this result uses the same shooting method as in Theorem 3.2. For $\varepsilon > 0$ small enough we construct the unique solution u of

$$\begin{cases} -\mathrm{div}(|\nabla u|^{p-2} \nabla u) + u|u|^{q-1} = 0 & \text{in } K(1,\infty), \\ u = \varepsilon g & \text{on } \bar{S} \cup K_S(1,\infty), \\ \limsup_{|x| \to +\infty} |x|^{p/(q+1-p)} u(x) < +\infty, \end{cases} \tag{3.23}$$

where g is defined in (3.4) and $\varepsilon g(x) \leq \gamma_{N,p,q} |x|^{-p/(q+1-p)}$; $u > 0$ in $K(1,\infty)$

and $u(x) \leq \gamma_{N,p,q} |x|^{-p/(q+1-p)}$. An important point is to prove that u does not decrease too quickly and this is obtained by showing that for $\rho > 0$ small enough the function $v(x) = \rho |x|^{-p/(q+1-p)} \omega_1(x/|x|)$ satisfies

$$\begin{cases} -\mathrm{div}\,(|\nabla v|^{p-2} \nabla v) + v^q \leq 0 & \text{in } K(1,\infty) \\ v \leq u & \text{in } \bar{S} \cup K_S(1,\infty). \end{cases} \tag{3.24}$$

Hence $\rho\omega_1(x/|x|) \leq |x|^{p/(q+1-p)} u(x) \leq \gamma_{N,p,q}$ in $\overline{K(1,\infty)}$. As in Theorem 3.2 there exists a function u* such that

$$u_R(x) = R^{p/(q+1-p)} u(Rx)$$

converges to u* in the C^1_{loc} topology of $K(1,\infty) \cup K_S(1,\infty)$ for some sequence $R_\nu \to +\infty$ and u* satisfies (0.7). Moreover

$$u^*(Rx) = \tilde{\Sigma}_R\, u^*(x), \tag{3.25}$$

which implies that $\tilde{\Sigma}_R = R^{-p/(q+1-p)}$ and

$$u^*(x) = u^*(r,\sigma) = r^{-p/(q+1-p)} \tilde{\omega}(\sigma), \tag{3.26}$$

with $\tilde{\omega}(\sigma) = u^*(1,\sigma)$. The function ω is then defined on S^{N-1} as in (3.16).

In the case N = 2 we can be precise about the bifurcation configuration for equation (3.22) (see [28]).

THEOREM 3.6: Assume N = 2, p > 1, k is a positive integer and β_k is the positive root of (3.21). If $\beta_k < p/(q+1-p) \leq \beta_{k+1}$ there exist at least k C^1 functions ω_j, $1 \leq j \leq k$, defined on $\mathbb{R}$, with least period $2\pi/j$ satisfying (3.22) on $\mathbb{R}$.

REMARK 3.3: There are still many unsolved problems concerning singularities of (0.1) or (0.7). One of great interest would be the following. Assume $u \in C^1(\Omega')$ is a solution of (0.1) (or (0.7)) in Ω' such that

$$\lim_{x \to 0} |x|^{\beta_1} u(x) = 0. \tag{3.27}$$

Then $u(x)/\mu(x)$ admits a finite limit when $x \to 0$ (for the case $p = 2$ see [26] and [27]).

References

1. Ph. Benilan and H. Brezis: Nonlinear problem related to the Thomas-Fermi equation. Unpublished work. See also H. Brezis, Some variational problems of the Thomas-Fermi type in Variational Inequalities Complimentary conditions. R.W. Cottle, F. Gianessi and J.L. Lions, eds, Wiley-Interscience, New-York (1980), 53-73.
2. B. Bojarski and T. Iwaniec: p-harmonic equation and quasi-linear mappings. Sunderforschungsbereich 72, Universität Bonn (1983).
3. H. Brezis and E.T. Lieb: Long range atomic potentials in Thomas-Fermi theory, Comm. Math. Phys., 65 (1979), 231-246.
4. H. Brezis and L. Oswald: Singular solutions for some semilinear elliptic equations (to appear).
5. P.T. Chrusciel: Conformally minimal foliations of three dimensional Riemannian manifolds and the energy of the gravitational field. Ph.D. Thesis, Polish Acad. Sci. Inst. Th. Phys. (1986).
6. M. Dobrowolski: Nichtlineare eckenprobleme und finite elemente methode, Z.A.M.M., 64, (1984) 270-271.
7. L.C. Evans: A new proof of local $C^{1,}$ regularity for solutions of certain degenerate elliptic P.D.E., Jl. Diff. Equ., 50, (1982), 315-338.
8. R.H. Fowler: Further studies on Emden's and similar differential equ., Quart. Il. Math., 2, 1931, 259-288.
9. A. Friedman et L. Veron: Solutions singulières d'équations quasilinéaires elliptiques, Cr. Acad. Sci. Paris, t. 302 I (1986), 145-150.
10. A. Friedman and L. Veron: Singular solutions of some quasilinear elliptic equations, Arch. Rat. Mech. Anal. (to appear).
11. S. Kichenassamy and L. Veron: Singular solutions of the p-Laplace equation, Math. Ann. 275 (1986), 599-615.
12. I.N. Kroll: On the behaviour of the solutions of a quasilinear equation near null salient points of the boundary, Proc. Steklov Inst. Math., 125 (1973), 130-136.
13. I.N. Kroll and V.G. Mazja: The lack of continuity and Hölder continuity of the solutions of a certain quasi-linear equation, Sem. Math. V.A. Steklov Math. Inst., 14 (1969), 44-45.

14. J.L. Lewis: Regularity of the derivatives of solutions to certain degenerate elliptic equations, Ind. Univ. Math. Il., 32 (1983), 849-858.

15. Y.G. Reshetniak: Mappings with bounded deformations as extremals of Dirichlet type integrals, Sibirsk. Math. Zh., 9 (1966), 652-666.

16. J. Serrin: Local behaviour of solutions of quasilinear equations, Acta Math., 111 (1964), 247-302.

17. J. Serrin: Isolated singularities of solutions of quasi-linear equations, Acta Math., 113 (1965), 219-240.

18. A. Sommerfeld: Asymptotische integrazion der differenzialgleichung der Thomas-Fermischen atoms, Z. für Phys., 78 (1932), 283-308.

19. P. Tolksdorf: Regularity for a more general class of quasi-linear elliptic equations, Jl. Diff. Eq., 51 (1984) 126-150.

20. P. Tolksdorf: On the Dirichlet problem for quasilinear equations in domain with conical boundary points, Comm. P.D.E., 8 (1983), 773-817.

21. N.S. Trudinger: On Harnack type inequalities and their applications to quasilinear elliptic equations, Comm. Pure Appl. Math., 20 (1967), 721-747.

22. K. Uhlenbeck: Regularity for a class of nonlinear elliptic systems, Acta Math., 138 (1977) 219-240.

23. N.N. Ural'ceva: Degenerate quasilinear systems, L.O.M.I., 7 (1968), 184-222.

24. J.L. Vazquez: An a priori interior estimate for the solutions of a nonlinear problem representing weak diffusion, Nonlinear Anal., 5 (1981), 95-103.

25. J.L. Vazquez and L. Veron: Removable singularities of some strongly nonlinear elliptic equations, Manuscripta Math., 33 (1980), 129-144.

26. J.L. Vazquez and L. Veron: Isolated singularities of some semilinear elliptic equations, Jl. Diff. Equ., 60 (1985), 301-321.

27. L. Veron: Singular solutions of some nonlinear elliptic equations, Nonlinear Anal., 5 (1981), 225-242.

28. L. Veron: Anisotropic singularities of some degenerate elliptic equations (in preparation).

L. Veron
Département de Mathématiques
Faculté des Sciences
Parc de Gradnmont
F 37200 Tours, France.